W9-CMI-303

# Partial Differential Equations for Computational Science

David Betounes

# Partial Differential Equations for Computational Science

## With Maple® and Vector Analysis

CR-ROM
INCLUDED

WITHDRAWN

Springer TELOS

THE
ELECTRONIC
LIBRARY
OF
SCIENCE

David Betounes
Department of Mathematics
University of Southern Mississippi
Hattiesburg, MS 39406
USA

Library of Congress Cataloging-in-Publication Data
Betounes, David.
        Partial differential equations for computational science : with Maple
and vector analysis / David Betounes.
            p.   cm.
        Includes index.
        ISBN 0-387-98300-7 (hardcover : alk. paper)
            1. Differential equations.    Partial—Numerical Solutions.
        2. Vector analysis.    3. Maple (Computer file)    I. Title.
        QA377.B53    1997
        515'.353    dc21                              97-26381

Printed on acid-free paper.

Production managed by Steven Pisano; manufacturing supervised by Joe Quatela.
Typeset by The Bartlett Press, Marietta, GA.
Printed and bound by Hamilton Printing Co., Rensselaer, NY.
Printed in the United States of America.

9  8  7  6  5  4  3  2  1

ISBN 0-387-98300-7 Springer-Verlag New York Berlin Heidelberg    SPIN 10635166

TELOS, The Electronic Library of Science, is an imprint of Springer-Verlag New York. Its publishing program encompasses the natural and physical sciences, computer science, mathematics, economics, and engineering. All TELOS publications have a computational orientation to them, as TELOS' primary publishing strategy is to wed the traditional print medium with the emerging new electronic media in order to provide the reader with a truly interactive multimedia information environment. To achieve this, every TELOS publication delivered on paper has an associated electronic component. This can take the form of book/diskette combinations, book/CD-ROM packages, books delivered via networks, electronic journals, newsletters, plus a multitude of other exciting possibilities. Since TELOS is not committed to any one technology, any delivery medium can be considered. We also do not foresee the imminent demise of the paper book, or journal, as we know them. Instead we believe paper and electronic media can coexist side-by-side, since both offer valuable means by which to convey information to consumers.

The range of TELOS publications extends from research level reference works to textbook materials for the higher education audience, practical handbooks for working professionals, and broadly accessible science, computer science, and high technology general interest publications. Many TELOS publications are interdisciplinary in nature, and most are targeted for the individual buyer, which dictates that TELOS publications be affordably priced.

Of the numerous definitions of the Greek word "telos," the one most representative of our publishing philosophy is "to turn," or "turning point." We perceive the establishment of the TELOS publishing program to be a significant step forward towards attaining a new plateau of high quality information packaging and dissemination in the interactive learning environment of the future. TELOS welcomes you to join us in the exploration and development of this exciting frontier as a reader and user, an author, editor, consultant, strategic partner, or in whatever other capacity one might imagine.

*TELOS, The Electronic Library of Science*
*Springer-Verlag New York, Inc.*

**THE ELECTRONIC LIBRARY OF SCIENCE**

### TELOS Diskettes

Unless otherwise designated, computer diskettes packaged with TELOS publications are 3.5″ high-density DOS-formatted diskettes. They may be read by any IBM-compatible computer running DOS or Windows. They may also be read by computers running NEXTSTEP, by most UNIX machines, and by Macintosh computers using a file exchange utility.

In those cases where the diskettes require the availability of specific software programs in order to run them, or to take full advantage of their capabilities, then the specific requirements regarding these software packages will be indicated.

### TELOS CD-ROM Discs

For buyers of TELOS publications containing CD-ROM discs, or in those cases where the product is a stand-alone CD-ROM, it is always indicated on which specific platform, or platforms, the disc is designed to run. For example, Macintosh only; Windows only; cross-platform, and so forth.

### TELOSpub.com (Online)

Interact with TELOS online via the Internet by setting your World-Wide-Web browser to the URL: *http://www.telospub.com.*

The TELOS Web site features new product information and updates, an online catalog and ordering, samples from our publications, information about TELOS, data-files related to and enhancements of our products, and a broad selection of other unique features. Presented in hypertext format with rich graphics, it's your best way to discover what's new at TELOS.

TELOS also maintains these additional Internet resources:

*gopher://gopher.telospub.com*
*ftp://ftp.telospub.com*

For up-to-date information regarding TELOS online services, send the one-line e-mail message:

*send info*

to: *info@TELOSpub.com.*

# Preface

This textbook is an introduction to the subject of partial differential equations (PDEs) and, as the title indicates, is oriented towards the needs of students in *computational science*. Such students are presently in some degree program which emphasizes the computational aspects of a traditional area of science or engineering and, indeed, may be an interdisciplinary program which combines several traditional areas of study into one with a strong component of modern computing. Such is the program here at my university, where the Departments of computer science, mathematics, and physics offer a Ph.D. in Scientific Computing with a core of coursework in all three departments. Thus in teaching our introductory graduate course in PDEs, over the last three years, it was typical to have students from physics and computer science in the same class with our mathematics grad students. From this situation, this book, its content, and methodology, evolved into its present form.

Having said this, I should also mention that many of our grad students in the traditional *pure* math program take, and benefit from, the course for which this book was designed, and I would think this would be the case at other universities. In certain cases, this book could be used in an advanced undergraduate course. The Chapters 1–8 and Chapter 12 have their origins in material developed for our undergraduate PDE course.

The philosophy behind this book is to provide an elementary, yet sophisticated, introduction to the subject–one which requires a minimum of mathematical prerequisites and yet discusses many difficult aspects of the theory. There are only a few theorems and proofs, and these are only outlines of what could be made rigorous with the proper tools from functional analysis and differential geometry. They are for pedagogical purposes and for providing a proper logical framework to the material. The goal is to have material that is accessible to first year graduate students not only in mathematics but also in other departments and programs as well. The only prerequisites are a good undergraduate background in calculus, differential equations, and linear algebra.

There are several themes which occur throughout the book. These are:

**(Use of the Computer)**    While this book is for computational scientists (and others) it is *not* about computer techniques and numerical methods. Rather, throughout, the computer is used as a pedagogical tool for understanding and enhancing the material. As such, the emphasis is on using the computer for visualization and interpretation of the solutions to PDEs, and the examples and exercises are specifically designed to be implemented with a computer algebra system (CAS). To make the examples concrete, we have chosen *Maple* as the CAS.

**(Use of Vector Analysis)**    Vector analysis, or more properly differential geometry, is an essential tool in the study of PDEs and the text illustrates why this is so throughout. Prior knowledge of topics from vector analysis, like Gauss's divergence theorem, is not assumed and Appendix A is provided as a background for such topics. This Appendix can either be assigned for independent study or parts of it can be covered as part of the course.

**(The Flow)**    The concept of the flow generated by a vector field is central to many areas of mathematics and naturally arises in the study of PDEs via heat flow lines, electric field lines, and fluid flow lines. Indeed the flow is essential in the development of the fluid flow equations and the equations for waves in elastic solids. The idea of the flow is easily motivated by these topics in the study of PDEs and is easily visualized by use of various software plotting packages.

**(Weak Solutions of PDEs)**    In a first course, students should have some exposure to abstract methods dealing with existence and uniqueness of solutions of PDEs. In Chapter 10, the Galerkin method is applied to the nonlinear system of Navier-Stokes equations and the same theme is repeated at the end of Chapter 11, where the Ritz-Galerkin method is applied to the linear system of Navier equations (for waves in elastic solids). Chapter 13 is devoted entirely to such methods as applied to Poisson's equation and the heat equation. The exercises to this chapter illustrate how these abstract methods lead to the finite element method for numerically solving PDEs.

**(The Physics Behind the PDEs)**    The book provides derivations and motivations of the underlying physics, for the heat equation, Maxwell's equations, the Navier-Stokes equations, and the wave equation (however, much of this for Maxwell's equations is more appropriately left to a physics course in electricity and magnetism). These derivations and motivations are essential for computational scientists (and many others), who will in their ensuing courses become immersed in the numerical aspects of PDEs and will need to exploit the special understanding of these PDEs that comes from knowing their physical origins. There is a limit of the extent to which these physical aspects can be fully discussed in a semester or two and with this in mind I have tried to arrange this material so that the amount of time spent on it can vary. Some is relegated to the exercise sets and for reference I have included some details from the theory of continuum mechanics in Appendix B.

The material in the book and some reasons for its order of presentation are as follows.

**Chapter 1** establishes the notation and basic concepts: gradient, curl, divergence, Laplacian, and covariant derivative operators, scalar fields, vector fields, various identities, and the flow generated by a vector field. An overview of systems of PDEs and the particular types to be studied in the text is given.

**Chapters 2-8** are devoted to a detailed study of the heat equation and one of the principal analytic techniques for solving certain PDEs: (generalized) Fourier series. The study of Laplace's equation and boundary value problems occurs in these chapters as well, since they naturally arise in the study of the heat equation. These chapters also involve the use of the computer to plot the complicated series solutions and study their evolution over time. This is an excellent way to motivate the time and labor spent in calculating a particular series solution to a heat problem, since the series is virtually impossible to interpret by hand. The material in Chapters 2-8 could comprise a large portion of any introductory course and is for the most part elementary. After establishing the ideas and algorithms for one and two dimensional heat problems, the three dimensional case in Chapter 8 could be assigned for independent study. Chapter 8 provides for the student (i) a review (summary) of all the material from the 1 and 2 dimensional cases, (ii) further scientific visualization tools, and (iii) independent projects (exercises). An elementary discussion of the theory of existence and uniqueness for the heat equation and related topics is deferred to Chapter 13, since some instructors will not want to cover this aspect of the material at this point in the course.

**Chapter 9** is centered on Maxwell's equations and is designed as a change of pace, or respite, from the routine of calculating series solutions of PDEs. Additionally it serves to introduce the student to a nice example of a *system* of PDEs and gives them a chance to use partial differential operators and see some standard identities. Since the theory for Maxwell's equations is quite deep and complicated and comprises the subject matter in many standard physics courses, the chapter concentrates mainly on the topic of potentials for the electomagnetic field and a few structural aspects of these equations. This chapter seems to be the appropriate place to introduce (in the exercises), for the first time, some elementary material on generalized functions (distributions).

**Chapter 10** discusses the Navier-Stokes equations (and the Euler equations as a special case) for incompressible fluids and serves as a good example of a *nonlinear* system of PDEs. While analytic techniques for dealing with these equations are generally lacking, there are enough special cases (like potential flows) to make the theory concrete and keep the students interested. In particular, the basic planar flows in Exercise set 10.3.2 are illustrated in many fluid mechanics texts, but viewing these static drawings is no match for the experience of generating the flow lines dynamically with a computer. The inability to write down analytic solutions to very simple fluid flow problems serves to motivate the theoretical approach outlined in Section 10.5 (the Galerkin method for existence of weak solutions) and also to demonstrate that such situations are common in the study of PDEs. Some aspects of this chapter may require reference to Appendix B on continuum mechanics, depending on the instructor's depth of presentation.

**Chapter 11** returns to the series representation techinique as it applies to the study of solutions of the wave equation. The wave equations for a vibrating string and membrane are derived by linearizing the equations of motion for these continua, as described in Appendix B. The discussion for the string is given in detail, while the membrane and solid cases have most of their details relegated to the exercise sets. The inability to derive analytic solutions to the Navier equation, which governs waves in elastic solids, is again (as in Chapter 10) used as a vehicle to motivate a modern, theoretical approach to the existence and uniqueness of solutions (the Ritz-Galerkin method).

**Chapter 12** returns to the analytic solution of the heat equation on planar domains and covers the the case where the domain has a circular type of geometry: a disk or an annulus. This serves to introduce the student to one of several special cases where a change of coordinates will allow a series representation in terms of Sturm-Liouville eigenfunctions (in this case, functions related to Bessel functions). I have found that covering this material immediately after Chapter 8 can be a real *drag* on the students, whereas at some later point in the course, this material can serve to reinforce the ideas behind the series technique. In addition, the material is ideally suited to being learned with the aid of a CAS (computer algebra system). The eigenfunctions and eigenvalues are made concrete when visualized through their plots and numerical calculations of tables of values. The integrals giving the Fourier-Bessel coefficients for the series representations are easily calculated numerically, where closed form expressions are impossible.

**Chapter 13** rounds out the text with an elementary discussion of the theory concerning the existence and uniqueness of solutions to the heat equation and Poisson's equation. Weak solutions, weak versions of BVP's and IBVPs, Dirichlet forms, the Lax-Milgram lemma, and the Ritz-Galerkin method are all discussed. The chapter gives as much of the essence of the theory as possible without relying on any background in functional analysis or differential geometry.

**Appendix A** provides an elementary introduction to vector analysis and the differential geometry of curves and surfaces.

**Appendix B** is a brief introduction to the topic of Continuum mechanics. This should provide a guide to the concepts and notation for this old and venerable subject, which seems to have never found a place in the mathematics or physics curriculum.

**Appendix C** serves as a short reference manual for *Maple*. Soon most undergraduates in the sciences will have had much experience with one or more of the CAS's (*Maple, Mathematica, Matlab*, etc.) and this appendix will be superfluous. My specfic intents with designing how *Maple* is integrated into the text were: (1) keep the *Maple* code as unobtrusive as possible for readability and to allow for non-*Maple* users to use the book and (2) defer most work with the computer until Chapter 5 in order to afford the students the time to digest the concepts and methods presented for the 1-d heat equation.

**CD-ROM:** The CD contains a large amount of electronic material designed to actively supplement the text. In essence it is an extension of most of the examples from the text that require the computer and *Maple* for their solution and visualization. An example in the text will, for instance, describe and analyze the nature of heat flow in a metal plate, or of wave motion on a membrane (drum head). But, for reasons of pedagogy and readability, there is a limit to the amount of time devoted to the example. This is where the CD picks up. For more in-depth analysis of the temperature evolution or the spatial vibration, one should work through the tutorials, exercises, and extended projects on the CD. The CD also serves as an excellent way to learn more about *Maple*'s syntax, programming features, and graphical capabilities.

You will gain the most benefit by using each worksheet interactively, as Maple was designed to be used. Thus you should move through the worksheet by executing the commands in succession. I have tried to design the worksheets with appropriately placed, empty, execution groups ($>$), so that the user will be able to read all the text and output between each nonblank execution group. Some scrolling (like back scrolling) may be necessary, but generally you can proceed by just hitting enter.

If you have the plots executed in a separate window (rather than "inline" in the worksheet) you will be able to control the plot outputs and compare with the original in the worksheet. If you have little experience with *Maple*, this will help you learn how to control the way a plot looks by adjusting the options in the plot window. If you adjust the options on the plot that is already pasted in, then you will lose the comparative value of the worksheet.

The exercises on the worksheet encourage you to make changes to the worksheet and re-execute in order to answer certain questions and generally study a given example in more detail. Any changes and alterations to the worksheet can be saved using Save As from the File menu. The original worksheet on the CD-ROM is read only, of course, and so cannot be written over. Thus, you will always have a copy of the original. Many of the exercises in the book require some *Maple* code similar to that on these worksheets, and the enterprising student will have already discovered that the bits and pieces of code on these worksheets can be excised and place in new worksheets for convenient use, more of less as a template.

### System Requirements and CD-ROM Table of Contents:

The accompanying CD-ROM contains files and programs that are executable on a variety of platforms. All files ending in .txt are readable across all platforms. It is recommanded to begin using the CD-ROM by first reading the legal.txt, cdreadme.txt, and cdtoc.txt.

It is recommended that users have at least 16MB of RAM installed on their computers in order to obtain the full educational value out of the worksheets. Naturally, 32MB of RAM will provide even more assurance that all of the programs will work properly, including those worksheets containing very large graphic files.

The primarly files on the CD-ROM are the 41 *Maple* worksheets (.mws extensions for *Maple* Release 4 and .ms extensions for Release 3). The cdtoc.txt outlines these (organized by the corresponding chapter in the book) and the other auxiliary files on the CD, with a brief description of their contents.

Also included on the accompanying CD-ROM is a *Maple* demo version for Macintosh and Windows users who have not invested in some version of the *Maple* software. The demo version of *Maple* has all the look and feel of the complete *Maple* application but, of course, has essential elements missing. With the demo version, you can open any of the worksheets on the CD-ROM with .mws extensions and scroll through them, reading the content and viewing the graphics as you go. You can manipulate the graphics (two- and three-dimensional plots) and execute some of the *Maple* commands. However most of the *Maple* commands will not work because the functionality of the demo was purposely limited. Thus, you can only read through the worksheets. To get the full benefit of these worksheets, you will need a full version of the *Maple* software.

legal.txt        Copyright, permissions, and disclaimers.

cdtoc.txt        The complete table of contents for the CD-ROM, including explanations on the given chapter *Maple* files.

cdreadme.txt     Explanations about the CD-ROM and installing the *Maple* demo.

MAPLEdemo        Contains *Maple* demo software for Macintosh and Windows users who do not have *Maple* software.

MapleR4files    Contains 22 *Maple* worksheets for *Maple* Release 4. These files end in .mws

MapleR3files    Contains 19 *Maple* worksheets for *Maple* Release 3. These files end in .ms

CH5.MWS and CH5.MS (Chapter 5: Computational Analysis)
    Example 5.1: 1-d heat flow
    Example 5.2: 1-d heat flow
    Example 5.4: Gibb's phenomenon
    Theory and history of Fourier series
    *Maple* topics: procedures, plot options, `fsolve` and `solve`.

CH6A.MWS and CH6A.MS (Chapter 6: Two-Dimensional Heat Flow)
    Example 6.1: Heat flow in a square plate
    Example 6.2: Heat flow in a square plate
    *Maple* topics: scaling axes, `orientation`, `thickness`, and `linecolor` in `plot3d`, the sequence command, arrays.

CH6B.MWS and CH6B.MS (Chapter 6: Two-Dimensional Heat Flow)
    Example 6.3: A semi-homogeneous heat flow in a square plate
    *Maple* topics: `color` and $z$-hue in `plot3d`, animations with the options $z$-hue and `contour`.

CH6C.MWS and CH6C.MS (Chapter 6: Two-Dimensional Heat Flow)
    Example 6.4: A square plate with nonhomogeneous BC's
    Gibbs phenomenon
    Odd/even periodic extensions for functions of two variables

CH7A.MWS and CH7A.MS (Chapter 7: Boundary Value Problems)
    Example 7.1: A Dirichlet problem with continuous BCs
    Example 7.3: A Dirichlet problem with discontinuous BCs
    Gibbs phenomenon, isotherms
    *Maple* topic: `implicitplot`.

CH7B.MWS and CH7B.MS (Chapter 7: Boundary Value Problems)
    Example 7.4: A Dirichlet problem with discontinuous BCs
    *Maple* topics: contours, plotting heat flow lines with DEplot, options (`obsrange`, `color`, `arrows`, `dirgrid`) in DEplot, plotting flows dynamically with a special procedure called DDEplot

CH7C.MWS and CH7C.MS (Chapter 7: Boundary Value Problems)
>Example 7.5: A Neumann problem for a square plate
>Example 7.8 and 7.9: Mixed BVPs for a square plate, plotting heat flow lines

CH8A.MWS and CH8A.MS (Chapter 8: Three-Dimensional Heat Flow)
[Note: Computers with less memory use CH8A1.MWS, CH8A2.MWS, CH8A3.MWS.]
>Example 8.2: A mixed BVP for the unit cube
>3-d visualization via slices, isotherms, and heat flow plots
>*Maple* topics: `implicitplot3d`, `DEplot3d`, options for `DEplot3d`, plotting flows dynamically with a special procedure called `DDEplot3d`

CH8B.MWS and CH8B.MS (Chapter 8: Three-Dimensional Heat Flow)
>Example 8.3: A heat IBVP for the unit cube
>Visualization, via slices and temperature profiles/spacetime surfaces along given line segment.

CH8C.MWS and CH8C.MS (Chapter 8: Three-Dimensional Heat Flow)
>Example 8.4: A nonhomogeneous heat IBVP for the unit cube
>Evolution toward the steady state
>Isotherms of the steady state

CH10A.MWS and CH10A.MS (Chapter 10: Fluid Mechanics)
>History of the Euler equations for a nonviscous fluid
>Examples: an infinite row of vortices and Karman's vortex sheet
>*Maple* topic: a special procedure called `DDEplot`

CH10B.MWS and CH10B.MS (Chapter 10: Fluid Mechanics)
>Example: flow out of a channel
>Exercise: the pitot tube
>*Maple* topic: conformal plots

CH10C.MWS and CH10C.MS (Chapter 10: Fluid Mechanics)
>Joukowski airfoils (cambered and uncambered)
>Flow past airplane wings

CH11A.MWS and CH11A.MS (Chapter 11: Waves in Elastic Materials)
>Example 11.2: A traveling triangular wave on a string
>Example 11.3: A traveling wave and an oscillating boundary
>Reversibility of the motion, animations, spacetime surface

CH11B.MWS and CH11B.MS (Chapter 11: Waves in Elastic Materials)
> Example 11.4: A traveling parabolic wave on a membrane
> Example 11.5: A wave on a membrane with two free
> boundaries
> Animations

CH11C.MWS and CH11C.MS (Chapter 11: Waves in Elastic Materials)
> Example: A traveling circular wave on a large membrane

CH12A.MWS and CH12A.MS (Chapter 12: The Heat IBVP in Polar
Coordinates)
> Example 12.1: A heat IBVP for a disk
> Bessel functions of the first and second kinds, computing
> zeros of Bessel functions and coefficients for Fourier-
> Bessel series
> *Maple* topics: `save`, `read`, `type`, `int`, `Int`, `cylinderplot`,
> piecewise defined functions, animations in polar
> coordinates

CH12B.MWS and CH12B.MS (Chapter 12: The Heat IBVP in Polar
Coordinates)
> Example 12.2: A heat IBVP ofr an annulus
> Calculation of eigenvalues and Fourier-Bessel coefficients
> numerically
> *Maple* topics: `diff`, `unapply`, and inert forms

CH12C.MWS and CH12C.MS (Chapter 12: The Heat IBVP in Polar
Coordinates)
> Example 12.3: A Dirichlet problem for a disk with
> piecewise defined boundary data
> Example 12.4: A Dirichlet problem for a disk with cosine
> wave boundary data
> Plotting flows in polar coordinates
> *Maple* topic: `spacecurve`

# Contents

# 1 Introduction

In this first chapter we present some of the notation and nomenclature connected with the study of partial differential equations (PDEs) and consider systems of partial differential equations in general. In the ensuing chapters we will focus on some particular systems of partial differential equations and discuss classical techniques for solving them as well as modern methods for proving existence and uniqueness of solutions.

Many of the partial differential equations we study (especially systems of partial differential equations like Maxwell's equations and the Navier–Stokes equations) are commonly formulated in terms of the notation and concepts from vector analysis, and so we discuss some of this first. Currently, vector anlaysis has been subsumed under the fields of analysis on manifolds and differential geometry, and indeed, we could formulate everything in terms of differential forms/tensor fields and exterior differential systems as perhaps the preferred modern viewpoint. To do so would require more background from the student, and for our goals here, the nineteenth century vector analysis is just as adequate. Many of the results from vector analysis are very useful for the theory of partial differential equations, and we have included some of them in Appendix A for your reference.

## 1.1. Vector Analysis: Some Basic Notions

The classical approach to vector analysis deals with two types of fields — scalar fields and vector fields — and several types of differential operators — the gradient, curl, divergence, covariant derivative, and Laplacian operators — that act on these fields.

For us, a scalar $a$ is a real number $a \in \mathbb{R}$, and a vector $v$ is a 3-tuple of real numbers, $v \in \mathbb{R}^3$:

$$v = (v_1, v_2, v_3).$$

Such a vector is often identified with a directed line segment in 3-dimensional space. The standard operations of vector addition, $v + w$, and multiplication of a vector by a scalar, $av$, have geometric interpretations as well. You should also recall the important operations on vectors of dot product, $v \cdot w$, the result of which is a scalar, and of cross product, $v \times w$, the result of which is a vector:

$$v \cdot w = v_1 w_1 + v_2 w_2 + v_3 w_3, \tag{1.1}$$

$$v \times w = (v_2 w_3 - v_3 w_2, v_3 w_1 - v_1 w_3, v_1 w_2 - v_2 w_1) \tag{1.2}$$

$$= \begin{vmatrix} \mathbf{i} & \mathbf{j} & \mathbf{k} \\ v_1 & v_2 & v_3 \\ w_1 & w_2 & w_3 \end{vmatrix}. \tag{1.3}$$

The dot product is fundamental in computing the magnitude, or length, $|v|$ of a vector, as well as the angle $\theta$ between vectors:

$$|v| = (v \cdot v)^{1/2} = (v_1^2 + v_2^2 + v_3^2)^{1/2},$$

$$v \cdot w = |v||w| \cos \theta.$$

The cross product gives a vector $v \times w$ that is perpendicular to both $v$ and $w$, and has magnitude $|v \times w| =$ the area of the parallelogram spanned by $v$ and $w$. Recall also that $u \cdot (v \times w) = \pm$ the volume of the parallelepiped determined by the vectors $u$, $v$, and $w$. This is known as the scalar triple product of these vectors, with the $\pm$ sign determined by the relative orientations of the vectors.

The terminology *scalar field* and *vector field* comes from way back in the history of physics (the 1800s or so). Thus in modern terms, a scalar field is just a function $u = u(x, y, z)$ of three variables (the spatial variables) defined on some domain (subset) $\Omega$ of $\mathbb{R}^3$, with range in $\mathbb{R}$. Symbolically, $u : \Omega \subseteq \mathbb{R}^3 \to \mathbb{R}$. One thinks of $u$ as assigning to each point $(x, y, z)$ in $\Omega$ a corresponding scalar $u(x, y, z)$. Thus looking at the region $\Omega$ in space and conceiving of actually labeling each point of $\Omega$ with the corresponding scalar (number), one obtains a picture consisting of a field of scalars, hence the name scalar field, for $u$. For example, in physical terms, $u(x, y, z)$ could represent the temperature at the point $(x, y, z)$. Often, the scalar fields of interest depend on a fourth variable, the time $u = u(x, y, z, t)$, and are called time-dependent scalar fields.

Analogously, a vector field is a function $E : \Omega \subseteq \mathbb{R}^3 \to \mathbb{R}^3$ that assigns to each point $(x, y, z)$ in $\Omega$ a vector $E(x, y, z)$. Thus

$$E(x, y, z) = \left( E^1(x, y, z), E^2(x, y, z), E^3(x, y, z) \right)$$

is the representation of $E$ in terms of its components $E^1$, $E^2$, and $E^3$, each of which is itself a scalar field. Pictorially, one can conceive of plotting the vector

$E(x, y, z)$ with initial point at $(x, y, z)$, and with $(x, y, z)$ varying over $\Omega$, one obtains a whole field of vectors emanating from the various points of $\Omega$. This is illustrated hypothetically in Figure 1.1. A typical physical example is when $E$ represents a static electric field on $\Omega$, so that $E(x, y, z)$ is the the electric force (per unit charge) at the point $(x, y, z)$. As before, our vector fields sometimes depend on the time $E = E(x, y, z, t)$, and in this case they are often referred to as time-dependent vector fields. It is more or less traditional not to make a big distinction here, and to let the scalar and vector fields be understood as static (not depending on $t$) or time-dependent from the context or example at hand.

You should recall the notions of continuity and differentiability of scalar fields from calculus. For a scalar field $u$, we will use interchangeably the standard notations for its partial derivatives. Thus with respect to the variable $x$, we use $\partial u / \partial x = u_x$ for the scalar field defined by

$$\frac{\partial u}{\partial x}(x, y, z) = \lim_{h \to 0} \frac{u(x + h, y, z) - u(x, y, z)}{h},$$

for $(x, y, z) \in \Omega$ (assuming that this partial derivative exists). The notation for the higher-order partial derviatives is also as usual; for example,

$$u_{xy} = \frac{\partial^2 u}{\partial y \partial x}.$$

Partial derivatives of a vector field $E = (E^1, E^2, E^3)$ are defined naturally by applying the partial dervative operator to each component. For example,

$$\frac{\partial E}{\partial x} = \left( \frac{\partial E^1}{\partial x}, \frac{\partial E^2}{\partial x}, \frac{\partial E^3}{\partial x} \right),$$

$$\frac{\partial^2 E}{\partial y \partial x} = \left( \frac{\partial^2 E^1}{\partial y \partial x}, \frac{\partial^2 E^2}{\partial y \partial x}, \frac{\partial^2 E^3}{\partial y \partial x} \right).$$

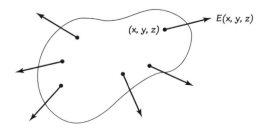

Figure 1.1.    Visualization of the vector field $E$.

There are four basic differential operators (besides the partial derivatives) connected with scalar and vector fields. Three of them, the gradient, curl, and divergence operators, are closely related. The fourth is the covariant derivative operator (with respect to a given vector field). These operators are defined as follows:

### The gradient

For a scalar field $u$, the gradient of $u$ is the vector field denoted by either grad($u$) or $\nabla u$ and defined by

$$\text{grad}(u) = \nabla u \equiv \left( \frac{\partial u}{\partial x}, \frac{\partial u}{\partial y}, \frac{\partial u}{\partial z} \right).$$

### The curl

For a vector field $E = (E^1, E^2, E^3)$, the curl of $E$ is the vector field denoted by curl($E$), or by $\nabla \times E$, and defined by

$$\text{curl}(E) = \nabla \times E \equiv \left( \frac{\partial E^3}{\partial y} - \frac{\partial E^2}{\partial z}, \frac{\partial E^1}{\partial z} - \frac{\partial E^3}{\partial x}, \frac{\partial E^2}{\partial x} - \frac{\partial E^1}{\partial y} \right)$$

$$= \begin{vmatrix} \mathbf{i} & \mathbf{j} & \mathbf{k} \\ \frac{\partial}{\partial x} & \frac{\partial}{\partial y} & \frac{\partial}{\partial z} \\ E_1 & E_2 & E_3 \end{vmatrix}.$$

### The divergence

For a vector field $E$, the divergence of $E$ is the scalar field denoted by div($E$), or by $\nabla \cdot E$, and defined by

$$\text{div}(E) = \nabla \cdot E \equiv \frac{\partial E^1}{\partial x} + \frac{\partial E^2}{\partial y} + \frac{\partial E^3}{\partial z}.$$

The notation grad, curl, div for the three basic operators is used mainly by mathematicians, while the alternative notation $\nabla$, $\nabla\times$, $\nabla\cdot$ for the same operators is preferred for the most part by physicists. This latter notation is very suggestive if one wishes to think of $\nabla$ as a vector; namely, just write symbolically, $\nabla = \left( \frac{\partial}{\partial x}, \frac{\partial}{\partial y}, \frac{\partial}{\partial z} \right)$.

### Geometric interpretation of the gradient

We will have immediate need of the geometrical significance of the gradient $\nabla u$ of a scalar field $u$. There are two aspects of this, which you probably studied in calculus, and each is connected with the chain rule:

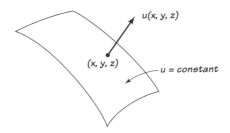

Figure 1.2. The gradient at $(x, y, z)$ is perpendicular to the level surface through $(x, y, z)$.

(1) The gradient vector $\nabla u(x, y, z)$ is perpendicular to the level surface of $u$ through the point $(x, y, z)$. See Figure 1.2. To see this, recall one version of the chain rule: If $\alpha : I \to \mathbb{R}^3$ is a curve in $\mathbb{R}^3$ that lies in the domain of $u$, then (assuming that $u$ and $\alpha$ are differentiable) the derivative of the composite function is given by

$$\frac{d}{dt}[u \circ \alpha](t) = \nabla u(\alpha(t)) \cdot \alpha'(t),$$

for all $t$ in the interval $I$. Specializing the curve $\alpha$ to a curve on the level surface through $(x, y, z)$, we can easily prove this assertion about this gradient (exercise).

(2) The gradient vector $\nabla u(x, y, z)$ points in the direction of greatest increase in $u$ at $(x, y, z)$. To make this more precise, recall the notion of *directional derivative* of $u$ at $(x, y, z)$ in the direction of $v = (v_1, v_2, v_3)$, where $v$ is any unit vector ($|v| = 1$). The curve

$$\alpha(t) = (tv_1 + x, tv_2 + y, tv_3 + z)$$

parametrizes a straight line through $(x, y, z)$ in the direction of $v$, and the composite function $f(t) \equiv u(\alpha(t))$ records the values of $u$ along this line. The rate of change $f'(0)$ gives, by definition, the directional derivative $\nabla_v u(x, y, z)$. By the chain rule, this reduces to

$$\nabla_v u(x, y, z) \equiv \frac{d}{dt}[u \circ \alpha](0) = \nabla u(\alpha(0)) \cdot \alpha'(0)$$

$$= \nabla u(x, y, z) \cdot v$$

$$= |\nabla u(x, y, z)| \cos \theta.$$

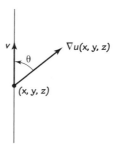

Figure 1.3.    The directional derivative $\nabla_v u$ of $u$ in the direction $v$.

Here $\theta$ is the angle between the vectors $\nabla u(x, y, z)$ and $v$. This last identity is illustrated in Figure 1.3 and shows that the greatest directional derivative $\nabla_v u(x, y, z)$ occurs when $\theta = 0$, i.e., when $v$ is in the direction of the gradient vector $\nabla u(x, y, z)$. This establishes the second assertion. We mention here that the notation $\nabla_v u$ is meant to suggest the relationship between the directional derivative and the covariant derivative (introduced next). Also, if $n = (n_1, n_2, n_3)$ is a unit normal vector to some surface, then

$$\nabla_n u = \nabla u \cdot n = \frac{\partial u}{\partial n}$$

is an alternative notation that is commonly used; it is called a *normal derivative*. Generally, we will just use $\nabla u \cdot n$ for the directional derivative of $u$ in the direction $n$.

(3) What has been said above in (1) and (2) about the gradient of a function of three variables also applies to a function of two variables. Thus suppose $u = u(x, y)$ is a function of two variables. Its gradient

$$\nabla u = \left( \frac{\partial u}{\partial x}, \frac{\partial u}{\partial y} \right)$$

is perpendicular to each level curve

$$u(x, y) = c$$

of $u$ and points in the direction of greatest increase of $u$.

### The covariant derivative

Suppose

$$v = v(x, y, z) = \left(v^1(x, y, z), v^2(x, y, z), v^3(x, y, z)\right)$$

is a vector field on $\mathbb{R}^3$, and $f = f(x, y, z)$ is a scalar field. Then the *covariant derivative of $f$ along $v$* is the scalar field

$$\nabla_v f = \nabla f \cdot v = v^1 \frac{\partial f}{\partial x} + v^2 \frac{\partial f}{\partial y} + v^3 \frac{\partial f}{\partial z}. \tag{1.4}$$

In addition to covariant derivatives of scalar fields, we also need to take covariant derivatives of vector fields. Thus, supposing $E = E(x, y, z) = \left(E^1(x, y, z), E^2(x, y, z), E^3(x, y, z)\right)$ is another vector field, then the *covariant derivative of $E$ along $v$* is the vector field $\nabla_v E$ defined by

$$\nabla_v E \equiv \left(\nabla_v E^1, \nabla_v E^2, \nabla_v E^3\right) \tag{1.5}$$

$$= E'v \tag{1.6}$$

$$= v^1 \frac{\partial E}{\partial x} + v^2 \frac{\partial E}{\partial y} + v^3 \frac{\partial E}{\partial z}. \tag{1.7}$$

Here and in the sequel we are using the prime "*′*" to denote the derivative of vector-valued functions. Its operation on a vector field $E$ yields the corresponding *Jacobian matrix*

$$E' = \begin{bmatrix} E^1_x & E^1_y & E^1_z \\ E^2_x & E^2_y & E^2_z \\ E^3_x & E^3_y & E^3_z \end{bmatrix}. \tag{1.8}$$

Equation (1.6) follows from the definition in equation (1.5), since the Jacobian matrix $E'$ has rows $\nabla E^1$, $\nabla E^2$, $\nabla E^3$, and the product in (1.6) is that of the matrix $E'$ applied to the vector $v$. The other expression for the covariant derivative $\nabla_v E$, given in (1.7), follows from properties of matrix multiplication and the fact that the columns of the Jacobian matrix $E'$ are $\partial E/\partial x$, $\partial E/\partial y$, and $\partial E/\partial z$.

In the case when $v$, $E$, $f$ depend on the time $t$, the definitions of the covariant derivatives given above still apply if we consider $t$ as fixed.

## Exercises 1.1

1. (*The Laplacian*) Show that for any scalar field $u$ and any vector field $E$, one has the identities

$$\text{curl}(\nabla u) = 0,$$

$$\text{div}(\text{curl}(E)) = 0,$$

$$\text{div}(\nabla u) = \frac{\partial^2 u}{\partial x^2} + \frac{\partial^2 u}{\partial y^2} + \frac{\partial^2 u}{\partial z^2}.$$

This last identity gives rise to an important operator, called the *Laplacian* and denoted by $\nabla^2$, since $\text{div}(\nabla u) = \nabla \cdot (\nabla u)$, or also by $\Delta$. Thus, the Laplacian of the scalar field $u$ is the scalar field

$$\nabla^2 u \equiv \frac{\partial^2 u}{\partial x^2} + \frac{\partial^2 u}{\partial y^2} + \frac{\partial^2 u}{\partial z^2} = u_{xx} + u_{yy} + u_{zz}.$$

2. Another useful identity comes from a repeated application of the curl operator. Namely, show that for any vector field $A$,

$$\text{curl}(\text{curl}(A)) = \nabla(\text{div}(A)) - \nabla^2 A.$$

Here $A = (A^1, A^2, A^3)$ and $\nabla^2 A \equiv (\nabla^2 A^1, \nabla^2 A^2, \nabla^2 A^3)$.

3. (a) Show by direct computation that

$$\nabla^2 \left( \frac{1}{r} \right) = 0,$$

where $r \equiv (x^2 + y^2 + z^2)^{1/2}$. This says that $1/r$ is a solution of Laplace's equation $\nabla^2 u = 0$. We will study this PDE in more detail later.

   (b) Show by direct computation that the function

$$u = g(x, y, z, t) \equiv (2at)^{-3/2} \exp\left( -\frac{x^2 + y^2 + z^2}{4at} \right)$$

satisfies the heat equation

$$\frac{\partial u}{\partial t} = a\nabla^2 u.$$

4. Prove the following identities. Here $f$, $g$ are scalar fields and $u$, $v$, $w$ are vector fields. Note that identities (a)–(f) exhibit how the various operators behave with respect to products, much like the *product rule* for ordinary derivatives. Also, (e) is a special case of (f), and you might want to use it in proving (f).

(a) $\mathrm{div}(fv) = \nabla f \cdot v + f\,\mathrm{div}(v)$.

(b) $\mathrm{curl}(fv) = \nabla f \times v + f\,\mathrm{curl}(v)$.

(c) $\nabla_v(fw) = (\nabla_v f)w + f\nabla_v w$.

(d) $\nabla_v(u \cdot w) = \nabla_v u \cdot w + u \cdot \nabla_v w$.

(e) $\frac{\partial}{\partial x_i}(u \times w) = \frac{\partial u}{\partial x_i} \times w + u \times \frac{\partial w}{\partial x_i}$.

(f) $\nabla_v(u \times w) = (\nabla_v u) \times w + u \times (\nabla_v w)$.

(g) $\nabla_{fv} w = f\nabla_v w$.

(h) $\mathrm{curl}(u \times w) = \nabla_w u - \nabla_u w + \mathrm{div}(w)u - \mathrm{div}(u)w$.

(i) $\nabla_v v = \nabla(\frac{|v|^2}{2}) + \mathrm{curl}(v) \times v$.

(j) $\mathrm{div}(u \times w) = \mathrm{curl}(u) \cdot w - u \cdot \mathrm{curl}(w)$.

(k) $\mathrm{div}(\nabla f \times \nabla g) = 0$.

5. Prove that $\nabla u(x, y, z)$ is perpendicular to the level surface of $u$ that passes through $(x, y, z)$.

6. Show that each of the following functions satisfies the heat equation $u_t = \alpha^2 \nabla^2 u$:

   (a) $u = e^{-14\alpha^2\pi^2 t} \sin \pi x \cos 3\pi y \sin 2\pi z$.

   (b) $u = e^{29\alpha^2\pi^2 t} \cosh 3\pi x \sinh 2\pi y \cosh 4\pi z$.

   (c) $u = \exp(29\alpha^2\pi^2 t + \pi(3x + 2y + 4z))$.

   (d) $u = x^2 + y^2 - 2z^2 - 3x - 5y + 6z + 1$.

   (e) $u = \cos 3\pi x \sin 4\pi y \sinh 5\pi z$.

7. Laplace's equation

$$\nabla^2 u = 0$$

is an important example of a partial differential equation, one that we will study in detail later. You perhaps noted in the last exercise the general fact that any (time-independent) solution $u$ of Laplace's equation is also a solution of the heat equation $u_t = \alpha^2 \nabla^2 u$. Show that each of the following functions is a solution of Laplace's equation:

   (a) $u = 4z^2 y - x^2 y - y^3$.

   (b) $u = x^3 - 3xy^2$.

   (c) $u = x^2 - y^2$.

   (d) $u = \ln(x^2 + y^2) + x - x(x^2 + y^2)^{-1}$

8. In addition to scalar fields and vector fields, many areas of modern mathematics and physics employ the notion of a tensor field. This concept probably first arose in continuum mechanics, which uses the notion of the stress tensor $T$ associated with the continuum. The stress tensor is a typical example of what is known as a second rank *tensor field* on $\mathbb{R}^3$. Such a tensor field $T = T(x, y, z)$ is, by definition, a matrix-valued function of three variables:

$$T(x, y, z) = \begin{bmatrix} T^{11}(x, y, z) & T^{12}(x, y, z) & T^{13}(x, y, z) \\ T^{21}(x, y, z) & T^{22}(x, y, z) & T^{23}(x, y, z) \\ T^{31}(x, y, z) & T^{32}(x, y, z) & T^{33}(x, y, z) \end{bmatrix}.$$

Technically, $T : \Omega \to \mathcal{M}_3$ is a map defined on some domain of $\mathbb{R}^3$ with values in the set $\mathcal{M}_3$ of $3 \times 3$ matrices. Note that the $(i, j)$th entry of $T$, i.e., $T^{ij}$, is a scalar field, and viewed like this, the tensor field

$$T = T(x, y, z) = \{T^{ij}(x, y, z)\}_{i,j=1,2,3} \tag{1.9}$$

is just a matrix of scalar fields. On the other hand, denoting the three rows of $T$ by $T^1, T^2, T^3$, we can view $T$ in terms of its rows as

$$T = T(x, y, z) = (T^1(x, y, z), T^2(x, y, z), T^3(x, y, z)), \tag{1.10}$$

and each row is a vector field. In terms of its rows, the divergence of $T$ is defined to be

$$\operatorname{div}(T) = \left(\operatorname{div}(T^1), \operatorname{div}(T^2), \operatorname{div}(T^3)\right). \tag{1.11}$$

Likewise, the covariant derivative of $T$ is defined in terms of its rows:

$$\nabla_v T = (\nabla_v T^1, \nabla_v T^2, \nabla_v T^3). \tag{1.12}$$

Based on this, prove the following identities. For vector fields $v, w$,

(a) $\nabla_v(Tw) = (\nabla_v T)w + T(\nabla_v w)$.

(b) $\operatorname{tr}(v') = \operatorname{div}(v)$ (here tr denotes the trace of a matrix).

(c) $\operatorname{div}(\operatorname{div}(v)I) = \nabla \operatorname{div}(v)$.

(d) $\operatorname{div}(v') = \nabla^2 v$.

(e) $\operatorname{div}(v'^*) = \nabla \operatorname{div}(v)$, where $*$ denotes the transpose operation on matrices.

## 1.2.    General Systems of PDEs

Here we look at the general form for a system of PDEs, briefly and from a elementary point of view, i.e., without jet bundles, which afford the proper modern setting for an advanced formulation and study of PDEs. We do not intend to do much with the general theory and will concentrate on some very particular PDEs in the rest of the book. However, you should see how these fit into the general picture.

We start with simple scalar PDEs of low order and work up to systems of PDEs of arbitrary order.

From an elementary point of view the general *first-order scalar PDE* for a single unknown function $u = u(x, y, z)$ is an equation involving $u$, its first order partials, and given functions of $x, y, z$. This is written symbolically as

**First-Order PDE**

$$F(x, y, z, u, u_x, u_y, u_z) = 0. \tag{1.13}$$

Here $F = F(x, y, z, u, p, q, r)$ is a function of seven variables, and if we wish to be more precise, we could state that:

(1) The *domain* of $F$ (as a subset of $\mathbb{R}^7$) has the form: $J^1 = \Omega \times I \times \mathbb{R}^3$, where $\Omega \subseteq \mathbb{R}^3$ is an open set and $I$ is an open interval.

(2) A *solution* of equation (1.13) is a scalar field $\psi$ defined and differentiable on a domain $\Omega_0 \subseteq \Omega$ such that $\psi(x, y, z) \in I$, and

$$F\left(x, y, z, \psi(x, y, z), \frac{\partial \psi}{\partial x}(x, y, z), \frac{\partial \psi}{\partial y}(x, y, z), \frac{\partial \psi}{\partial z}(x, y, z)\right) = 0,$$

for all $(x, y, z) \in \Omega_0$.

For example, the PDE

$$uu_x + zu_y = 0$$

is a first-order scalar PDE, and one solution of this is

$$u = \psi(x, y, z) = xz(y - 1)^{-1}.$$

The form of the general *second-order scalar PDE* is analogous to the first-order case, except that now partials up to the second order occur. If we restrict our attention to the two-variable case ($x$ and $y$), the general form is

**Second-Order PDE**

$$F(x, y, u, u_x, u_y, u_{xx}, u_{xy}, u_{yy}) = 0.$$

To allow for a larger number of variables than $x$, $y$, $z$, and for more than one function $u$, we find it more convenient to use subscripts and vector notation. Thus for $n$ variables $x_1, \ldots, x_n$ we use the notation

$$x = (x_1, \ldots, x_n),$$

and if the PDE is a system of equations, each involving the scalar functions $u^1 = u^1(x), \ldots, u^p = u^p(x)$ of $n$ variables, then we let

$$u(x) = (u^1(x), \ldots, u^p(x)).$$

You can also think of $u$ as a vector-valued function defined on some domain $\Omega$ of Euclidean $n$-space: $u : \Omega \subseteq \mathbb{R}^n \to \mathbb{R}^p$. Then, a $k$th-*order system of PDEs* has the general form

**$k$th-Order PDE**

$$F^r(x, u, u^{(1)}, u^{(2)}, \ldots, u^{(k)}) = 0, \tag{1.14}$$

for $r = 1, \ldots, q$. This is a system of $q$ equations for the $p$ unknown scalar functions $u^a = u^a(x)$, $a = 1, \ldots, p$. For the partial derivatives we have used the notation

$$u^{(1)} = \left\{ \frac{\partial u^a}{\partial x_i} \right\}_{i=1,\ldots,n}^{a=1,\ldots,p},$$

$$u^{(2)} = \left\{ \frac{\partial^2 u^a}{\partial x_i \partial x_j} \right\}_{i \leq j}^{a=1,\ldots,p},$$

$$\vdots$$

$$u^{(k)} = \left\{ \frac{\partial^k u^a}{\partial x_{i_1} \cdots x_{i_k}} \right\}_{i_1 \leq \cdots \leq i_k}^{a=1,\ldots,p}.$$

There is a special case of the general $k$th-order system (1.14) that has been well studied and is by all means the simplest type of PDE. This is known as a $k$th-*order linear system* and has the form

### *k*th-Order Linear PDE

$$\sum_{i_1 \leq \cdots \leq i_k} A_{i_1 \cdots i_k}(x) \frac{\partial^k u}{\partial x_{i_1} \cdots \partial x_{i_k}} + \cdots + \sum_{i=1}^{n} A_i(x) \frac{\partial u}{\partial x_i} + A(x)u + b(x) = 0. \quad (1.15)$$

Here $b(x) \in \mathbb{R}^q$, and the coefficients of $u$ and its partial derivatives are matrices: $A(x), A_i(x), \ldots, A_{i_1 \cdots i_k}(x)$ are $q \times p$ matrices. Usually, one only considers the case where $u$ is a scalar function ($p = 1$) (and there is only one equation, $q = 1$), in which case the above coefficients (and $b(x)$ as well) are scalars. The $k$th-order linear system is called *homogeneous* if $b(x) = 0$ for all $x$. In the case where the coefficients do not depend on $x$, the system is called a *constant-coefficient* system. Linear homogeneous systems have the following important property, which is easily verified:

### Superposition Principle

If $u_1, \ldots, u_m$ are solutions of a given homogeneous, linear PDE, then so is

$$u \equiv c_1 u_1 + \cdots + c_m u_m,$$

for any choice of constants $c_1, \ldots, c_m$.

As we have mentioned, there is an enormous amount of theory that has been developed for the general nonlinear system of PDEs (8), as well as for the general linear system (1.15). From time to time throughout this book we will mention various aspects of this theory, with citations for further study.

## 1.3.   The Main Examples of PDEs

There is a large number of very particular PDEs that have arisen historically in physics and the other sciences and that have been studied and analyzed by a variety of techniques. In this book we concentrate on a few of the the most prominent of these PDEs. These will be:

**The Heat Equation:**   $u_t - \alpha^2 \nabla^2 u = F$.

**Poisson's Equation:**   $\nabla^2 u = F$.

**The Wave Equation:**   $u_{tt} - c^2 \nabla^2 u = F$.

In these PDEs, $\alpha$ and $c$ are constants and $F$ is a given function. Each of these equations is a second-order linear PDE that is non-homogeneous when $F \neq 0$.

Note that the homogeneous version ($F = 0$) of Poisson's equation is also called Laplace's equation. These three equations are the main examples of what are known as *parabolic, elliptic*, and *hyperbolic* PDEs respectively. There are more general versions of these PDEs that model more complicated physical situations (like the heat equation with convection), but we will concentrate on these basic versions for the most part.

**Maxwell's Equations**

$$\operatorname{div}(E) = 4\pi\rho,$$

$$\operatorname{curl}(E) + \frac{1}{c}\frac{\partial B}{\partial t} = 0,$$

$$\operatorname{curl}(B) - \frac{1}{c}\frac{\partial E}{\partial t} = \frac{4\pi}{c}J,$$

$$\operatorname{div}(B) = 0.$$

In these equations, $\rho$ and $J = (J^1, J^2, J^3)$ are given scalar and vector fields respectively, while $E = (E^1, E^2, E^3)$ and $B = (B^1, B^2, B^3)$ are the unknowns. Thus Maxwell's equations actually comprise a system of eight first-order linear PDEs for the six (unknown) component functions $E^1, E^2, E^3, B^1, B^2, B^3$. Written out without using vector notation, Maxwell's equations are

$$E_x^1 + E_y^2 + E_z^3 = 4\pi\rho,$$

$$E_y^3 - E_z^2 + c^{-1}B_t^1 = 0,$$

$$E_z^1 - E_x^3 + c^{-1}B_t^2 = 0,$$

$$E_x^2 - E_y^1 + c^{-1}B_t^3 = 0,$$

$$B_y^3 - B_z^2 - c^{-1}E_t^1 = 4\pi c^{-1}J^1,$$

$$B_z^1 - B_x^3 - c^{-1}E_t^2 = 4\pi c^{-1}J^2,$$

$$B_x^2 - B_y^1 - c^{-1}E_t^3 = 4\pi c^{-1}J^3,$$

$$B_x^1 + B_y^2 + B_z^3 = 0.$$

It is good (for pedagogical reasons) to exhibit explicitly the eight PDEs comprising Maxwell's equations as we have just done, but the preferred version of the PDEs is the vector version, not just because it is more compact, but primarily because it makes the study of the structure of Maxwell's equations easier.

The unknowns $E$ and $B$ in Maxwell's equations represent an electric field and magnetic field respectively. There are actually many different forms of Maxwell's equations, depending on the physical conditions, and the above version we've chosen is a basic one that will illustrate most of the features involved in such a system of linear PDEs.

## The Navier–Stokes Equations

$$\frac{\partial \rho}{\partial t} + \nabla_v \rho = 0,$$

$$\rho \left[ \frac{\partial v}{\partial t} + \nabla_v v \right] = \rho F - \nabla p + \mu \nabla^2 v,$$

$$\operatorname{div}(v) = 0.$$

This is a *nonlinear* second order system of PDEs that models the motion of a viscous fluid. The $\rho$ is a scalar field representing the mass density of the fluid, $p$ is the pressure function, and $v = (v^1, v^2, v^3)$ is a vector field representing the velocity of the fluid. In the second equation, $\mu$ is a constant (viscosity constant), and when $\mu = 0$, the Navier–Stokes equations reduce to what are known as the *Euler equations*. Written out fully, the Navier–Stokes equations comprise a system of five PDEs:

$$\rho_t + \rho_x v^1 + \rho_y v^2 + \rho_z v^3 = 0,$$

$$\rho[v_t^1 + v_x^1 v^1 + v_y^1 v^2 + v_z^1 v^3] = \rho F^1 - p_x + \mu \nabla^2 v^1,$$

$$\rho[v_t^2 + v_x^2 v^1 + v_y^2 v^2 + v_z^2 v^3] = \rho F^2 - p_y + \mu \nabla^2 v^2,$$

$$\rho[v_t^3 + v_x^3 v^1 + v_y^3 v^2 + v_z^3 v^3] = \rho F^3 - p_z + \mu \nabla^2 v^3,$$

$$v_x^1 + v_y^2 + v_z^3 = 0,$$

## Linear Elasticity Equations

$$\rho \frac{\partial^2 u}{\partial t^2} = \rho F + \mu \nabla^2 u + (\lambda + \mu) \nabla \operatorname{div}(u).$$

Here $u = (u^1, u^2, u^3)$ is a time-dependent vector field, $\rho$ is given, and $\lambda, \mu$ are constants. The linear elasticity equations model the propagation of vibrational waves in elastic solids. These equations are also known as the *Navier equations*.

## 1.4.   The Flow Generated by a Vector Field

We end the Introduction with a discussion of a very important topic: the flow generated by a vector field. This concept, besides being essential in many areas of mathematics, like ordinary differential equations and differential geometry is instrumental in formulating the theories of continuum mechanics and fluid mechanics. It also serves to visualize the heat flow lines in heat problems and the field lines in electricity and magnetism.

In a sense, the study of flows generated by vector fields is nothing other than the study of solutions of systems of ordinary differential equations (ODEs). See [AM 78], [Arn 78], [Per 91], [Ver 90]. Any such system, in normal form, can be reduced to a corresponding first-order system, and this latter system is one that is associated with an appropriate vector field. The discussion here is for vector fields on domains in $\mathbb{R}^n$, although dimensions $n = 2$ and $n = 3$ will be the usual applications.

To allow for a larger number of variables than $x, y, z$, we again find it convenient to use subscripts:

$$x = (x_1, \ldots, x_n)$$

for a point $x \in \mathbb{R}^n$.

Suppose that $v$ is a time-dependent vector field

$$v(x, t) = \left( v^1(x, t), \ldots, v^n(x, t) \right)$$

on a region $\Omega$ in $\mathbb{R}^n$ with interval of times $I$, that is, $(x, t) \in \Omega \times I$. The only requirement is that $v^i$ and its partials $\partial v^i / \partial x_j$ be continuous functions. For convenience, we assume that the interval of times has the form $I = (-T, T)$, with $0 < T \leq \infty$.

Associated to each such vector field $v$ is a system of $n$ ordinary differential equations

$$x'_1 = v^1(x_1, \ldots, x_n, t),$$

$$\vdots \tag{1.16}$$

$$x'_n = v^n(x_1, \ldots, x_n, t),$$

where the prime "$'$" denotes differentiation with respect to $t$. A *solution* of the system (1.16) is, by definition, a differentiable, vector-valued function $\gamma : J \to \mathbb{R}^n$ (i.e., a smooth curve in $\mathbb{R}^n$), with $J \subseteq I$, that satisfies

(1) $\gamma(t)$ lies in $\Omega$ for all $t$ in $J$.

(2) The component functions $\gamma_1, \ldots, \gamma_n$ satisfy

$$\gamma_1'(t) = v^1\,(\gamma_1(t), \ldots, \gamma_n(t),\, t)\,,$$

$$\vdots \tag{1.17}$$

$$\gamma_n'(t) = v^n\,(\gamma_1(t), \ldots, \gamma_n(t),\, t)\,,$$

for all $t$ in $J$.

It will often be convenient to write the condition (1.17) in vector form:

$$\gamma'(t) = v(\gamma(t), t), \tag{1.18}$$

where $\gamma(t) = (\gamma_1(t), \ldots, \gamma_n(t))$.

The solutions of the system (1.16) are also known as *integral curves* of the vector field $v$, and equation (1.18) says geometrically that each integral curve has its tangent vector at a given point coincide with the value of $v$ at the point. The system (1.16) is called a 1st order, nonautonomous system of DE's in normal form.

**Theorem 1.1 (Existence and Uniqueness Theorem)**    *Assume that $v^i$ and $\partial v^i/\partial x_j$, for $i, j = 1, \ldots, n$, are continuous functions on $\Omega \times I$. Then for each $p$ in $\Omega$ and $t_0$ in $I$, there is a unique solution $\gamma : J \to \mathbb{R}^n$ of the system (1.18) with $t_0 \in J$ and*

$$\gamma(t_0) = p. \tag{1.19}$$

Generally, the existence and uniqueness theorem does not guarantee that the interval $J$ on which the solution $\gamma$ is defined is the same as $I$. However, it does give us a maximum interval on which the solution exists (we take $t_0 = 0$ from now on):

**Corollary 1.1 (Maximum Interval of Existence)**    *For each $p \in \Omega$, there is an interval $I_p \subseteq I$ and a solution $\gamma_p : I_p \to \mathbb{R}^n$ of the system (1.16) that satisfies*

$$\gamma_p(0) = p$$

*and that also has the property that $I_p$ contains the interval of definition of any other solution that passes through $p$ at time $t = 0$.*

This leads directly to the definition of the flow:

**Definition 1.1 (The Flow)**   Let

$$\mathcal{D} = \{ (x, t) \mid x \in \Omega, \text{ and } t \in I_x \}.$$

The flow $\phi$ generated by $v$ is the function $\phi : \mathcal{D} \to \Omega$ defined by

$$\phi(x, t) = \gamma_x(t), \tag{1.20}$$

for $x \in \mathcal{D}$. Here $\gamma_x$ denotes the maximal integral curve that passes through $x$ at time 0 (see the above corollary). Otherwise stated, for each $x \in \Omega$, one has

$$\frac{\partial \phi}{\partial t}(x, t) = v\left(\phi(x, t), t\right), \tag{1.21}$$

$$\phi(x, 0) = x, \tag{1.22}$$

for all $t \in I_x$.

In terms of components,

$$\phi(x, t) = \left(\phi^1(x, t), \ldots, \phi^n(x, t)\right).$$

The notation $\phi(x, t)$ explicitly indicates that the integral curve depends on the initial condition $x$. It is possible to show that this dependence on the initial conditions is smooth in the sense that $\phi$, as a vector-valued function of four variables, is differentiable.

**Example 1.1**   For ease of visualization, consider the case $n = 2$ and suppose the vector field $v$ is

$$v(x_1, x_2) = (-\pi \cos \pi x_1 \sinh \pi x_2, -\pi \sin \pi x_1 \cosh \pi x_2),$$

defined for all $(x_1, x_2)$, but for pedagogical reasons we restrict $(x_1, x_2) \in \Omega = [0, 1] \times [0, 1]$. The corresponding system of DE's that gives rise to the flow is

$$x_1' = -\pi \cos \pi x_1 \sinh \pi x_2,$$

$$x_2' = -\pi \sin \pi x_1 \cosh \pi x_2.$$

A picture of various integral curves (flow lines) for this system is shown in Figure 1.4. As we shall see later, the flow lines in this example represent the paths of heat flow (from hot to cold) in a square metal plate.

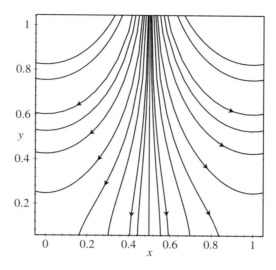

Figure 1.4.  A few of the flow lines generated by the vector field $v$ from Example 1.1.

**Remark**  In doing the exercises, here and elsewhere, you will need a computer. You can draw the phase portraits (pictures of the flow lines) in *Maple* if you wish, but at present this software package lacks the feature of exhibiting each flow line as it is being drawn. Thus you will not get a feeling for the dynamics of the flow. *Maple* does have, however, the full power of symbolic computation and programming capabilities that are lacking in most DE plotting software. *Phaser*, a computer software package for plotting flows dynamically, is good and simple but lacks *Maple*'s versatility (see [Ko 89]). The CD-ROM accompanying the text also has some special *Maple* code called DDEplot which will plot the flow lines dynamically.

The flow map $\phi$ is more than just a convenient notation for the integral curves of the system of DE's associated to $v$. It is rather a good conceptual tool for describing the kinematics of motion when $v$ is interpreted as the velocity vector field for, say, a fluid circulating in a tank. The discussion of this is as follows. We first introduce an alternative notation for the flow $\phi$:

$$\phi_t(x) \equiv \phi(x, t). \tag{1.23}$$

It is best to think of the case $n = 2$ or $n = 3$ in the following interpretation.

Suppose we take a point $x$ in $\Omega$ and follow the motion of a particle of the fluid that has this position at time zero ($\phi_0(x) = x$). Then the position of this particle at time $t$ is $\phi_t(x)$. Thus, over the time interval from 0 to $t$, the fluid particle traces out a portion of the integral curve as shown in Figure 1.5. Otherwise stated, one

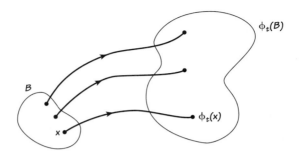

Figure 1.5.   The flow lines arising from the flow. These give rise to the motion and deformation of the part $B$ of $\Omega$ by the flow map over the interval of time from 0 to $t$.

views the reference point as *flowing* along the integral curve from the point $x$ to the point $\phi_t(x)$ (impelled by the motion). More generally, if we take a subregion $B$ of $\Omega$ and a time interval $J$ such that $B \times J \subseteq \mathcal{D}$, then applying the map $\phi_t$ to $B$ gives a new subregion,

$$\phi_t(B) \equiv \{\, \phi_t(x) \mid x \in B \,\}. \tag{1.24}$$

We can view each point $x$ in $B$ as flowing along its integral curve and ending up at $\phi_t(x)$ at time $t$. Thus the portion $B$ of the fluid moves and deforms into $\phi_t(B)$ under the action of the flow over the time interval from 0 to $t$. This is illustrated in Figure 1.5.

### Exercises 1.4

1. Use a computer to draw the phase portrait (family of integral curves), on the given domain $\Omega$, for each of the following vector fields:

   (a) (*Heat flow in a square plate*)

   $$v(x, y) = (-2\pi \cos 2\pi x \sinh 2\pi y, \ -2\pi \sin 2\pi x \cosh 2\pi y),$$

   on $\Omega = [0, 1] \times [0, 1]$.

   (b) (*A perturbed pendulum*)

   $$v(x, y, t) = (y, \ -\sin x + 0.2 \sin(2x - 0.1t)),$$

   on $\Omega = [-10, 10] \times [-5, 5]$. Here $x$ represents the angular displacement of a pendulum from equilibrium and $y$ its angular velocity. Plot the phase portrait in the $x$-$y$ plane (no $t$-axis). Each integral curve traces out an intricate pattern. If $t$ runs for a long time, does the pattern repeat?

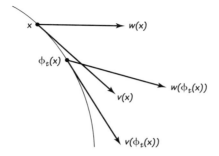

Figure 1.6.   Geometric significance of the covariant derivative $\nabla_v w$ of a vector field $w$ along a vector field $v$.

(c) (*Linear vector field*)

$$v(x, y) = (3x - 2y, -2x),$$

on $\Omega = [-10, 10] \times [-10, 10]$.

(d) (*Linear vector field*)

$$v(x, y) = (x - 2y, \ 2x + y),$$

on $\Omega = [-10, 10] \times [-10, 10]$.

(e) (*Predator–prey model*)

$$v(x, y) = (x - xy, \ xy - y),$$

on $\Omega = [0, 6] \times [0, 6]$. The system of differential equations for this vector field models the population fluctuations over time of two species: $x$ is the prey, $y$ the predators.

(f) (*Rossler's equations*)

$$v(x, y, z) = (-y - z, \ x + 0.2y, \ 0.2 + z(x - 4.2)),$$

on $\Omega = [-10, 10] \times [-10, 10] \times [-10, 10]$. Use several views. Try initial points $(0.5, 0.5, 0)$, $(1, 1, 0)$, $(2, 2, 0)$ and let $t$ go from 0 to 200.

2. For vector fields $v$, $w$, interpret the covariant derivative $\nabla_v w$ as the rate of change of $w$ along the flow $\phi$ generated by $v$. Specifically, show that

$$\left. \frac{d}{dt} w(\phi_t(x)) \right|_{t=0} = (\nabla_v w)(x),$$

for $x \in \Omega$. See Figure 1.6.

# 2 Derivation of the Heat Equation

This chapter is devoted to a derivation of the heat equation (in three dimensions) from basic physical principles. Actually, there are several different versions of this, depending upon the underlying assumptions. The heat equation we derive here is the scalar PDE

$$u_t = \alpha^2 \nabla^2 u + F,$$

where $\alpha^2$ is a constant and $F = F(x, y, z, t)$ is a given function. By specializing the 3-D heat equation one obtains the 2-D and 1-D heat equations, which are rather artificial physical situations, but make perfectly good sense mathematically and afford some simpler PDEs to deal with.

## 2.1. Heat Flow: Fourier's Law

To derive the equations that model the heat flow in a region of space, we need first to discuss a few naive notions about what heat is. Heat is measured in calories, and for a gas (system of molecules) in equilibrium in a closed container (for example), heat is a measure of the average kinetic energy $m|v|^2/2$ of the molecules in the gas ($v$ is the velocity of a molecule). Thus hotter gases are composed of molecules with higher individual kinetic energies, i.e., higher velocities when the masses are the same, while cold gases have their molecules in relative slow motion. Adding heat to the system is a means of increasing the kinetic energies of the individual molecules. For substances like liquids and solids, where the molecules are more tightly bound together, the same ideas apply, except that the motion of the individual molecules is more localized; and for solids with a rigid crystalline structure, the heat in the system consists primarily of vibrational energy.

Relative hotness or coldness of a substance, say a gas, is measured in degrees and is recorded by immersing a thermometer in the gas, allowing the mercury in the thermometer to come to equilibrium with molecules in the gas, and reading the scale. We commonly use the same method to record the temperature of a human body (a very complex system of molecules, indeed). Units of temperature (degrees) and heat (calories) are related by the definition, *One calorie is the amount of heat required to increase the temperature of one gram of water (at one atmosphere of pressure) by one degree Celsius.*

Now consider some basic notions about how heat flows, say in a region of space $\Omega$ occupied by a uniform gas, with $u(x, y, z, t)$ giving the temperature at each point $(x, y, z)$ at time $t$. If the gas is in equilibrium, the velocities of the molecules in various parts of the region $\Omega$ are the same (on average), and the temperature is constant, that is, $u$ is constant in $\Omega$. When the gas is not in equilibrium, some subregions of $\Omega$ have molecules flying about with greater velocities than in other subregions; and by colliding with nearby slow-moving molecules, the faster-moving molecules transfer some of their velocity, and this action propagates from hot subregions (with fast-moving molecules) to cold regions (with slow-moving molecules). In conventional terms, *heat flows from hot to cold.* (This is so despite the sensation that it flows the other way when you step on the cold ceramic tile of the bathroom floor with your warm bare feet in the morning.) In mathematical terms, $\nabla u$ is the direction of greatest increase of temperature (i.e., it points toward the hotter regions), and so $-\nabla u$ gives the direction of greatest heat flow at each point in the region $\Omega$.

**Definition 2.1 (Heat Flux Vector)**   If $u = u(x, y, z, t)$ is a temperature distribution on the region $\Omega$, then the (time-dependent) vector field

$$v = -\nabla u$$

is called the *heat flux vector field* for $u$. The flow generated by the heat flux vector field $v$ gives the heat flow lines.

Based on the above ideas, one can quantify the amount of heat that flows across a given surface $S$ in $\Omega$. This requires a natural postulate known as Fourier's law. Thus suppose $S$ is a planar surface in the region $\Omega$, small enough so that $\nabla u$ is constant (same direction and length) at all points of $S$. For a given unit vector $n$ perpendicular to $S$, one postulates

**Fourier's Law**

$$\{\text{heat flux across } S \text{ in the direction } n\} = -\kappa(\nabla u \cdot n)A. \qquad (2.1)$$

Here $A$ is the area of the planar surface $S$, and $\kappa \geq 0$ is the thermal conductivity of the gas (or other material), measured in units cal/(cm-sec-deg). One assumes that $\kappa \geq 0$ and that it generally varies from point to point in $\Omega$ and depends on the time as well. (However, the simplest case is when $\kappa$ is constant.) Note also that heat flux is measured in units cal/sec. Figure 2.1 illustrates geometrically why the minus sign is included on the right-hand side of Fourier's law. This will be important for us later.

Fourier's law above is stated for the simplest possible situation: a planar surface $S$ and $\nabla u$ constant on $S$. The more general situation of a curved surface $S$ (and no assumptions on $u$) requires a surface integral to express Fourier's law. If you do not know anything about surface integrals (as well as line and volume integrals), you can review the material in Appendix A if you wish. We can reason intuitively about the desired surface integral as follows. Let $\mathcal{F}$ be the flux (calories per second) of heat across $S$ at time $t$. By dividing $S$ up into small, almost planar, areas where $\nabla u$ is constant, then using the above law and adding up all the contributions to the total flux, one arrives at the following postulate:

**General Version of Fourier's Law**

$$\mathcal{F} = -\int_S \kappa(\nabla u \cdot n)\,dA. \tag{2.2}$$

This expression gives the heat flux, or net amount of heat that flows across $S$ from one side to another each second (the heat may be flowing in at some places and flowing out at others). The proper interpretation of $\mathcal{F}$ depends on the unit normal to the surface. For example, if the closed surface $S$ in Figure 2.2 is such that the unit normal $n$ is directed "outward" as shown, then $\mathcal{F}$ gives the (net) amount of *heat loss* each second by the region inside $S$. Note: Figure 2.2(a) depicts the situation where $\nabla u$ points toward the inside of $S$ at all points on $S$ (i.e., the inside of $S$ is hotter than the exterior). In this case, $-\kappa \nabla u \cdot n > 0$ at all points, and so $\mathcal{F} > 0$, giving a heat loss as expected, since heat is flowing out of the region bounded by $S$. Figure 2.2(b) shows the case where $\nabla u$ points outward at all points on $S$ (it is hotter outside $S$), and so $-\kappa \nabla u \cdot n < 0$ everywhere,

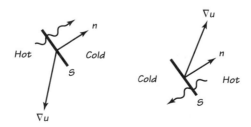

Figure 2.1.    Illustration of Fourier's law.

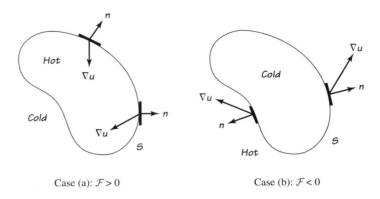

Case (a): $\mathcal{F} > 0$                    Case (b): $\mathcal{F} < 0$

Figure 2.2.    The heat flux $\mathcal{F}$ out of $B$.

giving that $\mathcal{F} < 0$. Thus we interpret a negative value of $\mathcal{F}$, the heat loss to the interior, as actually a heat gain to the interior. The two cases exhibited in Figure 2.2 are rather special, since in general, $\nabla u$ will be pointing toward the interior at some points on $S$ and toward the exterior at other points. Thus the interior region bounded by $S$ will be losing heat across the boundary at certain points and gaining heat across the boundary at other points.

## 2.2.   The Heat Equation (Without Convection)

The heat equation is a PDE for $u = u(x, y, z, t)$ describing how the temperature in the region $\Omega$ changes over time. In addition to Fourier's law, there are two other assumptions we need to make. The first is that for a small region with volume $dV$ in which the temperature is constant, the total heat inside the region at time $t$ is proportional to the mass times the temperature times the volume; i.e., total heat $= c\rho u \, dV$. Here $\rho = \rho(x, y, z, t)$ is the density function of the gas (or other substance) with units gm/cm$^3$, which is usually taken to be constant. Also, $c$ is the specific heat of the material, with units cal/(g-deg), and is usually constant as well. More generally, for an arbitrary subregion $B$ of $\Omega$, the total heat inside this region at time $t$ is given by the volume integral

$$\mathcal{H} = \int_B c\rho u \, dV. \tag{2.3}$$

A final assumption involves the possibility that heat is being generated in various regions of $\Omega$, called *heat sources*, or is possibly being extracted, from regions called *heat sinks*. Thus letting $f = f(x, y, z, t)$ represent the density function for this internally supplied/extracted heat, one assumes that

$$\mathcal{E} = \int_B f \, dV \tag{2.4}$$

gives the total heat added/subtracted from a subregion $B$ at time $t$ due to external sources/sinks. When $\mathcal{E} > 0$, it is viewed as a (net) amount of heat generated inside $B$ at time $t$, while when $\mathcal{E} < 0$, it is viewed as an amount of heat extracted from the region.

Using these concepts, we analyze the flow of heat in $\Omega$ as follows. Suppose we focus on an arbitrary subregion $B$ with bounding surface $S$ as shown in Figure 2.2. At each instant in time, the heat content inside $B$ is $\mathcal{H}$, and the increase in this heat content in going over to the next instant in time should be due to the heat gained due to flux inward across the bounding surface $S$ plus the heat added by sources inside $B$. Thus in terms of the above notation, one has $d\mathcal{H}/dt = -\mathcal{F} + \mathcal{E}$. In terms of the defining integrals, this equation explicitly is

$$\frac{d}{dt} \int_B c\rho u \, dV = \int_S \kappa (\nabla u \cdot n) dA + \int_B f \, dV \tag{2.5}$$

$$= \int_B \mathrm{div}(\kappa \nabla u) dV + \int_B f \, dV. \tag{2.6}$$

To get equation (2.6) from equation (2.5) we have used Gauss's divergence theorem to convert the surface integral over $S$ into a volume integral over $B$ (see Appendix A). It is important to note that the above equation for the rate of change of heat inside $B$ assumes that no heat enters or leaves $B$ due to convection. For example, if the region $\Omega$ is composed of a fluid *at rest* with various regions at different temperatures, then the above equations are valid. However, if the fluid is in motion, circulating inside $\Omega$, then there is convection; namely, the very motion of the fluid will convey the heat from one region to the next (see Problem 3 in Exercises B.1, Appendix B).

Next, let us rewrite equation (2.6) by carrying the derivative $d/dt$ under the integral sign on the left-hand side to get the equation $\int_B \partial(c\rho u)/\partial t \, dV = \int_B [\mathrm{div}(\kappa \nabla u) + f] dV$. Now, since $B$ is an arbitrary subregion of $\Omega$, we can take any point in $\Omega$, enclose it by smaller and smaller subregions $B$, and still have this equation hold. Thus in the limit, as volume$(B) \to 0$, the equation turns into the PDE

$$\frac{\partial}{\partial t}(c\rho u) = \mathrm{div}(\kappa \nabla u) + f. \tag{2.7}$$

This is the general form of the heat equation without convection. We can specialize this to obtain a simpler PDE by assuming that the heat capacity $c$, the material density $\rho$, and the conductivity $\kappa$ are all constants. Then equation (2.7) reduces to $c\rho u_t = \kappa \nabla^2 u + f$. Here we have used the identity $\text{div}(\nabla u) = \nabla^2 u$. Next, dividing through by $c\rho$ in the last equation gives the standard version of the heat equation:

**Heat Equation (Without Convection)**

$$u_t = \alpha^2 \nabla^2 u + F. \tag{2.8}$$

Here we have relabeled things for convenience: $\alpha^2 \equiv \kappa/(\rho c)$ is a constant called the *diffusivity of the material*, and $F \equiv f/(\rho c)$ is a scalar function called the *heat source density*.

## 2.3.   Initial Conditions and Boundary Conditions

Our study of the heat equation will show that it (as well as most other PDEs) has infinitely many solutions. However, if we specify some additional conditions that the temperature function $u$ must satisfy, then we will see that $u$ is uniquely determined. These conditions arise naturally from the physics of the situation.

We are interested in the way the temperature of the substance in the region $\Omega$ changes over time, and it is traditional to begin our observations of the temperature at time $t = 0$. Thus $u = u(x, y, z, t)$ should be defined for all $(x, y, z)$ in $\Omega$ and all $t \geq 0$, and satisfy the heat equation for all such points and $t > 0$. The initial condition on $u$ just says that we know the initial temperature $u(x, y, z, 0)$ at all points $(x, y, z)$ in the region $\Omega$:

**Initial Condition (IC)**

$$u = \phi \qquad \text{in } \Omega, \text{ for } t = 0.$$

Here $\phi = \phi(x, y, z)$ is a given function with domain $\Omega$, called the initial temperature distribution function, and the above notation for the initial condition is just shorthand for the more precise statement that $u(x, y, z, 0) = \phi(x, y, z)$ for all $(x, y, z)$ in $\Omega$.

The boundary conditions on $u$ say something about what is happening with the temperature and/or the flux of heat along the boundary of $\Omega$ for times $t > 0$. For convenience of notation we will let $\partial\Omega$ denote the bounding surface of $\Omega$. We work up to the most general form of the boundary conditions by first looking at some special cases:

**Boundary Conditions (BCs)**

**(O) No boundary:** In the case where $\Omega$ has no boundary, i.e., $\Omega = \mathbb{R}^3$, then naturally there are no boundary conditions. The evolution of the temperature is determined entirely by the the initial condition and the heat equation.

**(I) Fixed temperature on the boundary:** This is the condition that for all times $t > 0$, the temperature has a specified value $u(x, y, z, t) = g(x, y, z, t)$ at all points on the boundary, that is, for all $(x, y, z)$ on $\partial\Omega$. Here $g = g(x, y, z, t)$ is a given function (which in most cases for us will not depend on $t$). The abbreviated notation for this condition will be

$$u = g \qquad \text{on } \partial\Omega, \text{ for } t > 0.$$

Note: In order that the temperature be held fixed at the values specified by $g$, it is understood that some mechanism must be provided for extracting/adding heat at the boundary whenever any heat flows over to/away from the boundary.

**(II) Insulated boundary:** Saying that the boundary of $\Omega$ is insulated means that no heat can escape from or enter into $\Omega$ by flux across its boundary. In mathematical terms this is expressed by

$$\nabla u \cdot n = 0 \qquad \text{on } \partial\Omega, \text{ for } t > 0.$$

Here $n$ is the unit normal to the surface $\partial\Omega$, and this BC says that $n$ and $\nabla u$ are perpendicular. Another way of saying this is that $\nabla u$ is directed parallel to the boundary (i.e., lies in the tangent plane to the surface $\partial\Omega$) at each point $(x, y, z)$ on $\partial\Omega$. Thus there is no flux of heat outward or inward across the boundary.

**(III) Newton's law of cooling:** Suppose that $g = g(x, y, z, t)$, for $(x, y, z)$ on $\partial\Omega$, is a given function specifying the temperature of the external environment at each point on the boundary of $\Omega$, at each time $t > 0$, and consider a boundary condition of the form

$$-\kappa \nabla u \cdot n = h(u - g) \qquad \text{on } \partial\Omega, \text{ for } t > 0.$$

Here $h \geq 0$ is the heat exchange coefficient, with units cal/(sec-deg-cm$^2$). The boundary condition here is just the mathematical expression of *Newton's law of cooling*: The flux of heat per unit area, $-\kappa \nabla u \cdot n$, at each boundary point is proportional to the difference, $u - g$, in temperatures at

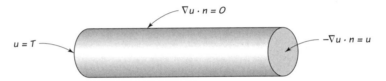

Figure 2.3.    Mixed boundary conditions.

the boundary. Thus the greater the difference in temperatures, the greater the flux of heat outward or inward across the boundary.

**(IV)  General boundary conditions:** The general type of boundary condition that we will use throughout the book is

$$\kappa \nabla u \cdot n + hu = g \qquad \text{on } \partial \Omega, \text{ for } t > 0.$$

Here $\kappa \geq 0, h \geq 0$ have the same meaning as above (the conductivity constant and heat exchange coefficient), and the assumption that these are nonnegative is motivated by the physics. From a mathematical point of view this assumption is not necessary, but its omission leads to unrealistic heat flows. Note that this general boundary condition contains BCs (I)–(III) above as special cases. Also note that $g$ and $h$ can vary from place to place on the boundary, and in many cases the boundary consists of several distinct pieces. Thus the general BC above really encompasses a whole variety of cases such as, for example, when parts of the boundary are insulated and other parts are held at a fixed temperature (see Figure 2.3).

## 2.4.  Initial-Boundary Value Problems

Now that we understand where the heat equation comes from — some of the physical and geometric ideas behind it — and have formulated some natural conditions for what happens initially and on the boundary, we need to explore, in the ensuing chapters, analytic methods for solving a given heat flow problem. The basic problem, then, is to determine the temperature function $u$ that satisfies

$$u_t = \alpha^2 \nabla^2 u + F \qquad \text{in } \Omega, \text{ for } t > 0,$$

$$\kappa \nabla u \cdot n + hu = g \qquad \text{on } \partial \Omega, \text{ for } t > 0,$$

$$u = \phi \qquad \text{in } \Omega, \text{ for } t = 0.$$

This is called an *initial-boundary value problem* (IBVP). The analytic solution of this problem, i.e.,, a description of the function $u$, can be quite involved unless the geometry of the region $\Omega$ is simple and the BCs are not too complicated. Even knowing the exact analytic solution $u$, giving the temperature at each point in $\Omega$ at each time $t$, may not be the end of what we want to know about the heat flow. Often we wish also to determine if the temperature in $\Omega$ is evolving toward some eventual temperature distribution, thus coming to a state of equilibrium. In order to speak precisely about this, we make the following definition:

**Definition 2.2 (Steady States)**    A time-independent function $S = S(x, y, z)$ is called a *steady state* for an IBVP if it satisfies the PDE and the BCs of the problem. If a solution $u = u(x, y, z, t)$ of an IBVP has a limit as $t \to \infty$, then the *eventual temperature distribution* is the function $u_\infty$ defined by

$$u_\infty(x, y, z) = \lim_{t \to \infty} u(x, y, z, t).$$

Note that in order for an IBVP to have any steady states, the BCs must not depend on the time, and likewise, the internal heat source/sink density $F$ must be time-independent. A given IBVP need not have an eventual temperature distribution, but when it does, we would expect it to coincide with one of the steady states of the problem.

**Example 2.1**    Suppose $\Omega$ is a sphere that initially is zero degrees throughout, $\phi = 0$, and the boundary of $\Omega$ is held at a fixed temperature of 75 degrees. Then, assuming that there is no internal heat source/sink ($F = 0$), we would expect, based on experience, that the temperature in the sphere would eventually become a constant 75 degrees throughout. Thus $u_\infty = 75$, and this is also a steady state of the problem.

**Example 2.2**    Consider the IBVP

$$u_t = u_{xx} + 1 \qquad \text{for } 0 < x < 1, t > 0,$$

$$u(0, t) = 0 \qquad \text{for } t > 0,$$

$$u_x(1, t) = 0 \qquad \text{for } t > 0,$$

$$u(x, 0) = 0. \qquad \text{for } 0 < x < 1.$$

This is known as a 1-dimensional heat problem. In the next chapter we will discuss how the general 3-dimensional problem specializes to a 1-D problem in certain cases. We interpret this 1-D problem as governing the heat flow in a thin rod of unit length lying along the $x$-axis with left end at $x = 0$ and right end at $x = 1$. The BCs say that the left end is held at the fixed temperature 0 degrees and the right end is insulated. The IC says that initially the rod is uniformly at a temperature of 0 degrees. The 1 in the heat equation indicates that heat is being

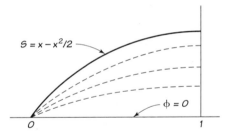

Figure 2.4.  A sketch of the initial temperature distribution $\phi$, several intermediate temperature distributions, and the steady state temperature distribution $S(x) = x - x^2/2$.

added at each point along the rod, increasing the temperature by 1 degree/sec at each point. So the rod is going to heat up. The right end is insulated (no heat can escape there) and any heat that flows over to the left end is extracted so as to maintain that end at 0 degrees. So what happens? Looking for a steady state $S$ by first solving the the differential equation $S_{xx} + 1 = 0$ and then trying to satisfy the BCs with the resulting $S$ leads us to the steady state

$$S(x) = x - \frac{x^2}{2}.$$

This is also the eventual temperature distribution for the problem. A sketch showing the evolution of the temperatures in the rod is displayed in Figure 2.4.

### Exercises 2.4

In this set of exercises, describe (based on your intuitive understanding of heat flow): (i) The steady states and eventual temperature distribution for the given IBVP, if any, and (ii) how the BCs figure into the determination of the heat flow (toward a steady state if there is one). For each of these problems the heat equation is

$$u_t = \alpha^2 \nabla^2 u + F,$$

and $\phi$ is the initial temperature distribution.

1. For the cylindrical rod shown in Figure 2.3, consider the following different situations:

   (a) $F = 0$ and $\phi = 0$. The left and right ends are held at temperatures $u = 0$ and $u = 1$, respectively, while the lateral surface of the cylinder is insulated.

   (b) $F = 0$ and $\phi = 0$. Each end is insulated, while the lateral surface is held at temperature $u = 1$.

   (c) $F = 1$ and $\phi = 0$. The BCs are the same as in (a) of this problem.

   (d) $F = 0$ and $\phi = 1$. The BCs are the same as in (a) of this problem.

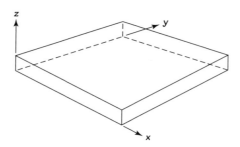

Figure 2.5.    A thin square plate.

2. For the thin square plate shown in Figure 2.5 with each side of unit length $L = 1$, consider the following situations:

(a) $F = 0$ and $\phi = 0$. The top and bottom are insulated, while each of the four sides is held at a fixed temperature $u = 1$.

(b) $F = 0$ and $\phi = 0$ The top and bottom are insulated, while each of the four sides is held at the following temperatures. A pair of opposite sides are held at temperatures 0 and 1, while the sides joining this pair are held at temperatures that vary linearly from 0 to 1.

(c) $F = 0$ and $\phi(x, y, z) = 1 - (x^2 + y^2)$. The top and bottom are insulated, as well as the four sides.

3. Interpret each of the following IBVPs (i.e.,, what is the geometry, what do the BCs mean, etc.?), and then do (i) and (ii) of the overall assignment as you did above. In each case $u = u(x, t)$ is a function of two variables $x$ and $t$ with $0 \le x \le 1$, and $t \ge 0$.

(a)

$$u_t = \alpha^2 u_{xx} \qquad \text{for } 0 < x < 1, t > 0,$$

$$u(0, t) = 0 \qquad \text{for } t > 0,$$

$$u(1, t) = 0 \qquad \text{for } t > 0,$$

$$u(x, 0) = \sin \pi x \qquad \text{for } 0 < x < 1.$$

(b)

$$u_t = \alpha^2 u_{xx} \qquad \text{for } 0 < x < 1, t > 0,$$

$$u_x(0, t) = 0 \qquad \text{for } t > 0,$$

$$u_x(1, t) = 0 \qquad \text{for } t > 0,$$

$$u(x, 0) = \sin \pi x \qquad \text{for } 0 < x < 1.$$

(c)

$$u_t = \alpha^2 u_{xx} + 1 \qquad \text{for } 0 < x < 1, t > 0,$$

$$u(0, t) = 0 \qquad \text{for } t > 0,$$

$$u(1, t) = 1 \qquad \text{for } t > 0,$$

$$u(x, 0) = x \qquad \text{for } 0 < x < 1.$$

(d) The PDE is

$$u_t = \alpha^2 u_{xx} \qquad \text{for } 0 < x < 1, t > 0,$$

and the initial condition is $u(x, 0) = 0$ for $0 < x < 1$. Investigate and compare what happens for the following four types of boundary conditions:

(1) $u(0, t) = 0 = u(1, 0)$,

(2) $u_x(0, t) = 0 = u(1, t)$,

(3) $u_x(0, t) = 0$, $-u_x(1, t) = u(1, t)$,

(4) $u_x(0, t) = 0 = u_x(1, t)$,

for all $t > 0$. Graph the steady states (if any).

(e) The PDE is

$$u_t = u_{xx} + \sin \pi x \qquad \text{for } 0 < x < 1, t > 0,$$

and the initial condition is $u(x, 0) = 0$ for $0 < x < 1$. Investigate and compare what happens for the following two types of boundary conditions:

(1) $u(0, t) = 0 = u(1, 0)$,

(2) $u(0, t) = 0 = u_x(1, t)$,

for all $t > 0$. Graph the steady states (if any).

# 3 The 1-D Heat Equation

We derive the 1-D heat flow problem as a special case (idealization) of the 3-D heat flow problem discussed in the last chapter. For this we make the following assumptions:

(1) The region $\Omega$ is a cylinder of length $L$ centered on the $x$-axis as shown in Fig. 3.1.

(2) The lateral surface (curved part) of the cylinder is insulated.

(3) The left end $(x = 0)$ and right end $(x = L)$, which are circular boundary surfaces, have BCs that do not depend on the $y$ and $z$ coordinates of points on these surfaces.

(4) The initial temperature distribution $\phi$ does not depend on $y$ and $z$; i.e., $\phi = \phi(x)$ is a function of $x$ only (and thus the initial temperature is constant on each circular cross section of the cylinder).

Under these assumptions, we look for a solution of the IBVP that is a function of $x$ and $t$ only:

$$u = u(x, t).$$

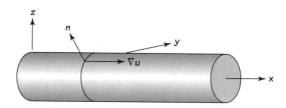

Figure 3.1. A cylindrical rod. The insulated BC $\nabla u \cdot n = 0$ on the lateral surface is automatically satisfied if $u = u(x, t)$ does not depend on $y$ and $z$.

In this case, one has $\nabla u = (u_x, 0, 0)$, so the direction of greatest increase of $u$ is always parallel to the $x$-axis. This immediately gives that $\nabla u \cdot n = 0$ on the lateral surface of the cylinder, and thus the insulated boundary condition is automatically satisfied (see Figure 3.1). The boundary conditions at the left and right ends of the cylinder specialize to:

(a) On the left end, the outward-directed unit normal is $n = (-1, 0, 0)$, and so $\kappa \nabla u \cdot n = -\kappa u_x$ everywhere on this part of the boundary.

(b) On the right end $n = (1, 0, 0)$, and so $\kappa \nabla u \cdot n = \kappa u_x$ everywhere there.

Finally, note that $\nabla^2 u = u_{xx}$, since $u$ does not depend on $y$ or $z$. Thus the general IBVP reduces to what is known as the 1-dimensional IBVP:

**1-D Heat Problem (IBVP)**

$$u_t = \alpha^2 u_{xx} + F \qquad \text{for } 0 < x < L, t > 0, \quad (3.1)$$

$$-\kappa_1 u_x(0, t) + h_1 u(0, t) = g_1(t) \qquad \text{for } t > 0, \qquad\qquad (3.2)$$

$$\kappa_2 u_x(L, t) + h_2 u(L, t) = g_2(t) \qquad \text{for } t > 0, \qquad\qquad (3.3)$$

$$u(x, 0) = \phi(x) \qquad \text{for } 0 < x < L. \qquad\qquad (3.4)$$

Some terminology connected with the above IBVP is that the heat equation is called *homogeneous* when $F \equiv 0$, and the boundary conditions are called *homogeneous BCs*, when $g_1 \equiv 0$ and $g_2 \equiv 0$.

It will take several chapters to develop all the techniques needed to construct the solution of this general 1-D heat problem. It is convenient to work up to the general case by examining the simpler cases first. The overall plan for this is as follows:

- The case where *both* the heat equation and the BCs are homogeneous will be covered in this chapter (this is the *homogeneous problem*).

- The case where the BCs are homogeneous but the heat equation is not will be covered in the Chapter 4 (this is the *semi-homogeneous problem*). That chapter will also show that this case actually suffices to solve the general case. This is done by introducing a technique of transforming a heat problem with non-homogeneous BCs into one with homogeneous BCs.

What we learn here in solving 1-D heat problems will lead directly to corresponding techniques for solving 2-D and 3-D heat problems in the later chapters.

There is a special assumption that will make our work on the 1-D heat problem a lot simpler. It is:

**Special assumption:**    Assume that the length of the cylinder (metal rod) is $L = 1$.

We assume this throughout our discussions, and later (in Exercises 4.3), we show how to deal with the case of arbitrary length $L$. The underlying principles for solving heat problems are the same regardless of the length, and since the length $L$ appears in the analytic solution of each problem, we can simplify the notation by taking $L = 1$ initially.

## 3.1.    The Homogeneous Problem

In this section we look at a method called *separation of variables* for solving the following special case of the heat flow problem when both the PDE and the BCs are homogeneous. Thus we consider solving

$$u_t = \alpha^2 u_{xx}, \tag{3.5}$$

$$-\kappa_1 u_x(0, t) + h_1 u(0, t) = 0, \tag{3.6}$$

$$\kappa_2 u_x(1, t) + h_2 u(1, t) = 0, \tag{3.7}$$

$$u(x, 0) = \phi(x). \tag{3.8}$$

Because the PDE and BCs are homogeneous, one can readily verify that the *superposition principle* holds: If $u$ and $v$ satisfy the PDE and the BCs, then so does $au + bv$ for any choice of constants $a$ and $b$. From this principle it follows, more generally, that if $u_1, u_2, \ldots, u_N$ are solutions of the homogeneous heat equation and boundary conditions, then likewise $\sum_{n=1}^{N} u_n$ is also a solution. The separation of variables technique gives a means of generating an infinite sequence of solutions, $\{u_1, u_2, \ldots\} = \{u_n\}_{n=1}^{\infty}$. Then by the superposition principle, the sum of all these solutions,

$$u(x, t) \equiv \sum_{n=1}^{\infty} u_n(x, t),$$

gives a solution of the PDE + BCs as well. The definition of $u(x, t)$ above by means of the series is heuristic. Questions concerning the convergence of the series and the interchangeability of series summation and partial derivatives (which are necessary to verify that $u$ actually is a solution) will not be discussed here. Such questions, originating in the works of Fourier, Euler, and D'Alembert, led to the development of Fourier analysis, as well as other areas of modern analysis. For us it suffices to handle the series solution in a formal way, knowing that under the appropriate assumptions, the series of functions $\sum_{n=1}^{\infty} u_n$ does actually converge in some sense to a solution $u$.

## 3.2.   Separation of Variables

The separation of variables technique is based on looking for solutions $u$ of the heat equation $u_t = \alpha^2 u_{xx}$, that have the special form

$$u(x, t) = X(x)T(t),$$

namely a product of a function of $x$ times a function of $t$. Substituting this in the heat equation gives

$$X(x)T'(t) = \alpha^2 X''(x)T(t).$$

Separating the parts of this equation that depend on $t$ from those parts that depend on $x$, one arrives at

$$\frac{T'(t)}{\alpha^2 T(t)} = \frac{X''(x)}{X(x)}.$$

Since the first ratio depends only on $t$, while the second ratio depends only on $x$, and these ratios are equal, it must be the case that each ratio is constant:

$$\frac{T'(t)}{\alpha^2 T(t)} = \frac{X''(x)}{X(x)} = k,$$

for all $x$ and $t$. Here $k$ is called the *separation constant*, or *eigenvalue*. Our arguments show that if $u = XT$ is to satisfy the heat equation $u_t = \alpha^2 u_{xx}$, then $X$ and $T$ must satisfy the ordinary differential equations

$$X'' = kX, \tag{3.9}$$

$$T' = k\alpha^2 T. \tag{3.10}$$

Now, the general solution of the first differential equation depends on the value of the separation constant, and so we consider three cases:

(1) ($k = 0$) In this case the general solution is $X(x) = Ax + B$, where $A$ and $B$ are arbitrary constants.

(2) $(k = \lambda^2)$ This is the case where we are assuming that $k$ is a positive constant, and for convenience we write it as the square of another postive constant $\lambda$. The general solution of the differential equation is $X(x) = Ae^{\lambda x} + Be^{-\lambda x}$.

(3) $(k = -\lambda^2)$ In this case the general solution of the differential equation is $X(x) = A \sin \lambda x + B \cos \lambda x$, with $A$ and $B$ arbitrary constants.

The general solution of the differential equation for $T$ is $T(t) = C \exp(k\alpha^2 t)$, where $C$ is an arbitrary constant, and $k = 0, \lambda^2$, or $-\lambda^2$ in the respective cases. Note, however, that when we take products of the $X$'s and $T$, the arbitrary constants $A, B, C$ can be combined and relabeled to give just two arbitrary constants. For example, in the first case above, $T(t) = C$ (since $k = 0$), and so $u = XT = (Ax + B)C = ACx + BC = ax + b$. Thus if we do this in all three cases, we obtain the following types of solutions to the heat equation:

$$u(x, t) = Ax + B, \tag{3.11}$$

$$u(x, t) = e^{\lambda^2 \alpha^2 t}(Ae^{\lambda x} + Be^{-\lambda x}), \tag{3.12}$$

$$u(x, t) = e^{-\lambda^2 \alpha^2 t}(A \sin \lambda x + B \cos \lambda x). \tag{3.13}$$

The method of separation of variables thus gives infinitely many solutions of the heat equation arising from the above three forms by allowing the constants $A, B$, and $\lambda$ to range over all possible values (with the one restriction that $\lambda > 0$).

Having derived the solutions (3.11)–(3.13) based on the separation of variables argument, one can now verify by direct calculation that each of these does indeed satisfy the heat equation $u_t = \alpha^2 u_{xx}$.

## 3.3.   Sturm–Liouville Problems

Now consider the problem of determining which of the above solutions (parametrized by the infinitely many values of $A, B, \lambda$) satisfy the boundary conditions as well. Of course, in a trivial sense, the choice of $A = 0 = B$ gives $u \equiv 0$, the identically zero function, which clearly satisfies the BCs (since they are homogeneous). This is known as the *trivial* solution. To obtain nontrivial solutions of the BCs (i.e., with not both $A$ and $B$ zero) requires, as we now argue, further restrictions on the arbitrariness of $\lambda$. It turns out that $\lambda$ must be restricted to a certain infinite sequence of values $\lambda_1, \lambda_2, \lambda_3, \ldots$ in order for the above types of solutions of the heat equation also to satisfy the boundary conditions. To see this, just substitute each of the above types into the BCs and look at what these conditions say. Here's how the work goes.

First note that in general, since $u = XT$, one has $u_x = X'T$. Thus the BCs (3.6)-(3.7) reduce to

$$\left[-\kappa_1 X'(0) + h_1 X(0)\right] T(t) = 0, \tag{3.14}$$

$$\left[\kappa_2 X'(1) + h_2 X(1)\right] T(t) = 0. \tag{3.15}$$

Now, these equations must hold for all $t > 0$, and so this necessarily gives the following conditions on $X$:

**Boundary Conditions on $X$**

$$-\kappa_1 X'(0) + h_1 X(0) = 0, \tag{3.16}$$

$$\kappa_2 X'(1) + h_2 X(1) = 0. \tag{3.17}$$

Thus we seek to determine solutions $X$ of the differential equation $X'' = kX$, for each of the cases $k = 0, \lambda^2, -\lambda^2$, that also satisfy the above boundary conditions. This is known as a *Sturm–Liouville (SL) problem*. Here, the Sturm–Liouville problem arises from trying to use the separation of variables method to solve the heat equation + BCs. More generally, when using separation of variables to solve other PDEs + BCs, one naturally gets a corresponding Sturm–Liouville problem. This led Sturm and Liouville to consider the most general case:

**The General Sturm–Liouville Problem**

Find those functions $X$ that satisfy the differential equation

$$[p(x)X'(x)]' + q(x)X(x) = k\omega(x)X(x) \tag{3.18}$$

and the above boundary conditions on $X$. Here the functions $p$, $q$, and $\omega$ are given functions. The nonzero functions $X$ that satisfy the DE + BCs are called *eigenfunctions*. Note that the zero function $X(x) = 0$, for all $x$, always satisfies the PDE + BCs. This trivial solution is excluded from being an eigenfunction. The corresponding values of $k$ for which the SL problem has nontrivial solutions are called the *eigenvalues* of the problem, since in the three cases $k = 0, \lambda^2, -\lambda^2$ one often calls $\lambda$ an eigenvalue, even though, strictly speaking, $k$ is the eigenvalue.

We will not discuss the general Sturm–Liouville problem (see [Pin 91]), but rather just concentrate on the special case

$$X''(x) = kX(x),$$

together with the general boundary conditions on $X$. In the general theory this is the case where $p(x) = 1, q(x) = 0$, and $\omega(x) = 1$, and this is the simplest possible Sturm–Liouville DE. Before discussing how to solve this Sturm–Liouville problem in general, let us look at some particular examples of it.

**Example 3.1   (Fixed Temperature at Both Ends)**   For the heat problem with both ends held at fixed temperature zero, $u(0, t) = 0 = u(1, t)$, the corresponding BCs on $X$ are $X(0) = 0 = X(1)$. In the three cases for the different types of $X$, we get:

(1) $(k = 0)$ For $X(x) = Ax + B$, the BCs just say that

$$B = 0, \tag{3.19}$$

$$A + B = 0, \tag{3.20}$$

which gives only the trivial solution $A = 0 = B$.

(2) $(k = \lambda^2)$ For $X(x) = Ae^{\lambda x} + Be^{-\lambda x}$, the BCs are

$$A + B = 0, \tag{3.21}$$

$$Ae^{\lambda} + Be^{-\lambda} = 0. \tag{3.22}$$

This a system of equations for $A$ and $B$, and we know from linear algebra that such a system has a nontrivial solution iff the determinant of the coefficient matrix is zero, which in this case gives the equation $e^{-\lambda} - e^{\lambda} = 0$. Rearranging this gives $e^{2\lambda} = 1$, which says that $\lambda = 0$. Since we are assuming that $\lambda > 0$, this says that there are no nontrivial solutions in this case either.

(3) $(k = -\lambda^2)$ For $X(x) = A \sin \lambda x + B \cos \lambda x$, the BCs are

$$B = 0, \tag{3.23}$$

$$A \sin \lambda + B \cos \lambda = 0. \tag{3.24}$$

Thus $B = 0$ and $A \sin \lambda = 0$. If we want a nontrivial solution in this case, we must require $\sin \lambda = 0$ (since otherwise $A = 0$). This forces $\lambda$ to be a multiple of pi: $\lambda = n\pi$ for $n = 0, \pm 1, \pm 2, \ldots$.

From all of this we get that only functions of the form $X(x) = A \sin n\pi x$ are solutions of the SL problem. For $n = 0$ we get the trivial solution $X = 0$, so we rule out that value of $n$. Also, since $A \sin(-n\pi x) = -A \sin n\pi x$, we might as well just take $n = 1, 2, 3, \ldots$. In summary, the eigenvalues are $k = -\lambda_n^2$, where

$$\lambda_n = n\pi,$$

and the corresponding eigenfunctions are

$$X_n = \sin n\pi x,$$

for $n = 1, 2, 3, \ldots$. The graphs of the first four eigenfunctions are shown in Figure 3.2 Notice how the geometry of the graphs clearly exhibits that the eigenfunctions satisfy the boundary conditions $X(0) = 0$ and $X(1) = 0$.

The time-dependent part $T$ that goes with $X_n$ is easily seen to be given by $T_n(t) = \exp(-n^2\pi^2\alpha^2 t)$. By the superposition principle, we get that the function

$$u(x, t) = \sum_{n=1}^{\infty} A_n e^{-n^2\pi^2\alpha^2 t} \sin n\pi x \tag{3.25}$$

is a solution of the PDE + BCs for any choice of the coefficients $A_1, A_2, A_3, \ldots$. As we will see shortly, this arbitrariness in the coefficients $\{A_n\}_{n=1}^{\infty}$ will allow us to satisfy the initial condition with a solution of this form.

**Example 3.2   (Fixed Temperature + Specified Flux)**   Consider the heat problem with boundary conditions $u(0, t) = 0$ (fixed temperature of zero at the left end) and $u_x(1, t) + u(1, t) = 0$ (flux in the outward direction $-u_x(1, t) = u(1, t)$ equal to the temperature there). The corresponding boundary conditions on $X$ are

$$X(0) = 0, \tag{3.26}$$

$$X'(1) + X(1) = 0. \tag{3.27}$$

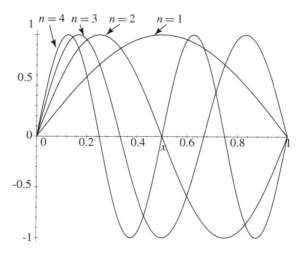

Figure 3.2.   Graphs of the eigenfunctions $X_n(x) = \sin n\pi x$ for $n = 1, 2, 3, 4$.

Dividing into the three cases, based on whether the separation constant is zero, positive, or negative, gives

(1) $(k = 0)$ For $X(x) = Ax + B$, one has $X'(x) = A$. The boundary conditions are

$$B = 0, \tag{3.28}$$

$$A + (A + B) = 0. \tag{3.29}$$

But this just gives the trivial solution $A = 0$, $B = 0$.

(2) $(k = \lambda^2)$ Here $X(x) = Ae^{\lambda x} + Be^{-\lambda x}$, and so $X'(x) = \lambda Ae^{\lambda x} - \lambda Be^{-\lambda x}$. Thus the BCs on $X$ are

$$A + B = 0, \tag{3.30}$$

$$(1 + \lambda)e^{\lambda}A + (1 - \lambda)e^{-\lambda}B = 0. \tag{3.31}$$

The determinant of the coefficient matrix for this system is $D = (1 - \lambda)e^{-\lambda} - (1 + \lambda)e^{\lambda}$. Thus we have nontrivial solutions in this case iff $D = 0$, or equivalently,

$$e^{2\lambda} = \frac{1 - \lambda}{1 + \lambda}. \tag{3.32}$$

The only solution of this equation is $\lambda = 0$ (this can be seen by graphing the functions $f(\lambda) = e^{2\lambda}$ and $g(\lambda) = (1 - \lambda)/(1 + \lambda)$ and determining where they intersect). Since we are assuming that $\lambda$ is positive, we see that there are no nontrivial solutions in this case.

(3) $(k = -\lambda^2)$ In this case $X(x) = A \sin \lambda x + B \cos \lambda x$, and so $X'(x) = \lambda A \cos \lambda x - \lambda B \sin \lambda x$. Therefore, the boundary conditions on $X$ are

$$B = 0, \tag{3.33}$$

$$(\lambda \cos \lambda + \sin \lambda)A + (\cos \lambda - \lambda \sin \lambda)B = 0. \tag{3.34}$$

Because $B = 0$, the second equation reduces to $(\lambda \cos \lambda + \sin \lambda)A = 0$. Hence we get a nontrivial solution iff $\lambda$ satisfies the equation $\lambda \cos \lambda + \sin \lambda = 0$, or equvialently,

$$\tan \lambda = -\lambda. \tag{3.35}$$

There are infinitely many solutions $\lambda_1, \lambda_2, \lambda_3, \ldots = \{\lambda_n\}_{n=1}^{\infty}$ to equation 3.35. This can be seen by graphing the functions $y = \tan \lambda$ and $y = -\lambda$.

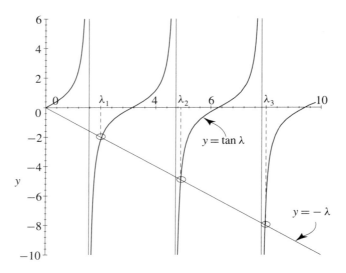

Figure 3.3.    Graphical solutions of $\tan \lambda = -\lambda$.

See Figure 3.3. The infinitely many intersections of these two graphs give the solutions of equation (3.35) as shown. While there is no exact formula for $\lambda_n$ in terms of $n$, one can compute the approximate values of the $\lambda_n$ by various numerical methods. Table 3.1 gives the first five of these values. *Note:* It is suggested from the graph as well as the numbers in the table that for large $n$, the approximation

$$\lambda_n \approx (2n - 1)\pi/2$$

holds. Thus by the above work, for each such value $\lambda_n$, we get a nontrivial solution $B = 0$, $A \neq 0$ of the Sturm–Liouville problem of the form $X(x) = A \sin \lambda_n x$.

Table 3.1.    Numerical values of the roots $\lambda_1, \lambda_2, \lambda_3, \ldots$ of $\tan \lambda = -\lambda$ and the approximations $(2n - 1)\pi/2$ to these roots.

| $n$ | $\lambda_n$ | $(2n - 1)\pi/2$ |
|---|---|---|
| 1 | 2.028757839 | 1.570796327 |
| 2 | 4.913180439 | 4.712388981 |
| 3 | 7.978665712 | 7.853981635 |
| 4 | 11.08553841 | 10.99557429 |
| 5 | 14.20743673 | 14.13716694 |

In summary then, we have found for this example that the eigenvalues are $k = -\lambda_n^2$, where $\{\lambda_n\}_{n=1}^{\infty}$ are the solutions of $\tan \lambda = -\lambda$, and the corresponding eigenfunctions are:

$$X_n(x) = \sin \lambda_n x,$$

for $n = 1, 2, 3, \ldots$. The graphs of the first four of these eigenfunctions are shown in Figure 3.4. Notice that these graphs appear to be geometrically right for functions that satisfy the boundary conditions $X(0) = 0$ and $X'(1) + X(1) = 0$. This latter BC can be written as

$$X'(1) = -X(1),$$

and it is interpreted geometrically as saying that the slope of the tangent line to $X$ at $x = 1$ is the negative of the value of $X$ at $x = 1$.

Thus for this homogeneous heat flow problem, the superposition principle gives us that the function

$$u(x, t) = \sum_{n=1}^{\infty} A_n e^{-\lambda_n^2 \alpha^2 t} \sin \lambda_n x \qquad (3.36)$$

$$= A_1 e^{-4\alpha^2 t} \sin 2x + A_2 e^{-24\alpha^2 t} \sin 4.9x + \cdots$$

is a solution of the PDE + BCs for any choice of constants $A_1, A_2, \ldots$.

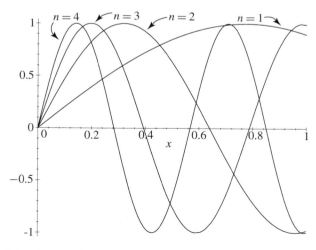

Figure 3.4.   Graphs of the eigenfunctions $X_n(x) = \sin \lambda_n x$, for $n = 1, 2, 3, 4$.

The above examples illustrate some of the details that arise when solving the Sturm–Liouville problem connected with the heat equation. However, it should be noted that these examples are not indicative of what the solutions (eigenvalues and eigenfuctions) of Sturm–Liouville problems look like in other situations. The above examples both give nontrivial solutions only in case (3) ($k = -\lambda^2$), and the eigenfunctions are sine functions. The exercises given later will show that for certain choices of boundary conditions:

- One obtains nontrivial solutions of the Sturm-Liouville problem from case (1), *the zero eigenvalue case.*

- Even in case (3) the solutions can involve both sine and cosine functions.

To give some idea of what can be expected in general, we conclude with a brief discussion of the solution of the *general* Sturm–Liouville problem for $X'' = kX$.

## 3.3.1.  Solutions of $X'' = kX$ with General BCs

Guided by the above examples, we look at the problem of finding solutions of the DE that satisfy

$$-\kappa_1 X'(0) + h_1 X(0) = 0, \tag{3.37}$$

$$\kappa_2 X'(1) + h_2 X(1) = 0. \tag{3.38}$$

Dividing into the three canonical cases, one obtains the following results:

(1) ($k = 0$) Here $X(x) = Ax + B$, and so $X'(x) = A$. Consequently, the boundary conditions on $X$ are

$$-\kappa_1 A + h_1 B = 0, \tag{3.39}$$

$$(\kappa_2 + h_2)A + h_2 B = 0. \tag{3.40}$$

This system of equations for $A$ and $B$ will have a nontrivial solution if and only if the determinant of the coefficent matrix is zero, i.e., if and only if

$$\kappa_1 h_2 + \kappa_2 h_1 + h_1 h_2 = 0. \tag{3.41}$$

Thus unless the given coefficients $\kappa_1$, $h_1$, $\kappa_2$, $h_2$ satisfy this equation, we find that $X(x) = Ax + B$ satisfies the BCs only in the trivial case $A = B = 0$. Therefore, this type of SL problem does *not* have 0 as an eigenvalue. For example, fixing the temperature at each end at zero degrees, $X(0) = 0 = X(1)$, gives only the trivial solution $X = 0$ in this case.

Now, $\kappa_1, h_1$ are nonnegative and not both zero (since then there would be no boundary condition). Likewise for $\kappa_2, h_2$. Thus the only type of BCs for which (3.41) holds is $h_1 = 0 = h_2$. This is the BC where each end is insulated

$$X'(0) = 0, \ X'(1) = 0,$$

and we easily see that $X(x) = A$, where $A$ is any nonzero constant, is a nontrivial solution. Thus this is the *only,* type of Sturm–Liouville boundary conditions that has zero as an eigenvalue. The corresponding eigenfunction we take to be

$$X_0(x) = 1.$$

(2) $(k = \lambda^2)$ This case is left as an exercise: Show that there are no nontrivial solutions of the BCs of the form $X(x) = Ae^{\lambda x} + Be^{-\lambda x}$.

(3) $(k = -\lambda^2)$ In this case $X$ has the form $X(x) = A \sin \lambda x + B \cos \lambda x$, and so $X'(x) = \lambda A \cos \lambda x - \lambda B \sin \lambda x$. Substituting these into the boundary conditions gives

$$-\kappa_1 \lambda A + h_1 B = 0, \qquad (3.42)$$

$$\kappa_2(\lambda A \cos \lambda - \lambda B \sin \lambda) + h_2(A \sin \lambda + B \cos \lambda) = 0. \qquad (3.43)$$

To analyze these equations further, we divide into two cases:

(3a) $(h_1 \neq 0)$ In this case, solving the first equation for $B$ gives $B = \lambda \kappa_1 A / h_1$. Substituting this in the second equation and rearranging a bit gives

$$A \left[ (\kappa_2 h_1 + \kappa_1 h_2)\lambda \cos \lambda + (h_1 h_2 - \kappa_1 \kappa_2 \lambda^2) \sin \lambda \right] = 0.$$

Now, if $A = 0$, then $B = 0$ too, and we get the trivial solution. However, if $A \neq 0$, then in the above equation, the expression in the brackets must be zero. Rewriting this condition slightly, we arrive at

$$\tan \lambda = \frac{(\kappa_1 h_2 + \kappa_2 h_1)\lambda}{\kappa_1 \kappa_2 \lambda^2 - h_1 h_2}. \qquad (3.44)$$

One can show, using graphical methods, that there are infinitely many solutions $\{\lambda_n\}_{n=1}^{\infty}$ of this equation. The corresponding functions $X$ (taking into account that $B = \kappa_1 \lambda A / h_1$) simplfes to $X =$

$A(h_1 \sin \lambda x + \kappa_1 \lambda \cos \lambda x)/h_1$. Thus deleting the arbitrary constant $A/h_1$ gives the eigenfunctions in this case:

$$X_n(x) = h_1 \sin \lambda_n x + \kappa_1 \lambda_n \cos \lambda_n x. \tag{3.45}$$

In the above derivation of the eigenvalue equation (3.44), we divided by $\kappa_1 \kappa_2 \lambda^2 - h_1 h_2$, and this quantity can be zero when $\kappa_1 = 0 = h_2$ (and only in this case, since we're assuming that $h_1 \neq 0$, and not both $\kappa_2$, $h_2$ are zero). When this is so, it is easy to see that the eigenvalue equation is

$$\cos \lambda = 0,$$

and so the eigenvalues are $\lambda_n = (2n - 1)\pi/2$, and the corresponding eigenfunctions $X_n(x) = \cos((2n - 1)\pi x/2)$.

(3b) ($h_1 = 0$) In this case necessarily $\kappa_1 \neq 0$ (otherwise the first BC is automatically satisfied). But this forces $A = 0$ (see equation (3.42)). Then equation (3.43) reduces to

$$B [-\lambda \kappa_2 \sin \lambda + h_2 \cos \lambda] = 0.$$

Since we want a nontrivial solution $B \neq 0$, we choose $\lambda$ so that the expression in the brackets is zero. This gives one of the following equations for the eigenvalues: Either

$$\cot \lambda = \frac{\kappa_2}{h_2}\lambda, \tag{3.46}$$

when $h_2 \neq 0$, or, when $h_2 = 0$,

$$\sin \lambda = 0. \tag{3.47}$$

Each of these equations has infinitely many solutions, $\{\lambda_n\}_{n=1}^{\infty}$, and the corresponding eigenfunctions are

$$X_n(x) = \cos \lambda_n x. \tag{3.48}$$

The above general discussion shows that the Sturm-Liouville problem that arises from the homogeneous heat equation + homogeneous BCs always has infinitely many negative eigenvalues, sometimes has 0 as an eigenvalue, and never has positive eigenvalues. The above examples, as well as the exercises, illustrate these aspects of the general theory.

## Exercises 3.3 (Homogeneous Heat Problems)

1. In this exercise consider the 1-dimensional homogeneous heat problem

$$u_t = \alpha^2 u_{xx},$$

$$-\kappa_1 u_x(0, t) + h_1 u(0, t) = 0,$$

$$\kappa_2 u_x(1, t) + h_2 u(1, t) = 0,$$

$$u(x, 0) = \phi(x).$$

In each of the following parts to this exercise, the initial temperature distribution $\phi$ is arbitrary (unspecified), while several different sets of specific boundary conditions are given. In each case:

- Solve the corresponding Sturm–Liouville problem; i.e., find all the SL eigenfunctions $X_n$ and eigenvalues $\lambda_n$. In doing this, go through the three cases $k = 0, k > 0, k < 0$.

- Write down the general solution (in terms of a series) of the heat equation + BCs. This will have the form

$$u(x, t) = \sum_{n=0}^{\infty} A_n e^{-\lambda_n^2 \alpha^2 t} X_n(x),$$

where the $A_n$ are arbitrary constants. *Note*: If 0 is not an eigenvalue, omit the first term in the series and start the summation at $n = 1$.

- When the eigenvalues cannot be determined analytically, draw a graph to illustrate graphically how the eigenvalues are determined. Then calculate $\lambda_1, \lambda_2, \lambda_3, \lambda_4$ numerically to several decimal places of accuracy. Finally, write out the first four terms of the general series solution.

(a)
$$u_x(0, t) = 0 \qquad \text{for } t > 0,$$
$$u_x(1, t) = 0 \qquad \text{for } t > 0.$$

(b)
$$u(0, t) = 0 \qquad \text{for } t > 0,$$
$$u_x(1, t) = 0 \qquad \text{for } t > 0.$$

(c)
$$u_x(0, t) = 0 \qquad \text{for } t > 0,$$
$$u(1, t) = 0 \qquad \text{for } t > 0.$$

(d)

$$u_x(0, t) = 0 \qquad \text{for } t > 0,$$

$$u_x(1, t) + u(1, t) = 0 \qquad \text{for } t > 0.$$

(e)

$$-u_x(0, t) + u(0, t) = 0 \qquad \text{for } t > 0,$$

$$u_x(1, t) = 0 \qquad \text{for } t > 0.$$

(f)

$$-u_x(0, t) + u(0, t) = 0 \qquad \text{for } t > 0,$$

$$u(1, t) = 0 \qquad \text{for } t > 0.$$

(g)

$$-u_x(0, t) + u(0, t) = 0 \qquad \text{for } t > 0,$$

$$u_x(1, t) + u(1, t) = 0 \qquad \text{for } t > 0.$$

2. Find the eigenfunctions and eigenvalues for the SL problem

$$X'' = kX,$$

$$X(0) = 0,$$

$$X'(1) + aX(1) = 0.$$

Your answer should involve the parameter $a$, where it is assumed that $a \geq 0$. The eigenvalues will have to be determined graphically (when $a \neq 0$). What does your answer reduce to when $a = 0$?

3. Find the eigenfunctions and eigenvalues for the Sturm–Liouville problem

$$X'' = kX,$$

$$X'(0) + aX(0) = 0,$$

$$X(1) = 0.$$

Your answer should involve the parameter $a$, where it is assumed that $a \geq 0$. The eigenvalues will have to be determined graphically (when $a \neq 0$). What does your answer reduce to when $a = 0$? *Note:* The boundary condition at $x = 0$ does not arise in realistic heat flow problems (when $a > 0$) since it says that

$$X'(0) = -aX(0).$$

Interpreting this in terms of Newton's law of cooling, we see that the environment in which the left end of the rod is immersed is 0 degrees and that heat flows into the rod at this end if the temperature there is positive ($X(0) > 0$) and flows out of the rod there if the temperature is negative ($X(0) < 0$). This, of course, is an unrealistic physical situation.

4. For each of the problems you worked in Exercise 1 above:

   • Describe the boundary conditions in terms of heat flow and interpret how this is reflected in the graph of $u(x, t)$ at $x = 0$ and $x = 1$ (for each fixed $t$).

   • Use a computer to graph the first four SL eigenfunctions for the problem and comment on how the SL boundary conditions are reflected in the geometry of the graph.

5. For the case $k = \lambda^2$, the positive eigenvalue case, show that there are no nontrivial eigenfunctions, i.e., solutions $X$ of $X'' = -\lambda^2 X$, that satisfy the general BCs with $\kappa_1, \kappa_2, h_1, h_2$ non-negative. *Hint*: You may (or may not) find it easier to work this problem if you use

$$X(x) = A \sinh \lambda x + B \cosh \lambda x$$

as the general solution to the DE: $X'' = -\lambda^2 X$. This should make the discussion similar to that in the text for case (3).

6. Use the discussion in Section 3.31 and pertinent graphs to make an argument for the assertion that there are infinitely many eigenvalues $k = -\lambda_n^2$, $n = 0, 1, 2, \ldots$, with the $\lambda_n$'s forming an increasing sequence

$$\lambda_0 = 0 < \lambda_1 < \lambda_2 < \lambda_3, \ldots$$

that is unbounded: $\lim_{n \to \infty} \lambda_n = \infty$.

7. As we have seen, separation of variables in a 1-D heat problem leads to a Sturm–Liouville problem with BCs

$$-\kappa_1 X'(0) + h_1 X(0) = 0,$$

$$\kappa_2 X'(1) + h_2 X(1) = 0,$$

where the constants $\kappa_1, \kappa_2, h_1, h_2$ are, for physical reasons, assumed to be *nonnegative*. Because of this nonnegativity, we get two results: (i) 0 is an eigenvalue if and only if the boundary is insulated, and (ii) there are no positive eigenvalues. This happens in dimensions 2 and 3 as well, and the general result is in Exercises 13.1, problem 5.

From a mathematical standpoint (forgetting about the physics), one could consider the more general case where there are no restrictions on $\kappa_1, \kappa_2, h_1, h_2$. For example, Exercise 3 above seems perfectly reasonable from a mathematical viewpoint. In this exercise, make no assumptions on the constants (other than that $\kappa_1$ and $h_1$ are not both zero, and $\kappa_2$ and $h_2$ are not both zero), and do the following:

(a) Determine the eigenfunction and eigenvalues for the SL problem. Go through each case $k = 0$, $k = \lambda^2$, $k = -\lambda^2$ in detail.

(b) Describe the general solution of the corresponding heat problem and its limiting behavior.

# 4 Solution of the 1-D Heat Problem

In the last chapter, using separation of variables and Sturm–Liouville analysis, we have shown how to construct a series solution

$$u(x, t) = \sum_{n=0}^{\infty} A_n e^{-\lambda_n^2 \alpha^2 t} X_n(x) \tag{4.1}$$

that at least formally satisfies the homogeneous heat equation and homogeneous BCs. The series is constructed out of the Sturm–Liouville eigenfuctions $X_n$ and eigenvalues $\lambda_n$, together with arbitrary constants $A_n$. Note that we have started the series summation at $n = 0$, and we make the convention that $\lambda_0 = 0$, so that the zero[th] term is $A_0 X_0(x)$. However, as we have seen, not all heat flow problems have 0 as an eigenvalue. So we make the convention that $X_0(x) = 0$ (i.e. delete the zero[th] term from the series) when 0 is *not* an eigenvalue. We have also seen that the only physically realistic set of BCs for which 0 is an eigenvalue is the totally insulated case, and in this case $X_0(x) = 1$.

The series solution can be thought of as the *general solution* of the homogeneous heat flow problem, since it involves the arbitrary constants $\{A_n\}_{n=0}^{\infty}$. By specifying particular values for these constants, we can get a particular solution of the PDE + BCs that *also* satisfies the initial condition $u(x, 0) = \phi(x) =$ the initial temperature distribution in the rod. This is where *Fourier analysis* comes into the problem. In fact, the $A_n$'s are the (generalized) Fourier coefficients of $\phi$. It was in this exact context that Fourier in the 1800s originated the study of this large area of modern mathematical analysis. Thus if we take $t = 0$ in the above series representation of the temperature function, the IC becomes

$$\sum_{n=0}^{\infty} A_n X_n(x) = \phi(x). \tag{4.2}$$

The possiblity of representing a given function $\phi$ by such a series, called the (generalized) *Fourier series* for $\phi$, was the topic of much debate amoung mathematicians after it was introduced by Fourier. These debates led to more precise ideas on how a series of functions can converge (pointwise, uniformly, in mean square, etc.), but we will not discuss this aspect of Fourier analysis. Fourier also gave a formula (an integral formula) for calculating the constants $A_n$ in the series representation (4.2). In the general setting that we are dealing with, this formula is

$$A_n = L_n^{-1} \int_0^1 \phi(x) X_n(x) dx. \tag{4.3}$$

In this equation, $L_n$, the so-called normalizing constant, is given by

$$L_n = \int_0^1 [X_n(x)]^2 \, dx. \tag{4.4}$$

The reasoning that led Fourier to a formula like this for the $A_n$'s is based upon the orthogonality relations for the Sturm–Liouville eigenfunctions:

**Theorem 4.1**   *Let $\{X_n\}$ be the eigenfunctions for the Sturm–Liouville problem:*

$$[p(x)X'(x)]' + q(x)X(x) = k\omega(x)X(x), \tag{4.5}$$

$$-\kappa_1 X'(0) + h_1 X(0) = 0, \tag{4.6}$$

$$\kappa_2 X'(1) + h_2 X(1) = 0. \tag{4.7}$$

*Then for $n \neq r$, the eigenfuctions $X_n$ and $X_r$ are orthogonal:*

$$\int_0^1 X_n(x) X_r(x) \omega(x) dx = 0. \tag{4.8}$$

The proof of the theorem is left as an exercise. Equation (4.8) in the theorem expresses what is known as the *orthogonality relations* for the Sturm–Liouville eigenfunctions.

*Note:* For the particular Sturm–Liouville DE connected with the heat equation, $\omega(x) = 1$, $p(x) = 1$, and $q(x) = 0$, for all $x$. This will be the only case of interest to us here. Many of the formulas (like (4.3)–(4.4) above) that we use for this special case would have to be modified for the more general case.

Using the theorem, we can give a heuristic derivation of the integral formula (4.3) for the constants $A_n$ needed to represent a given function $\phi$ in terms of the series (4.2) of Strum–Liouville eigenfunctions. Thus assume that $\phi$ can be represented by

$$\phi(x) = \sum_{n=0}^{\infty} A_n X_n(x), \tag{4.9}$$

for some choice of constants $A_n$. If we multiply both sides of this equation by $X_r(x)$ and then integrate both sides from 0 to 1, we get

$$\int_0^1 \phi(x) X_r(x) dx = \int_0^1 \left( \sum_{n=0}^{\infty} A_n X_n(x) X_r(x) \right) dx$$

$$= \sum_{n=0}^{\infty} A_n \int_0^1 X_n(x) X_r(x) dx \tag{4.10}$$

$$= A_r \int_0^1 [X_r(x)]^2 dx \tag{4.11}$$

$$= A_r L_r. \tag{4.12}$$

To get equation (4.10) we had to interchange the order of integration and series summation. Equation (4.11) comes from using the orthogonality relations, and (4.12) comes from the definition of $L_k$. This shows that we get equation (4.3) for the constants $A_n$.

The constants $A_n$ used in the series representation (4.2) for $\phi$ (and calculated by formula (4.3)) are known as the *Fourier coefficients* for $\phi$ (with respect to the eigenfunctions $\{X_n\}$). It might be better to refer to the $A_n$'s as the *generalized Fourier coefficients* of $\phi$, since historically the terms Fourier coefficients and Fourier series have been used mainly in reference to the case when the $X_n$ are sine or cosine functions.

## 4.1.   Solution of the Homogeneous Heat IBVP

This brings us to the conclusion of the complete solution of the homogeneous heat IBVP. Namely, the solution is given by

$$u(x, t) = \sum_{n=0}^{\infty} A_n e^{-\lambda_n^2 \alpha^2 t} X_n(x), \tag{4.13}$$

where the eigenvalues $\lambda_n$ and eigenfuctions $X_n$ are determined by the separation of variables technique (or from Tables in Appendix D, if the BCs are among those listed there). The Fourier coefficients in (4.13) are calculated by

$$A_n = L_n^{-1} \int_0^1 \phi(x) X_n(x) \, dx, \tag{4.14}$$

where $L_n$, the normalizing constant, is computed by

$$L_n = \int_0^1 [X_n(x)]^2 \, dx. \tag{4.15}$$

The labor involved in writing down the solution (4.13) to any given problem is not too great. One must first determine the eigenvalues (which sometimes must be computed by numerical tecniques) and eigenfunctions. Then the computation of the integrals (4.14) is explicitly doable (if often tedious) for a variety of standard initial temperature distributions $\phi$. Note, too, that for some $\phi$'s the integrals are not computable in closed form, and numerical evaluation must be used.

The series representation (4.13) for the solution $u$ is useful for the theory, but in practice one is often content to use the approximation to $u$ afforded by the use of the the first few terms of the series. We saw in the last chapter that the eigenvalues increase to infinity, $\lim_{n\to\infty} \lambda_n = \infty$, and so when this divergence is particularly fast, the initial terms of the series do provide a good approximation to $u$. This is so because the exponential factors

$$e^{-\lambda_n^2 \alpha^2 t}$$

in the terms of the series are essentially negligible (close to zero) for all $t \geq 1$. It is important to note also that since

$$\lim_{t\to\infty} e^{-\lambda_n^2 \alpha^2 t} = 0,$$

for all $n$, the theoretical series representation (4.13) for the solution $u$ to any *homogeneous* heat IBVP (with physically realistic BCs) has the property that

$$\lim_{t\to\infty} u(x,t) = A_0. \tag{4.16}$$

This says that in the long run (i.e., for $t$ large), the temperature distribution $u(x,t)$ in the rod will approach a constant temperature distribution $A_0$ throughout the rod. Furthermore, if both ends of the rod are insulated, this constant temperature is given by

$$A_0 = \int_0^1 \phi(x) dx,$$

which is the average value of the initial temperature distribution $\phi$. However, if both ends of the rod are *not* insulated, then

$$A_0 = 0.$$

Thus we see that the homogeneous problem always has eventual temperature distributions that are constants.

**Example 4.1**   We return to Example 3.1 from the last chapter and complete the solution of this IBVP by computing the Fourier coefficients. The eigenfunctions are $X_n(x) = \sin n\pi x$, and so

$$L_n = \int_0^1 \sin^2 n\pi x \, dx = \frac{1}{2}, \tag{4.17}$$

for $n = 1, 2, \ldots$. Consequently, the Fourier coefficients are given by

$$A_n = 2 \int_0^1 \phi(x) \sin n\pi x \, dx, \tag{4.18}$$

for $n = 1, 2, \ldots$. A particular example of an initial temperature distribution for which these coefficients are most easily computed is

$$\phi(x) = T,$$

a constant temperature $T$. Then one sees that

$$A_n = 2T \int_0^1 \sin n\pi x \, dx = \frac{2T}{n\pi}(1 - \cos n\pi). \tag{4.19}$$

Note that all the even Fourier coefficients are zero ($A_{2k} = 0$), and the odd coefficients are

$$A_{2k-1} = \frac{4T}{(2k - 1)\pi}. \tag{4.20}$$

If we now substitute the computed values (4.19) into the series representation (3.25) for $u$ from Example 3.1, and reduce the summation to just a sum over all odd $n$, we get

$$u(x, t) = \sum_{k=1}^{\infty} \frac{4T}{(2k - 1)\pi} e^{-[(2k-1)\pi\alpha]^2 t} \sin(2k - 1)\pi x. \tag{4.21}$$

This, then, is the solution of the homogeneous heat equation $u_t = \alpha^2 u_{xx}$, with boundary conditions of holding both ends at fixed temperature 0, and initial constant $T$ temperature distribution throughout the rod. Note that the steady state to which this solution tends is $\lim_{t\to\infty} u(x,t) = 0$. See Figure 4.1. This steady state should be expected, since all the heat flows from within the rod towards each end (symmetrically), where it is extracted.

Note also that the above calculation gives Fourier a sine series representation of the constant function $\phi(x) = T$:

$$T = \sum_{k=1}^{\infty} \frac{4T}{(2k-1)\pi} \sin(2k-1)\pi x. \tag{4.22}$$

This is an interesting result in its own right, and it exhibits that the convergence of a Fourier series need not be *pointwise*. For example, if we substitute either endpoint $x = 0$ or $x = 1$ into the series on the right side of (4.22), we get a series of zeros, and so (4.22) reduces to the statement $T = 0$, which is absurd!! (unless we started with $T = 0$). However, one can show that for any other point $x, 0 < x < 1$, the series on the right side of (4.22) *does* converge to $T$ (regardless of which $x$ you choose!).

**Example 4.2**   As in the last example, we return to Chapter 3, Example 3.2, and compute the Fourier coefficients $A_n$ in equation (3.36). For this we assume that the initial temperature distribution is

$$\phi(x) = x.$$

We first compute the normalizing constants $L_n$. There are several ways we can express our answer because it involves the eigenvalues $\lambda_n$, and these eigenvalues,

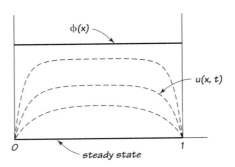

Figure 4.1.   Steady state and transient temperatures for Example 4.1

recall, satisfy

$$\tan \lambda_n = -\lambda_n.$$

Using this we find

$$
\begin{aligned}
L_n &= \int_0^1 \sin^2 \lambda_n x \, dx = \int_0^1 \frac{1}{2}(1 - \cos(2\lambda_n x)) dx \\
&= \frac{1}{2}\left(1 - \frac{1}{2\lambda_n} \sin 2\lambda_n\right) \\
&= \frac{1}{2}(1 + \cos^2 \lambda_n).
\end{aligned}
\tag{4.23}
$$

Next we compute the $A_n$ using integration by parts (and the above eigenvalue equation to simplify):

$$
\begin{aligned}
A_n &= L_n^{-1} \int_0^1 x \sin \lambda_n x \, dx \\
&= \frac{1}{L_n}\left(-\frac{\cos \lambda_n}{\lambda_n} + \frac{\sin \lambda_n}{\lambda_n^2}\right) \\
&= \frac{-2 \cos \lambda_n}{\lambda_n L_n}.
\end{aligned}
\tag{4.24}
$$

Thus, inserting the above results into the series representation (3.36) from Chapter 3, we get the following explicit representation of the solution of that heat IVBP:

$$
u(x, t) = \sum_{n=1}^{\infty} \frac{-4 \cos \lambda_n}{\lambda_n(1 + \cos^2 \lambda_n)} e^{-\lambda_n^2 \alpha^2 t} \sin \lambda_n x.
\tag{4.25}
$$

Again we see that the steady-state temperature distribution in the rod is identically 0 throughout. Our calculations also give us a (generalized) Fourier series representation of the function $\phi(x) = x$:

$$
x = \sum_{n=1}^{\infty} \frac{-4 \cos \lambda_n}{\lambda_n(1 + \cos^2 \lambda_n)} \sin \lambda_n x.
\tag{4.26}
$$

**Exercises 4.1**

1. (*Calculating Fourier coefficients*) This exercise deals with the topic of expanding a given function $\phi$ into a general Fourier series:

$$\phi(x) = \sum_{n=0}^{\infty} A_n X_n(x), \tag{4.27}$$

with respect to a given sequence $\{X_n(x)\}_{n=0}^{\infty}$ of Sturm–Liouville eigenfunctions. As we have seen, this amounts to computing the Fourier coefficients $A_n$ from the integral formula

$$A_n = L_n^{-1} \int_0^1 \phi(x) X_n(x) \, dx, \tag{4.28}$$

where $L_n$ is given by

$$L_n = \int_0^1 [X_n(x)]^2 \, dx. \tag{4.29}$$

The method of expanding a function $\phi$ into an eigenfunction series (4.27) is important and useful in many areas of mathematics, and we know how it arises in the solution of the heat equation IBVP. This method is also used for solving problems associated with other PDEs (like Laplace's equation and the wave equation). So the work you do in this exercise is *not* necessarily associated with the heat equation.

(a) The first thing to do is to compute the normalizing constants $L_n$ in (4.29) for each of the following choices of eigenfunctions:

 (1) $X_n(x) = \sin n\pi x, \quad n = 1, 2, \ldots$.

 (2) $X_n(x) = \cos n\pi x, \quad n = 0, 1, \ldots$.

 (3) $X_n(x) = \sin \frac{2n-1}{2}\pi x, \quad n = 1, 2, \ldots$.

 (4) $X_n(x) = \cos \lambda_n x, \quad n = 1, 2, \ldots$. Here the $\lambda_n$'s are the solutions of the equation: $\cot \lambda = \lambda$.

(b) Compute the Fourier coefficients $A_n$ in the following generalized Fourier series. Simplfy your answers as much as possible. In particular, use the identity $\cos n\pi = (-1)^n$. Also, explicitly write out the resulting Fourier series in terms of the $A_n$'s you compute.

 (1) $x = \sum_{n=1}^{\infty} A_n \sin n\pi x$.

 (2) $x = A_0 + \sum_{n=1}^{\infty} A_n \cos n\pi x$.

 (3) $x^2 = A_0 + \sum_{n=1}^{\infty} A_n \cos n\pi x$.

(4) $x - x^2 = \sum_{n=1}^{\infty} A_n \sin n\pi x.$

(5) $x - x^2 = A_0 + \sum_{n=1}^{\infty} A_n \cos n\pi x.$

(6) $x = \sum_{n=1}^{\infty} A_n \sin \frac{2n-1}{2} \pi x.$

(7) $1 = \sum_{n=1}^{\infty} A_n \cos \lambda_n x$ (where the $\lambda_n$'s are the solutions of $\cot \lambda = \lambda$).

(8) $5 \sin 2\pi x - 3 \sin 5\pi x = \sum_{n=1}^{\infty} A_n \sin n\pi x.$

(9) $3x^2 - 5x + 4 = A_0 + \sum_{n=1}^{\infty} A_n \cos n\pi x.$

(10) $\phi(x) = \sum_{n=1}^{\infty} A_n \sin n\pi x,$ where

$$\phi(x) = \begin{cases} 0 & \text{for } 0 \le x \le 1/2, \\ x - 1/2 & \text{for } 1/2 \le x \le 1. \end{cases} \qquad (4.30)$$

(11) $\phi(x) = \sum_{n=1}^{\infty} A_n \sin n\pi x,$ where

$$\phi(x) = \begin{cases} \sin 2\pi x & \text{for } 0 \le x \le 1/2, \\ 0 & \text{for } 1/2 \le x \le 1. \end{cases} \qquad (4.31)$$

(12) $\phi(x) = \sum_{n=1}^{\infty} A_n \sin n\pi x,$ where: $\phi(x) = H_{a,p,b}(x),$ and $H_{a,p,b}$ is the *triangular pulse function*

$$H_{a,p,b}(x) = \begin{cases} 0 & \text{for } 0 \le x < a, \\ (x-a)/(p-a) & \text{for } a \le x < p, \\ (x-b)/(p-b) & \text{for } p \le x < b, \\ 0 & \text{for } c \le x \le 1. \end{cases} \qquad (4.32)$$

Here we assume $0 \le a < p < b \le 1.$

(13) Suppose $0 \le a < b < c < a' < b' < c' \le 1$ and $k, k' > 0.$ Graph the functions $k H_{a,b,c}$ and $k H_{a,b,c} - k' H_{a',b',c'},$ and expand these functions in their Fourier sine series. (use the results from the last problem.)

2. (*Solving homogeneous heat IBVPs*) For each of the following homogeneous IBVPs (1) write out the exact solution in terms of a series; (2) graph the initial temperature distribution $\phi$, the eventual temperature distribution $\lim_{t \to \infty} u(x, t) = S(x)$, and sketch a few of the intermediate temperature distributions. *Note*: You may use previous results, the tables in Appendix D, and if you already know what the eigenfunctions/values are, do not go through the separation of variables routine. Also, the box: ⬛ indicates a problem that is studied computationally in the exercises in Chapter 5.

(a) (*Insulated boundaries*)

$$u_t = \alpha^2 u_{xx} \qquad\qquad \text{for } 0 < x < 1, t > 0,$$

$$u_x(0, t) = 0 \qquad\qquad \text{for } t > 0,$$

$$u_x(1, t) = 0 \qquad\qquad \text{for } t > 0,$$

$$u(x, 0) = x - x^2 \qquad\quad \text{for } 0 < x < 1.$$

(b) ⟿ (*Fixed temperature and insulated boundaries*)

$$u_t = \alpha^2 u_{xx} \qquad\qquad \text{for } 0 < x < 1, t > 0,$$

$$u(0, t) = 0 \qquad\qquad \text{for } t > 0,$$

$$u_x(1, t) = 0 \qquad\qquad \text{for } t > 0,$$

$$u(x, 0) = x \qquad\qquad \text{for } 0 < x < 1.$$

(c) ⟿ (*Insulated and specified flux*)

$$u_t = \alpha^2 u_{xx} \qquad\qquad\qquad \text{for } 0 < x < 1, t > 0,$$

$$u_x(0, t) = 0 \qquad\qquad\qquad \text{for } t > 0,$$

$$u_x(1, t) + u(1, t) = 0 \qquad\quad \text{for } t > 0,$$

$$u(x, 0) = 1 \qquad\qquad\qquad \text{for } 0 < x < 1.$$

(d) (*Fixed temperature at the boundaries*)

$$u_t = \alpha^2 u_{xx} \qquad\qquad\qquad\qquad \text{for } 0 < x < 1, t > 0,$$

$$u(0, t) = 0 \qquad\qquad\qquad\qquad \text{for } t > 0,$$

$$u(1, t) = 0 \qquad\qquad\qquad\qquad \text{for } t > 0,$$

$$u(x, 0) = 3 \sin 2\pi x - \sin 4\pi x \qquad \text{for } 0 < x < 1.$$

(e) (*Fixed temperature at the boundaries*)

$$u_t = \alpha^2 u_{xx} \qquad\qquad \text{for } 0 < x < 1, t > 0$$

$$u(0, t) = 0 \qquad\qquad \text{for } t > 0$$

$$u(1, t) = 0 \qquad\qquad \text{for } t > 0$$

$$u(x, 0) = \phi(x) \qquad\quad \text{for } 0 < x < 1,$$

where $\phi(x)$ is the function in equation (4.31) above.

(f) $\boxed{\rightsquigarrow}$ (*Fixed temperature at the boundaries*)

$$u_t = \alpha^2 u_{xx} \qquad\qquad \text{for } 0 < x < 1, t > 0$$

$$u(0, t) = 0 \qquad\qquad \text{for } t > 0$$

$$u(1, t) = 0 \qquad\qquad \text{for } t > 0$$

$$u(x, 0) = H_{1/4,1/2,3/4}(x) \qquad \text{for } 0 < x < 1.$$

See equation (4.32) for the definition of $H_{a,p,b}$.

3. Prove Theorem 4.1. Also, derive the more general formula for $A_n$ and $L_n$. (Formulas (4.3)–(4.4) are for the special case where $\omega = 1$, $p = 1$, and $q = 0$ in the Sturm–Liouville problem.)

## 4.2.   Transforming to Homogeneous BCs

We have just seen in the previous section how to formulate the complete solution of the 1-D heat IBVP

$$u_t = \alpha^2 u_{xx} + F, \tag{4.33}$$

$$-\kappa_1 u_x(0, t) + h_1 u(0, t) = g_1(t), \tag{4.34}$$

$$\kappa_2 u_x(1, t) + h_2 u(1, t) = g_2(t), \tag{4.35}$$

$$u(x, 0) = \phi(x) \tag{4.36}$$

in the *homogeneous* case, i.e., provided that

$$F = 0, \; g_1 = 0, \; g_2 = 0. \tag{4.37}$$

We would now like to remove these restrictions (4.37) and see how the solution of the *non-homogeneous* problem is obtained. The non-homogeneous problem is solved in two steps:

**Step 1:** Transform the non-homogeneous problem (4.33)–(4.35) into a new IBVP of the form

$$U_t = \alpha^2 U_{xx} + G, \tag{4.38}$$

$$-\kappa_1 U_x(0, t) + h_1 U(0, t) = 0, \tag{4.39}$$

$$\kappa_2 U_x(1, t) + h_2 U(1, t) = 0, \tag{4.40}$$

$$U(x, 0) = \psi(x) \tag{4.41}$$

for a new unknown function $U$. This new problem is called *semi-homogeneous* since it has homogeneous BCs. In general, we cannot expect the heat equation for $U$ also to be homogeneous after the transformation. Note that the $F$ changes to $G$, and the initial condition function $\phi$ will also change to $\psi$. See below for how this works.

**Step 2:** Solve the semi-homogeneous IBVP (4.38)–(4.41).

Step 1 is pretty simple and will be discussed in this section. The second step is more complicated, and the discussion of it will be deferred to the next section. In essence, what the two steps say is that if we know how to solve a semi-homogeneous heat IBVP, then we can construct the solution of the general non-homogeneous heat IBVP from it. This will be clear in a moment.

The idea behind Step 1, transforming to homogeneous BCs, is related to the topic of steady-state solutions. Here's how it goes. We look for a solution of the original non-homogeneous problem (4.33)–(4.36) that has the special form

$$u = S + U,$$

where $S$ has the form

$$S(x, t) = A(t)x + B(t), \tag{4.42}$$

and $U$ is a new unknown function. The functions $A$, $B$ are also unknown, but we we will end up choosing them such that $u$ satisfies its non-homogeneous BCs when $U$ satisfies the homogeneous BCs (4.39)–(4.40). Toward this end we substitute $u = S + U$ into the original BCs (4.34)–(4.35) to get

$$-\kappa_1 \left[ A(t) + U_x(0, t) \right] + h_1 \left[ B(t) + U(0, t) \right] = g_1(t), \tag{4.43}$$

$$\kappa_2 \left[ A(t) + U_x(1, t) \right] + h_2 \left[ A(t) + B(t) + U(1, t) \right] = g_2(t). \tag{4.44}$$

Inspecting this, we see that these BCs will be satisfied if we require $U$ to satisfy the BCs (4.39)–(4.40) and we require $A$, $B$ to satisfy

$$-\kappa_1 A(t) + h_1 B(t) = g_1(t), \tag{4.45}$$

$$(\kappa_2 + h_2) A(t) + h_2 B(t) = g_2(t). \tag{4.46}$$

This is a system of two linear equations for the two unknowns $A(t)$, $B(t)$, and it has a solution if the determinant of its coefficient matrix is not zero, i.e., if

$$-(\kappa_1 h_2 + \kappa_2 h_1 + h_1 h_2) \neq 0.$$

This is always the case unless both $h_1$ and $h_2$ are zero. We leave this case (where the flux is fixed at both ends) for the exercises, where a modification of the form of $S$ is suggested that will solve the problem. So we proceed under the assumption that not both $h_1$ and $h_2$ are zero. Then the above argument shows that an $S$ of the form $S = A(t)x + B(t)$ can always be chosen to satisfy conditions (4.45)–(4.46). This says, by the way, that $S$ satisfies the original BCs on $u$.

With $A$ and $B$ determined, we have $S$ determined, and there remains the determination of $U$. We have already imposed the BCs (4.39)–(4.40) on $U$. Since $U$ is related to $u$ by $u = S + U$, and we want $u$ to satisfy the PDE (4.33) and IC (4.36), we substitute $u = S + U$ into these and get

$$S_t + U_t = \alpha^2 U_{xx} + F, \tag{4.47}$$

$$S(x, 0) + U(x, 0) = \phi(x). \tag{4.48}$$

Since $S$ is a known function, we can rearrange (4.47)–(4.48) to get the transformed PDE (4.38) and IC (4.41), with the $G$, $\psi$ there given by

$$G(x, t) = F(x, t) - S_t(x, t), \tag{4.49}$$

$$\psi(x) = \phi(x) - S(x, 0). \tag{4.50}$$

**Summary:**    If in the BCs, $h_1, h_2$ are not both zero, then we can determine an $S$ of the form

$$S(x, t) = A(t)x + b(t)$$

that satisfies the BCs. Then the transformed problem (4.38)–(4.41) is a semi-homogeneous IBVP for $U$ with $G$ and $\psi$ given by (4.47)–(4.48). We use the method from the next section to find $U$, and then we get the original heat distribution function $u = S + U$. This then will completely solve the general non-homogeneous heat IBVP.

**Example 4.3**    Suppose we consider the general situation of fixing the temperature at the left and right ends of the rod at temperatures $c_1$ and $c_2$ respectively.

Thus we consider the IBVP

$$u_t = \alpha^2 u_{xx} \qquad \text{for } 0 < x < 1, t > 0,$$

$$u(0, t) = c_1 \qquad \text{for } t > 0,$$

$$u(1, t) = c_2 \qquad \text{for } t > 0,$$

$$u(x, 0) = 0 \qquad \text{for } 0 < x < 1.$$

Note that the initial temperature distribution throughout the rod is a constant zero degrees, and immediately thereafter the left and right ends are brought to temperatures $c_1, c_2$ and maintained at these temperatures thereafter. Assuming that $c_1 > 0, c_2 > 0$, the BCs then cause heat to flow into the rod and raise the temperature there. Let us see how this works out analytically.

We look for a function $S$ of the form

$$S(x, t) = A(t)x + B(t)$$

that satisfies the BCs. This means that

$$B(t) = c_1,$$

$$A(t) + B(t) = c_2.$$

Since $c_1, c_2$ are constants, so are $A, B$. Indeed, $B = c_1, A = c_2 - c_1$. Consequently, the function $S$ does not depend on $t$ either and is given by

$$S(x) = (c_2 - c_1)x + c_1.$$

Having this, we transform $u = S + U$ to get the new IBVP for $U$. Since $S_t = 0$ in this case, and $S_{xx} = 0$, the new PDE is still homogeneous, and of course the new BCs are the homogeneous versions of the old BCs (since $S$ was chosen to achieve this). Thus, $U$ must satisfy

$$U_t = \alpha^2 U_{xx} \qquad \text{for } 0 < x < 1, t > 0,$$

$$U(0, t) = 0 \qquad \text{for } t > 0,$$

$$U(1, t) = 0 \qquad \text{for } t > 0,$$

$$U(x, 0) = -(c_2 - c_1)x - c_1 \qquad \text{for } 0 < x < 1.$$

Recall that the solution of this homogeneous IBVP is

$$U(x, t) = \sum_{n=1}^{\infty} A_n e^{-\alpha^2 n^2 \pi^2 t} \sin n\pi x,$$

with the $A_n$'s being the Fourier sine coefficients of the initial condition function. In this case,

$$A_n = -2 \int_0^1 [(c_2 - c_1)x + c_1] \sin n\pi x \, dx$$

$$= -(c_2 - c_1) \frac{2(-1)^{n+1}}{n\pi} - c_1 \frac{2[1 - (-1)^n]}{n\pi}$$

$$= \frac{2[c_2(-1)^n - c_1]}{n\pi}.$$

Thus the complete solution of our original problem is

$$u(x, t) = (c_2 - c_1)x + c_1 + \sum_{n=1}^{\infty} A_n e^{-\alpha^2 n^2 \pi^2 t} \sin n\pi x. \qquad (4.51)$$

From this we see that the eventual temperature distribution in the rod is

$$\lim_{n \to \infty} u(x, t) = (c_2 - c_1)x + c_1 = S(x),$$

which coincides with the function we used to transform to homogeneous BCs. The graphs of the initial temperature distribution, several intermediate temperature distributions, and the final distribution $S$ are shown in Figure

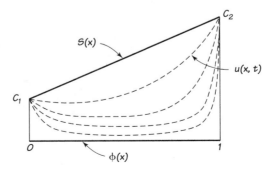

Figure 4.2.    Temperature distributions for Example 4.3.

4.2. Note that while the initial temperature distribution in this example is $\phi(x) = 0$, this does not affect the final temperature distribution $S$. Specifically, if we let $\phi(x)$ be an arbitrary function (choose your favorite one), then the solution will still be given by equation (4.52), except now the $A_n$'s will be given by

$$A_n = 2 \int_0^1 [\phi(x) - (c_2 - c_1)x - c_1] \sin n\pi x \, dx.$$

Regardless of what these numbers are, we still get that $U(x, t)$ decays to zero as $t$ tends to infinity.

**Example 4.4**    In this example let us also consider the case of maintaining specified temperatures at the boundary, but allowing one of these temperatures to vary with time. Specifically, consider the IBVP

$$u_t = \alpha^2 u_{xx} \qquad \text{for } 0 < x < 1, t > 0,$$

$$u(0, t) = 1 \qquad \text{for } t > 0,$$

$$u(1, t) = ae^{-t} \qquad \text{for } t > 0,$$

$$u(x, 0) = 0 \qquad \text{for } 0 < x < 1.$$

Now the function $S = Ax + B$ used to transform to homogeneous BCs must satisfy

$$B(t) = 1,$$

$$A(t) + B(t) = ae^{-t}.$$

Solving this gives $B = 1$ and $A(t) = ae^{-t} - 1$. Consequently,

$$S(x, t) = (ae^{-t} - 1)x + 1.$$

Looking at partial derivatives of this gives $S_t = -ae^{-t}x$, and of course $S_{xx} = 0$. When we transform, the new IBVP for $U$ is

$$U_t = \alpha^2 U_{xx} + ae^{-t}x \qquad \text{for } 0 < x < 1, t > 0,$$

$$U(0, t) = 0 \qquad \text{for } t > 0,$$

$$U(1, t) = 0 \qquad \text{for } t > 0,$$

$$U(x, 0) = -(a - 1)x - 1 \qquad \text{for } 0 < x < 1.$$

Thus, even though the original PDE was homogeneous, this transformed one is non-homogeneous. As a result, this IBVP for $U$ is semi-homogeneous, and we do not know how to solve this yet (see the next section).

**Example 4.5**    This example is related to the last one, except that the PDE is modified so that transforming to homogeneous BCs gives a homogeneous PDE. Thus, consider

$$u_t = \alpha^2 u_{xx} - ae^{-t} \qquad \text{for } 0 < x < 1, t > 0,$$

$$u(0, t) = 1 \qquad \text{for } t > 0,$$

$$u(1, t) = ae^{-t} \qquad \text{for } t > 0,$$

$$u(x, 0) = 0 \qquad \text{for } 0 < x < 1.$$

Since the BCs are the same as before, we have $S(x, t) = (ae^{-t} - 1)x + 1$. However, now the transformed problem is

$$U_t = \alpha^2 U_{xx} \qquad \text{for } 0 < x < 1, t > 0,$$

$$U(0, t) = 0 \qquad \text{for } t > 0,$$

$$U(1, t) = 0 \qquad \text{for } t > 0,$$

$$U(x, 0) = -(a - 1)x - 1 \qquad \text{for } 0 < x < 1.$$

Hence the solution of the original problem is

$$u(x, t) = (ae^{-t} - 1)x + 1 + \sum_{n=1}^{\infty} A_n e^{-\alpha^2 n^2 \pi^2 t} \sin n\pi x. \qquad (4.52)$$

Here the series part is the expression for $U(x, t)$, with the $A_n$'s the Fourier sine coefficients for $-(a - 1)x - 1$. Note that in this example there is no steady state, but there is an eventual (limiting) temperature distribution

$$\lim_{t \to \infty} u(x, t) = -x + 1.$$

This makes sense, because the problem is interpreted as follows: The BCs say that the left end is held at fixed temperature 1, while the right end is held at temperatures that gradually decrease to 0. The term $-ae^{-t}$ in the PDE indicates that heat is being extracted internally, the same amount at each point along the rod, and that this amount decreases over time.

**Comment:**    In transforming to homogeneous BCs, the $S$ function is sometimes a steady state, as in Example 4.3, but this is not always the case, as Example 4.5 shows.

## Exercises 4.2

1. In the following IBVPs, find a function $S = Ax + B$ that satisfies the non-homogeneous BCs and use it to transform to a new problem with homogeneous BCs. When the new IBVP has a homogeneous PDE, solve it, and then write out the solution $u = S + U$ of the original IBVP. Also, interpret your answer with graphs showing how the temperature distribution changes over time. Find steady states and eventual temperature distributions if there are any.
   *Note*: The box $\boxed{\leadsto}$ indicates a problem that is studied computationally in the Chapter 5 exercises.

   (a) $\boxed{\leadsto}$

   $$u_t = \alpha^2 u_{xx} \qquad \text{for } 0 < x < 1, t > 0,$$
   $$u(0, t) = 5 \qquad \text{for } t > 0,$$
   $$u(1, t) = 2 \qquad \text{for } t > 0,$$
   $$u(x, 0) = 0 \qquad \text{for } 0 < x < 1.$$

   (b)

   $$u_t = \alpha^2 u_{xx} \qquad \text{for } 0 < x < 1, t > 0,$$
   $$u(0, t) = 1 \qquad \text{for } t > 0,$$
   $$u_x(1, t) = 2 \qquad \text{for } t > 0,$$
   $$u(x, 0) = 0 \qquad \text{for } 0 < x < 1.$$

   (c)

   $$u_t = \alpha^2 u_{xx} \qquad \text{for } 0 < x < 1, t > 0,$$
   $$u_x(0, t) = -1 \qquad \text{for } t > 0,$$
   $$u(1, t) = 2 \qquad \text{for } t > 0,$$
   $$u(x, 0) = 0 \qquad \text{for } 0 < x < 1.$$

   (d) $\boxed{\leadsto}$

   $$u_t = \alpha^2 u_{xx} + x(1 + t)^{-2} \qquad \text{for } 0 < x < 1, t > 0,$$
   $$u(0, t) = 0 \qquad \text{for } t > 0,$$
   $$u(1, t) = t(1 + t)^{-1} \qquad \text{for } t > 0,$$
   $$u(x, 0) = \phi(x) \qquad \text{for } 0 < x < 1,$$

where $\phi(x) = x - x^2$. Also do this problem for $\phi(x) = 0$.

(e)

$$u_t = \alpha^2 u_{xx} + (1 - x)(1 + t)^{-2} \qquad \text{for } 0 < x < 1, t > 0,$$

$$u(0, t) = t(1 + t)^{-1} \qquad \text{for } t > 0,$$

$$u_x(1, t) = (1 + t)^{-1} \qquad \text{for } t > 0,$$

$$u(x, 0) = \phi(x) \qquad \text{for } 0 < x < 1,$$

where $\phi(x) = 0$. Also do this problem for $\phi(x) = x$.

(f)

$$u_t = \alpha^2 u_{xx} \qquad \text{for } 0 < x < 1, t > 0,$$

$$u(0, t) = 50 \qquad \text{for } t > 0,$$

$$u_x(1, t) + u(1, t) = 30 \qquad \text{for } t > 0,$$

$$u(x, 0) = 50 \qquad \text{for } 0 < x < 1.$$

(g)

$$u_t = \alpha^2 u_{xx} \qquad \text{for } 0 < x < 1, t > 0,$$

$$u_x(0, t) = 50 \qquad \text{for } t > 0,$$

$$u_x(1, t) + u(1, t) = 30 \qquad \text{for } t > 0,$$

$$u(x, 0) = 50x \qquad \text{for } 0 < x < 1.$$

2. As indicated in the text, the BCs where the flux is fixed at each end cannot be transformed to homogeneous functions by using a function of the form $S = A(t)x + B(t)$. In this problem show that something of the form

$$S(x, t) = A(t)x^2 + B(t)x$$

will transform the following IBVPs to ones with homogeneous BCs. In each case find the actual $S$ that works, do the transformation, and solve the total problem if possible.

(a) $\boxed{\rightsquigarrow}$

$$u_t = \alpha^2 u_{xx} - 6\alpha^2 \qquad \text{for } 0 < x < 1, t > 0,$$

$$u_x(0, t) = -2 \qquad \text{for } t > 0,$$

$$u_x(1, t) = 4 \qquad \text{for } t > 0,$$

$$u(x, 0) = 2x \qquad \text{for } 0 < x < 1.$$

(b)

$$u_t = \alpha^2 u_{xx} \qquad \text{for } 0 < x < 1, t > 0,$$

$$u_x(0, t) = 2 \qquad \text{for } t > 0,$$

$$u_x = ae^{-t} \qquad \text{for } t > 0,$$

$$u(x, 0) = 2x \qquad \text{for } 0 < x < 1.$$

## 4.3.  The Semi-Homogeneous Heat Problem

In this section we look at the method for solving the semi-homogeneous heat IBVP:

$$u_t = \alpha^2 u_{xx} + F, \qquad (4.53)$$

$$-\kappa_1 u_x(0, t) + h_1 u(0, t) = 0, \qquad (4.54)$$

$$\kappa_2 u_x(1, t) + h_2 u(1, t) = 0, \qquad (4.55)$$

$$u(x, 0) = \phi(x). \qquad (4.56)$$

The method is a quite natural extension of the one we used to solve the homogeneous IBVP in the first section of this chapter. Indeed, the method here must reduce to the one there when $F = 0$.

Thus we start by finding the Sturm–Liouville eigenvalues $\{\lambda_n\}$ and eigenfunctions $\{X_n\}$ associated to the problem, i.e., we solve first

$$X'' = kX, \qquad (4.57)$$

$$-\kappa_1 X'(0) + h_1 X(0) = 0, \qquad (4.58)$$

$$\kappa_2 X'(1) + h_2 X(1) = 0. \qquad (4.59)$$

If this is one of the standard Sturm–Liouville problems that we have discussed, then it suffices to look up the eigenvalues/eigenfunctions in the Table D.3 in Appendix D. Having these, we know that the *homogeneous* case, $F = 0$, of (4.53)–(4.56) has solution

$$u(x, t) = \sum_{n=0}^{\infty} A_n e^{-\lambda_n^2 \alpha^2 t} X_n(x), \qquad (4.60)$$

and so it seems reasonable in the semi-homogeneous case, $F \neq 0$, to look for a solution of the form

$$u(x, t) = \sum_{n=0}^{\infty} T_n(t) X_n(x). \tag{4.61}$$

Here the $T_n$'s are unknown functions of $t$ that we seek to determine. This we do by substituting the expression (4.61) for $u$ into the PDE (4.53) and IC (4.56). Doing this and using some Fourier analysis, we finally arrive at the following result:

**Theorem 4.2**   *Suppose F and $\phi$ are expanded in Fourier series with respect to the Sturm–Liouville eigenfunctions*

$$F(x, t) = \sum_{n=0}^{\infty} F_n(t) X_n(x), \tag{4.62}$$

$$\phi(x) = \sum_{n=0}^{\infty} A_n X_n(x), \tag{4.63}$$

*with the Fourier coefficients given by the usual integral formulas*

$$F_n(t) = L_n^{-1} \int_0^1 F(x, t) X_n(x) dx, \tag{4.64}$$

$$A_n = L_n^{-1} \int_0^1 \phi(x) X_n(x) dx. \tag{4.65}$$

*Further, suppose that for each $n = 0, 1, 2, \ldots$, we let $T_n$ be the solution of the initial value problem*

$$T_n' + \lambda_n^2 \alpha^2 T_n = F_n, \tag{4.66}$$

$$T_n(0) = A_n. \tag{4.67}$$

*Then the function u given by*

$$u(x, t) = \sum_{n=0}^{\infty} T_n(t) X_n(x) \tag{4.68}$$

*satisfies the semi-homogeneous heat IBVP (4.53)–(4.55).*

The proof of the theorem is left as an exercise (a rigorous proof would require quite a bit of advanced analysis, so just give the usual type of heuristic arguments). Theorem 4.2 gives us a recipe for the series solution

$$u(x, t) = \sum_{n=0}^{\infty} T_n(t) X_n(x) \tag{4.69}$$

of the semi-homogeneous problem. The steps to follow are:

(1) Find the Sturm–Liouville eigenfunctions $X_n$ and eigenvalues $\lambda_n$ associated with the problem.
  *Note*: In most cases now, we can simply look these up in Table D.2 in the appendix.

(2) Compute the Fourier coefficients $F_n(t)$ and $A_n$ for the internal heat source/sink function $F(x, t)$, and initial temperature distribution function $\phi(x)$. The $A_n$'s are always constants, but the $F_n(t)$'s in general will depend on $t$.

(3) For each $n$ find the function $T_n(t)$ by solving the initial value problem

$$T_n'(t) + \alpha^2 \lambda_n^2 T_n(t) = F_n(t),$$

$$T_n(0) = A_n.$$

Note that by using elementary techniques from an introductory differential equations course, you can express the solution of this initial value problem as:

$$T_n(t) = \left[ A_n + \int_0^t F_n(s) e^{a_n s} \, ds \right] e^{-a_n t}. \tag{4.70}$$

Here and in the sequel we use the notation

$$a_n = \alpha^2 \lambda_n^2$$

for the sake of convenience. Formula (4.70) will be the basic formula to use to calculate the $T_n$'s. Note in particular how this generalizes the homogeneous problem. In the homogeneous problem we found that $T_n(t) = A_n e^{-a_n t}$, and of course the above formula (4.70) reduces to this in the case that $F(x, t) = 0$.

(4) Put everything together in $u(x, t) = \sum_{n=0}^{\infty} T_n(t) X_n(x)$.

**Note:**   You must pay careful attention to whether zero is an eigenvalue in any problem you solve. If zero is *not* an eigenvalue, then the term $T_0(t)X_0(x)$ is *not* present in the series solution, since by convention $X_0 = 0$ in this case.

**Example 4.6**   Consider the following IBVP:

$$u_t = \alpha^2 u_{xx} + b \sin \pi x \qquad \text{for } 0 < x < 1, t > 0,$$

$$u(0, t) = 0 \qquad \text{for } t > 0,$$

$$u(1, t) = 0 \qquad \text{for } t > 0,$$

$$u(x, 0) = 1 \qquad \text{for } 0 < x < 1.$$

Here $b$ is a given constant (when $b = 0$ the problem is homogeneous). The corresponding Sturm–Liouville problem has BCs $X(0) = 0 = X(1)$, and so the eigenfunctions are $X_n(x) = \sin n\pi x$ and $\lambda_n = n\pi$, for $n = 1, 2, 3, \ldots$. Note that zero is not an eigenvalue for this problem. Since the internal heat source/sink $F(x) = \sin \pi x$ is actually the the first eigenfunction, its Fourier coefficients are easy to compute:

$$F_1 = b, \qquad F_n = 0$$

for all other $n$. The initial temperature distribution is $\phi(x) = 1$, and its Fourier coefficients are

$$A_n = \frac{2[1 - (-1)^n]}{n\pi}.$$

Consequently, computing the integrals in formula (4.70) gives

$$T_1(t) = \left[ A_1 + \int_0^t b e^{a_1 s} ds \right] e^{-a_1 t}$$

$$= \left[ A_1 + \frac{b e^{a_1 t} - b}{a_1} \right] e^{-a_1 t}$$

$$= A_1 e^{-a_1 t} + \frac{b - b e^{-a_1 t}}{a_1},$$

and for all other $n$

$$T_n(t) = A_n e^{-a_n t}.$$

Putting all of this together and using the value $a_n = \alpha^2 n^2 \pi^2$ gives the following solution of this problem:

$$u(x,t) = \left[ \frac{b - be^{-\alpha^2 \pi^2 t}}{\alpha^2 \pi^2} \right] \sin \pi x + \sum_{n=1}^{\infty} A_n e^{-\alpha^2 n^2 \pi^2 t} \sin n\pi x. \qquad (4.71)$$

The series solution of this semi-homogeneous problem also shows us what effect the internal heat source function $F(x) = b \sin \pi x$ has on the eventual temperature distribution in the rod:

$$S(x) = \lim_{t \to \infty} u(x,t) = \frac{b}{\alpha^2 \pi^2} \sin \pi x.$$

This seems reasonable, since the ends of the rod are maintained at a temperature of zero degrees, and the internal heat source generates heat in the rod, raising the temperature by $F(x) = b \sin \pi x$ degrees at the point $x$ for each instant in time. See Figure 4.3.

**Example 4.7**   As in the last example, let us use an $F$ that does not depend on $t$, so that the $F_n$'s are constants and the integral formula (4.70) for the $T_n$'s is as simple as possible (other than the case $F = 0$). Thus consider the IBVP

$$u_t = \alpha^2 u_{xx} + bx \qquad \text{for } 0 < x < 1, t > 0,$$

$$u(0,t) = 0 \qquad \text{for } t > 0,$$

$$u(1,t) = 0 \qquad \text{for } t > 0,$$

$$u(x,0) = 1 \qquad \text{for } 0 < x < 1.$$

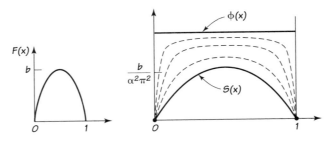

Figure 4.3.   Temperature profiles for Example 4.6.

This problem is identical to the last one, except that now $F(x) = bx$ is not a multiple of one of the eigenfunctions $X_n(x) = \sin n\pi x$ for the problem. Thus we must compute the $F_n$'s from scratch:

$$F_n = 2 \int_0^1 bx \sin n\pi x \, dx = \frac{2b(-1)^{n+1}}{n\pi},$$

for $n = 1, 2, 3, \ldots$. Now, the computation of the $T_n$'s is

$$T_n(t) = \left[ A_n + \int_0^t F_n e^{a_n s} ds \right] e^{-a_n t}$$

$$= \left[ A_n + F_n \frac{e^{a_n t} - 1}{a_n} \right] e^{-a_n t}$$

$$= \frac{F_n}{a_n} + \left[ A_n - \frac{F_n}{a_n} \right] e^{-a_n t}.$$

Having this, we multiply by $X_n(x) = \sin n\pi x$ and sum from 1 to $\infty$ to get the solution of this problem. In doing so, we separate out the terms that do not depend on $t$ from those that do (and also restore $a_n = \alpha^2 n^2 \pi^2$):

$$u(x, t) = \sum_{n=1}^{\infty} \frac{F_n}{\alpha^2 n^2 \pi^2} \sin n\pi x + \sum_{n=1}^{\infty} \left[ A_n - \frac{F_n}{\alpha^2 n^2 \pi^2} \right] e^{-\alpha^2 n^2 \pi^2 t} \sin n\pi x.$$

(4.72)

The latter series is the transient part of the temperature distribution because it dies out over time (goes to 0 as $t \to \infty$), and consequently, $u$ tends to the steady state

$$S(x) = \lim_{t \to \infty} u(x, t) = \sum_{n=1}^{\infty} \frac{F_n}{\alpha^2 n^2 \pi^2} \sin n\pi x. \qquad (4.73)$$

As usual, a figure depicting the graphs of the initial temperature distribution $\phi(x)$, several of the intermediate temperature distributions $u(x, t_1), u(x, t_2), \ldots$, and the eventual temperature distribution $S(x)$ would be helpful in fully understanding the answer. However, $S(x)$ as given by (4.73) would be rather hard to graph from just looking at the series. But it turns out that we can sum this series to get a closed-form formula for $S(x)$. This is not too hard to believe, since the series in (4.73) is just the Fourier series expansion of $S(x)$. Note that

$$bx = \sum_{n=1}^{\infty} F_n \sin n\pi x,$$

since the $F_n$'s are the Fourier coefficients of $F(x) = bx$. Here we have the inverse problem of determining a formula for $S(x)$ knowing only that its Fourier coefficients are $F_n/(\alpha^2 n^2 \pi^2)$. There is an easy way to solve this. The theory (which we have not explained here) tells us that $S$ must be a solution of the heat equation that does not depend on $t$, i.e. $S$ satisfies

$$0 = \alpha^2 S_{xx} + bx;$$

and also, $S$ must satisfy the BCs of the problem, i.e., $S(0) = 0 = S(1)$ (you can verify these things directly from the series representation for $S$). Thus to determine $S$, we first solve the above differential equation:

$$S_{xx} = -\frac{bx}{\alpha^2}.$$

Integrating once gives

$$S_x = -\frac{bx^2}{2\alpha^2} + A,$$

where $A$ is a constant of integration. A second integration gives

$$S = -\frac{bx^3}{6\alpha^2} + Ax + B.$$

These are all the steady states of the non-homogeneous heat equation in this problem. There are infinitely many, since $A$ and $B$ are arbitrary constants. Let us choose $A$, $B$ such that the above $S$ also satisfies the BCs of the problem. Substituting the expression for $S$ into $S(0) = 0$ and $S(1) = 0$ gives

$$B = 0,$$

$$-\frac{b}{6\alpha^2} + A + B = 0.$$

This gives $B = 0$ and $A = b/6\alpha^2$. Substituting these values back into the expression for $S$ and factoring gives the $S$ we are looking for:

$$S(x) = -\frac{b}{6\alpha^2} x(1 - x^2).$$

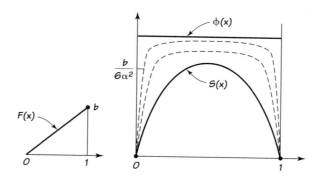

Figure 4.4.    The temperature distributions for Example 4.7.

Having found this, one could now directly compute the Fourier coefficients and verify that they are what we found above. The graphical analysis for this example is shown in Figure 4.4

## Exercises 4.3

In Exercises 1-8, solve the semi-homogeneous heat problems. You may use the Tables in Appendix D and any previous results we have found. Find the steady state $S$ and the eventual temperature distribution: $\lim_{t \to \infty} u(x, t) = u_\infty(x)$ (if any). If $S$ is given by a series, also find the closed form formula for $S$. Where possible draw graphs of $\phi$, $S$, and several intermediate temperature profiles.

*Note*: The box: $\boxed{\leadsto}$ indicates a problem that also occurs in Chapter 5.

1. $\boxed{\leadsto}$ *(Fixed-temperature BCs)*

$$u_t = \alpha^2 u_{xx} + 1 \qquad \text{for } 0 < x < 1, t > 0,$$

$$u(0, t) = 0 \qquad \text{for } t > 0,$$

$$u(1, t) = 0 \qquad \text{for } t > 0,$$

$$u(x, 0) = 1 \qquad \text{for } 0 < x < 1.$$

2. *(Fixed-temperature BCs)*

$$u_t = \alpha^2 u_{xx} + bxt \qquad \text{for } 0 < x < 1, t > 0,$$

$$u(0, t) = 0 \qquad \text{for } t > 0,$$

$$u(1, t) = 0 \qquad \text{for } t > 0,$$

$$u(x, 0) = 1 \qquad \text{for } 0 < x < 1.$$

3. (*Fixed-temperature BCs*)

$$u_t = \alpha^2 u_{xx} + b \sin \omega t \qquad \text{for } 0 < x < 1, t > 0,$$

$$u(0, t) = 0 \qquad \text{for } t > 0,$$

$$u(1, t) = 0 \qquad \text{for } t > 0,$$

$$u(x, 0) = 1 \qquad \text{for } 0 < x < 1.$$

4. (*Insulated BCs*)

$$u_t = \alpha^2 u_{xx} + 1 \qquad \text{for } 0 < x < 1, t > 0,$$

$$u_x(0, t) = 0 \qquad \text{for } t > 0,$$

$$u_x(1, t) = 0 \qquad \text{for } t > 0,$$

$$u(x, 0) = 1 \qquad \text{for } 0 < x < 1.$$

5. (*Insulated BCs*)

$$u_t = \alpha^2 u_{xx} + x^2 \qquad \text{for } 0 < x < 1, t > 0,$$

$$u_x(0, t) = 0 \qquad \text{for } t > 0,$$

$$u_x(1, t) = 0 \qquad \text{for } t > 0,$$

$$u(x, 0) = 1 \qquad \text{for } 0 < x < 1.$$

6. (*Mixed BCs*)

$$u_t = \alpha^2 u_{xx} + b \sin \omega t \qquad \text{for } 0 < x < 1, t > 0,$$

$$u_x(0, t) = 0 \qquad \text{for } t > 0,$$

$$u(1, t) = 0 \qquad \text{for } t > 0,$$

$$u(x, 0) = 1 \qquad \text{for } 0 < x < 1.$$

7. (*Mixed BCs*)

$$u_t = \alpha^2 u_{xx} + bx \qquad \text{for } 0 < x < 1, t > 0,$$

$$u(0, t) = 0 \qquad \text{for } t > 0,$$

$$u_x(1, t) = 0 \qquad \text{for } t > 0,$$

$$u(x, 0) = 1 \qquad \text{for } 0 < x < 1.$$

8. (*Mixed BCs*)

$$u_t = \alpha^2 u_{xx} + be^{-t} \qquad \text{for } 0 < x < 1, t > 0,$$

$$u(0, t) = 0 \qquad \text{for } t > 0,$$

$$u_x(1, t) = 0 \qquad \text{for } t > 0,$$

$$u(x, 0) = 1 \qquad \text{for } 0 < x < 1.$$

In each of the following exercises, solve the IBVP for $u$ by transforming to homogeneous boundary conditions $u = S + U$ and solving the resulting semi-homogeneous problem for $U$. Sketch the graphs of the initial, intermediate, steady state, and eventual temperature distributions $\phi$, $u(\cdot, t)$, $S$, $u_\infty$ (the latter two when they exist).

9. (*This was discussed in Example 4.4.*)

$$u_t = \alpha^2 u_{xx} \qquad \text{for } 0 < x < 1, t > 0,$$

$$u(0, t) = 1 \qquad \text{for } t > 0,$$

$$u(1, t) = ae^{-t} \qquad \text{for } t > 0,$$

$$u(x, 0) = 0 \qquad \text{for } 0 < x < 1.$$

10. $\boxed{\rightsquigarrow}$

$$u_t = \alpha^2 u_{xx} + 1 \qquad \text{for } 0 < x < 1, t > 0,$$

$$u(0, t) = 0 \qquad \text{for } t > 0,$$

$$u(1, t) = 1 \qquad \text{for } t > 0,$$

$$u(x, 0) = x \qquad \text{for } 0 < x < 1.$$

11. $\boxed{\rightsquigarrow}$

$$u_t = \alpha^2 u_{xx} \qquad \text{for } 0 < x < 1, t > 0,$$

$$u(0, t) = 0 \qquad \text{for } t > 0,$$

$$u(1, t) = \sin t \qquad \text{for } t > 0,$$

$$u(x, 0) = 1 \qquad \text{for } 0 < x < 1.$$

12.

$$u_t = \alpha^2 u_{xx} + mx \sin \omega t \qquad \text{for } 0 < x < 1, t > 0,$$

$$u(0, t) = 0 \qquad \text{for } t > 0,$$

$$u(1, t) = a \qquad \text{for } t > 0,$$

$$u(x, 0) = ax + b \qquad \text{for } 0 < x < 1.$$

13.

$$u_t = \alpha^2 u_{xx} \qquad \text{for } 0 < x < 1, t > 0,$$

$$u(0, t) = a \qquad \text{for } t > 0,$$

$$u(1, t) = b \qquad \text{for } t > 0,$$

$$u(x, 0) = ax \qquad \text{for } 0 < x < 1.$$

14. (*Steady states*) This problem deals with the determination of steady states $S$ and eventual temperature distributions $u_\infty$ for the semi-homogeneous IBVP

$$u_t = \alpha^2 u_{xx} + F(x),$$

$$-\kappa_1 u_x(0, t) + h_1 u(0, t) = 0,$$

$$\kappa_2 u_x(1, t) + h_2 u(1, t) = 0,$$

$$u(x, 0) = \phi(x).$$

*Note*: We are assuming that $F$ does not depend on $t$.

(a) Find all the steady states of this IBVP. Do this in two ways. In the first way, solve the steady state equation $0 = \alpha^2 S''(x) + F(x)$ by integrating twice and then choosing the constant of integration such that $S$ satisfies the BCs. Express the iterated double definite integral in the answer as a single definite integral (*Hint:* change the order of integration). In the second way use formulas (4.66) and (4.68) with the $T_n$'s constants).

(b) Explicitly write out the solution of the IBVP, separating out the transient part from the nontransient part. Determine under what conditions the eventual temperature distribution

$$u_\infty(x) = \lim_{t \to \infty} u(x, t)$$

exists.

(c) Assuming that $u_\infty$ exists, determine if it necessarily coincides with some steady state $S$ of the IBVP.

15. Prove Theorem 4.2. Also show how to get equation (4.70).

16. (*Heat flow in non-unit length rods*) All of your previous work on solving 1-D Heat IBVPs has been limited to the case where the rod has unit length $L = 1$. This exercise examines how things change when $L \neq 1$. Thus consider the general 1-D heat IBVP

$$u_t = \alpha^2 u_{xx} + F,$$

$$-\kappa_1 u_x(0, t) + h_1 u(0, t) = g_1(t),$$

$$\kappa_2 u_x(L, t) + h_2 u(L, t) = g_2(t),$$

$$u(x, 0) = \phi(x),$$

for $x \in [0, L]$ and $t > 0$. It is possible to use a change of scale transformation to transform this problem into one for a unit-length rod and use this to solve the original problem. Do not use this approach here, but rather just reexamine all the techniques used for the case $L = 1$ and see what changes occur for the general case (this will also provide a quick review of the subject).

(a) Since the problem is non-homogeneous, verify that the technique of transforming to homogeneous BCs can be used exactly as before. Namely, write $u = S + U$, choose $S$ of the form $S(x, t) = C(t)x^2 + A(t)x + B(t)$ to satisfy the BCs, and then get a new IBVP for $U$ that is semi-homogeneous. It is easy to see that $S$ now depends on the parameter $L$ (length of the rod).

(b) To complete the solution, examine how the solution of the semi-homogeneous problem is affected for the non-unit-length case. Thus look at solving

$$u_t = \alpha^2 u_{xx} + F,$$

$$-\kappa 1 u_x(0, t) + h_1 u(0, t) = 0,$$

$$\kappa_2 u_x(L, t) + h_2 u(L, t) = 0,$$

$$u(x, 0) = \phi(x).$$

Verify that the series

$$u(x, t) = \sum_{n=0}^{\infty} T_n(t) X_n(x)$$

is formally the solution of the problem. Here the $X_n$s and $T_n$s are essentially the same as before. Specifically, the $X_n$'s are the Sturm–Liouville eigenfunctions for the SL problem

$$X'' = kX,$$

$$-\kappa_1 X'(0) + h_1 X(0) = 0,$$

$$\kappa_2 X'(L) + h_2 X(L) = 0,$$

and they are determined by going through the cases $k = 0, k = \lambda^2, k = -\lambda^2$ to get the eigenvalues and eigenfunctions. The only change is that now the eigenvalues $\lambda_n$ and eigenfunctions $X_n$ will depend on the parameter $L$. The $T_n$'s are given (*defined*) by the same integral formula:

$$T_n(t) = \left[ A_n + \int_0^t F_n(s)e^{a_n s}\, ds \right] e^{-a_n t}.$$

Here

$$a_n = \alpha^2 \lambda_n^2,$$

and $F_n(t), A_n$ are the Fourier coefficents of $F(x, t), \phi(x)$ with respect to the eigenfunctions $X_n$:

$$F(x, t) = \sum_{n=0}^{\infty} F_n(t) X_n(x),$$

$$\phi(x) = \sum_{n=0}^{\infty} A_n X_n(x).$$

Verify, using heuristic arguments like those in equations (4.10)–(4.12), that these coefficients are given by the integral formulas

$$F_n(t) = L_n^{-1} \int_0^L F(x, t) X_n(x) dx,$$

$$A_n = L_n^{-1} \int_0^L \phi(x) X_n(x) dx,$$

where the normalizing constants $L_n$ are now given by

$$L_n = \int_0^L [X_n(x)]^2 dx.$$

Note that the arguments in equations (4.10)–(4.12) use the orthogonality property of the eigenfunctions; i.e., for $n \neq r$ the eigenfunctions $X_n$ and $X_r$

are orthogonal on the interval $[0, L]$:

$$\int_0^L X_n(x) X_r(x) dx = 0.$$

The proof of this is exactly the same as what you did in Exercise 15 above. In summary, observe that there are only three basic changes in dealing with a rod of length $L$:

(1) Due to the BCs involving the endpoint $x = L$, the function $S$ used to transform to homogeneous BCs will involve $L$ as a parameter.

(2) Due to the BCs in the associated Sturm–Liouville problem, the parameter $L$ enters into the expressions for the eigenfunctions and eigenvalues. So, · be careful at this stage of the solution.

(3) The integral formulas for the Fourier coefficients and normalizing constants will have $[0, L]$ as the interval of integration.

(c) Use what you learned in parts (a) and (b) to solve several of the heat IBVPs assigned for this section (especially Exercises 4, 9, and 13 if these were assigned) for rods of length $L = 10$.

# 5 Computational Analysis

The previous two chapters developed all the details for solving the general 1-D heat problem by means of a series of the form

$$u(x, t) = S(x, t) + T_0(t) + \sum_{n=1}^{\infty} T_n(t) X_n(x). \qquad (5.1)$$

These details involved (1) transforming to homogeneous BCs (this is where the $S$ comes in), resulting in a new, semi-homogeneous IBVP to solve; (2) solving the semi-homogeneous problem by determining the eigenfunctions $X_n$, the eigenvalues, and the amplitude functions $T_n$.

The series expression (5.1) for the solution to the heat IBVP is particularly suitable for interpretation by computer. So far, you have had experience with interpreting the solution manually by graphing the initial temperature distribution, graphing the steady state (if any), and using your understanding of what the PDE + BCs say to sketch some of the intermediate temperature distributions. Now however, we will use *Maple*, a computer algebra system (CAS), to get a more exact description of the series solution. Other CASs you might want to use are *Mathematica, MathCad, Macsyma,* and *Derive*. In addition to using the computer to plot an approximation to the temperature distribution curve $u(x, t)$ at time $t$, we will also use it for several related activities. One is finding the eigenvalues $-\lambda_n^2$ in the more complicated Sturm–Liouville problems where the $\lambda_n$'s are not known exactly except through a defining equation they must satisfy (e.g., $\tan \lambda = \lambda$). The other activity involves understanding Fourier analysis better in its own right, not necessarily connected with some heat problem.

## 5.1.  Plotting Solutions of Heat Problems

The best way to explain how to use mathematical software to augment our analysis of heat flow problems is just to work through a few examples. By the time you finish this section, you should appreciate the value of having series solutions to IBVPs, even though from a human point of view these series can be very complicated and the information they contain is not readily apparent without a computer.

**Example 5.1**    This first example is rather elementary, since it can be readily done by hand. However, in using the computer to assist you with any task, it is always best to check your methodology and algorithm on an example for which you know the answer. Thus we consider the IBVP

$$u_t = u_{xx} \qquad \text{for } 0 < x < 1, t > 0,$$

$$u(0, t) = 0 \qquad \text{for } t > 0,$$

$$u(1, t) = 0 \qquad \text{for } t > 0,$$

$$u(x, 0) = x \qquad \text{for } 0 < x < 1.$$

The interpretation of the problem is this: The rod is initially heated so that the temperature increases uniformly from 0 degrees at the left end to 1 degree at the right end. Immediately thereafter, both ends of the rod are brought to, and held at, 0 degrees temperature for the duration of the experiment. The heat will flow from the interior of the rod towards the ends, where it is extracted. Thus the eventual temperature distribution will be 0 degrees throughout the rod. Figure 5.1 shows a hand-drawn sketch of the initial temperature distribution $\phi(x) = x$, the final temperature distribution $S(x) = 0$, and some of the intermediate temperature curves. The exact series solution of this problem is

$$u(x, t) = \sum_{n=1}^{\infty} A_n e^{-n^2 \pi^2 t} \sin n\pi x,$$

where:

$$A_n = \frac{2(-1)^{n+1}}{n\pi}.$$

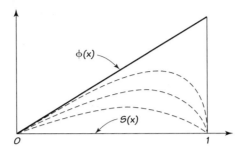

Figure 5.1.   Hand-drawn temperature curves.

The theory tells us that the sum of the first $k$ terms of the series,

$$u(x, t, k) = \sum_{n=1}^{k} A_n e^{-n^2\pi^2 t} \sin n\pi x,$$

will give an approximation to the actual temperature function:

$$u(x, t) \approx u(x, t, k).$$

Precise estimates on the closeness of this approximation and other details from numerical analysis are not discussed here. Suffice it to say that we expect the approximation $u(x, t, k)$ to be fairly good for large $k$ (with some exceptions, as we shall see).

The following *Maple* code illustrates how to produce graphs of the approximate temperature distributions $u(x, t, k)$ for various times $t$. First, for convenience, we define the Fourier coefficients $A_n$, and then we define $u(x, t, k)$ via a procedure statement with a do loop to perform the summation from 1 to $k$:

```
A := n -> 2*(-1)^(n+1)/(n*Pi);
u := proc(x,t,k) s:=0;
    for n from 1 to k do
    s := s + A(n)*exp(-n^2*Pi^2*t)*sin(n*Pi*x) od;
    end;
```

We can now use the plot command to plot any of the curves $u(x, t, k)$ for a given time $t$ and a given $k$. We begin with $t = 0$, in order to see how well the approximation matches the initial temperature distribution, whose graph is a simple straight line on the interval [0,1]. The approxmation is

$$x \approx u(x, 0, k) = \sum_{n=1}^{k} A_n \sin n\pi x,$$

which is just an attempt to approximate the graph of $\phi(x) = x$ with a sum of sine waves. We produce two separate *Maple* plots with the commands

```
plot(u(x,0,20),x=0...1);
plot(u(x,0,50),x=0...1);
```

These plots are shown in Figures 5.2–5.3. As you can readily see, the approximation with $k = 20$ is pretty poor, while the one with $k = 50$ is fairly good except near $x = 1$. This lack of closeness in the approximation near $x = 1$, which persists for all values of $k$, is known as the *Gibbs phenomenon*. (you should try some larger values of $k$ to observe this further). It arises essentially because of the discontinuity in the graph of $\phi$ when it is extended to an odd periodic function on all of $\mathbb{R}$.

From the above experimentation, we find that $k = 50$ produces a reasonable approximation $u(x, 0, 50)$ to $\phi(x) = x$ throughout most of the interval $[0, 1]$, and so we use $k = 50$ in plotting the intermediate temperature distributions $u(x, t, 50)$ for various times $t$. Choosing the times $t$ for the plots takes some experimentation too, since the temperature in the rod decays rather rapidly to $S(x) = 0$ throughout. We choose here $t = 0, 0.01, 0.1, 0.2$ for illustration. These are all plotted in the same picture by using the command:

```
plot({u(x,0,50),u(x,.01,50),u(x,.1,50),u(x,.2,50)},x=0..1);
```

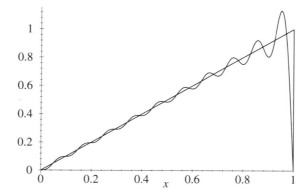

Figure 5.2.   Computer plot of $u(x, 0, 20)$, which is an approximation to the initial temperature curve (the straight line in the pictures).

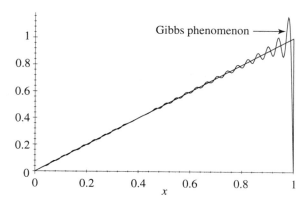

Figure 5.3.   Computer plot of $u(x, 0, 50)$, which is an approximation to the initial temperature curve (the straight line in the pictures).

The results are shown in Figure 5.4

The plots show how the Gibbs phenomenon vanishes for times $t > 0$. The initial discontinuity of having the temperature at $x = 1$ jump from being 1 at time zero to being 0 thereafter is present only in the initial curve. The plots also give some indication of how rapidly the temperature decays to zero. In theory, it takes infinitely long for the temperature to reach 0 degrees throughout, but on any practical scale this happens in a finite amount of time, which can be found experiementally (see the exercises).

**Remarks on *Maple*:**   Here are some general observations to bear in mind when learning and using *Maple* (or any computer algebra system).

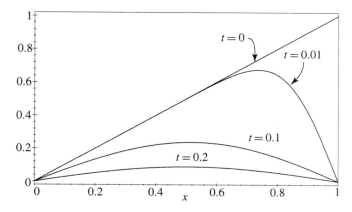

Figure 5.4.   Computer plots of $u(x, t, 50)$ for times $t = 0, 0.01, 0.1, 0.2$, which are approximations to the true temperature curves at these times.

(1) *Maple* has many features and commands that allow a given task to be done in different ways. It is best to learn a few basic commands well, rather than trying to master everything at once.

(2) *Maple*'s data structures and syntax are similar to those of many programming languages. It also has, built in, numerous commands, which are really algorithms, for performing certain mathematical operations. These commands usually consist of numerous lines of code and may have many exotic features in them. For example, the sum command consists of several pages of code, and we could have used this instead of the procedure with the do loop in the above example. For instance, we could have written

```
A := n -> 2*(-1)^(n+1)/(n*Pi);
u := (x,t,k)->
     sum(A(n)*exp(-n^2*Pi^2*t)*sin(n*Pi*x),n=1..k);
```

to accomplish the same thing. However, to do so seems rather silly. From a user's standpoint, the sum command may be more convenient, shorter, and give a prettier output, but from a computational standpoint it is equivalent to substituting about 50 lines of code (designed to do many tasks other than the simple one here) into your *Maple* worksheet. In certain situations, the overall execution time can take longer.

(3) As far as execution time goes, the code we used in the above example is fairly quick. But later, when we get to 3-D heat problems where triple sums require three nested do loops, we will need more efficient code. One thing that makes the above code inefficient is that the coefficients A(n) have to be recomputed every time the procedure proc(x,t,k) is executed. If there are 100 values of $x$ and 10 values of $t$ used in producing the plots, this means recomputing each $A(n)$ one thousand times. As an alternative to this, make A a one-dimensional array and compute each of its values just once. For example, the following code does this:

```
f := n -> 2*(-1)^(n+1)/(n*Pi);
A := array(1..200);
for n from 1 to 200 do A[n] := f(n) od:
u := proc(x,t,k) s:=0;
   for n from 1 to k do
   s := s + A[n]*exp(-n^2*Pi^2*t)*sin(n*Pi*x)
   od; end;
```

Notice several important things here. Now A is an array with 200 elements (choose a larger number if you think more $A_n$'s will be needed), and the $n$th element of A is referred to by A[n] with square brackets (as opposed to the parentheses in the function notation f(n)). Secondly, the do loop that assigns the values to A ends with a *colon*. This will suppress the output from being displayed on the terminal (you do not usually want to see a list of these 200 values).

**Example 5.2**  This example demonstrates better the need for computer assistance in the analysis of heat flow. The initial and final temperature distributions are easy enough to draw by hand, but the intermediate temperature distributions are essentially impossible to discern from the series solution without a computer. The IBVP is

$$u_t = u_{xx} - ae^{-t} \qquad \text{for } 0 < x < 1, t > 0,$$

$$u(0, t) = 1 \qquad \text{for } t > 0,$$

$$u(1, t) = ae^{-t} \qquad \text{for } t > 0,$$

$$u(x, 0) = 0 \qquad \text{for } 0 < x < 1,$$

and it is one we solved in the previous chapter (Example 4.5). The series solution is

$$u(x, t) = (ae^{-t} - 1)x + 1 + \sum_{n=1}^{\infty} A_n e^{-n^2\pi^2 t} \sin n\pi x,$$

where the $A_n$'s are the Fourier sine coefficients of $(1 - a)x - 1$:

$$A_n = 2(1 - a) \int_0^1 x \sin n\pi x \, dx - 2 \int_0^1 \sin n\pi x \, dx$$

$$= \frac{2[(1 - a)(-1)^{n+1} + (-1)^n - 1]}{n\pi}.$$

$$= \frac{-2[a(-1)^{n+1} + 1]}{n\pi}.$$

The interpretation of this IBVP is that the rod initially is at 0 degrees throughout and thereafter has its left end held at a constant 1 degree, while the right-end temperature decreases with time according to $ae^{-t}$. There is also a heat sink decreasing the temperature uniformly along the rod at a rate of $ae^{-t}$ at time $t$ (for large $t$ this is essentially negligible). From the series solution, we see that the eventual temperature distribution is

$$\lim_{t \to \infty} u(x, t) = -x + 1.$$

However, as mentioned, this information gives us no clue as to what the intermediate temperature distributions look like. Therefore, we execute the following *Maple* code to investigate the heat flow:

```
a:= 2;
A:= n -> -2*(a*(-1)^(n+1)+1)/(n*Pi);
u := proc(x,t,k) s:=0;
    for n from 1 to k do
    s := s + A(n)*exp(-n^2*Pi^2*t)*sin(n*Pi*x)
    od;
    s := (a*exp(-t)-1)*x + 1 + s;
end;
```

Notice that the code has the parameter *a* built into it, with value 2 assigned at the outset. Later on, after investigating the heat flow for the $a = 2$ case, we can reassign the value of *a* to any other desired value. We now plot the approximate solutions $u(x, t, k)$ for various times *t* using the *Maple* command

```
plot({u(x,.05,100),u(x,.1,100),u(x,.2,100),u(x,.4,100),
u(x,.5,100),u(x,.6,100),u(x,.7,100),u(x,.8,100),
u(x,.9,100),u(x,1,100),u(x,5,100)},x=0..1);
```

The number of terms ($k = 100$) in the series that we have chosen to use, as well as the particular time values, resulted from some initial experimentation. The resulting plots are shown in Figure 5.5.

As you can see from the picture, it takes about 5 seconds to go from the initial temperature distribution ($\phi(x) = 0$) to the final (steady-state) temperature distribution ($S(x) = -x + 1$). At times $t = 0.05, 0.1, 0.2$, the temperature distributions are somewhat parabolic in shape. By time $t = 0.4$ these have flattened out into essentially straight-line temperature distributions. This can also be explained from inspection of the series expression for $u(x, t)$, which we abbreviate by

$$u(x, t) = (ae^{-t} - 1)x + 1 + w(x, t).$$

Note that each of the summands in $w(x, t)$ contains a factor of the form $e^{-n^2\pi^2 t}$, $n = 1, 2, ...$, and each of these is considerably smaller than $ae^{-t}$ for *t* large. Otherwise stated, for large *t* we have the approximation

$$u(x, t) \approx (ae^{-t} - 1)x + 1.$$

The graph of the function on the right side (for fixed *t*) is just a straight line. For $t = 0.4, ..., 5$, these are essentially the straight lines you see in Figure 5.5.

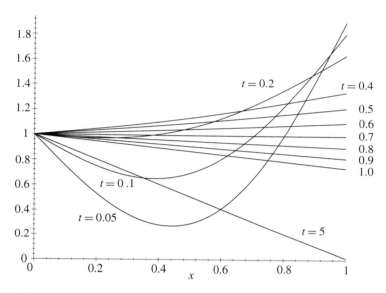

Figure 5.5.   Plots of the temperature distributions $\phi(x) = 0$ (initial), $S(x) = -x + 1$ (final), and intermediate ones for selected times $t$.

## 5.2.   Sturm–Liouville Problems

The first complicated Sturm–Liouville problem we encountered in the text was

$$X'' = kX,$$

$$X(0) = 0,$$

$$X'(1) + X(1) = 0,$$

and we found that the eigenfunctions are $X_n(x) = \sin \lambda_n x$ and the eigenvalues are $k_n = -\lambda_n^2$. The $\lambda_n$'s were found to be the solutions (roots) of the equation

$$\tan \lambda = -\lambda.$$

The determination of the approximate values of the $\lambda_n$'s required two steps: (1) sketch the graphs of $y = \tan \lambda$ and $y = -\lambda$ and read off, from the points of intersection, the rough first estimates of $\lambda_1, \lambda_2, \ldots$, and (2) improve these estimates by using a bisection method (trial and error) on a hand-held calculator. This manual process is relatively quick and easy.

To accomplish the same thing using mathematical software is entirely similar, but step (2) is somewhat quicker. Using *Maple*, step (2) amounts to just issuing a command like

```
fsolve(tan(x)=-x,x, 1..2.5);
```

This should return a value of 2.028757838 for $x$. Note that in the above command, x,1..2.5 indicates that you want *Maple* to search for a solution $x$ with $x \in [1, 2.5]$. Thus while the above command is quick and efficient, you will need to have some idea about where all the solutions lie so that the search interval can be specified. This is where step (1) comes in. To have the computer plot $\tan x$ and $-x$ is simple enough, but because of the vertical asymptotes in $\tan x$, you should specify a range of $y$ values that are acceptable, for example,

```
plot({tan(x), -x}, x=0..15,y=15..15);
```

This produces the plot shown in Figure 3.3 back in Chapter 3. As you know, the plot of $y = \tan x$ can be rather inaccurate near its vertical asymptotes, and this can cause inaccuracies in the results from the fsolve command. In order to circumvent this problem, observe that the equation we are trying to solve, $\tan x = -x$, is the same as $\sin x = -x \cos x$, or more simply $\sin x + x \cos x = 0$. For convenience, we let

$$f(x) = \sin x + x \cos x,$$

and consider finding the roots of the equation $f(x) = 0$. We did not use this approach when working this problem by hand because graphing $f(x)$ is more difficult for us. For the computer it's actually easier, since this function has no asymptotes. To do this, we use the *Maple* code

```
f:= x -> sin(x) + x*cos(x);
plot(f(x),x=0..15);
```

Figure 5.6 shows the resulting plot, which is quite a bit more satisfactory than Figure 3.3 for determining the intervals in which the solutions lie.

From this graphical analysis we can discern the intervals in which the first five eigenvalues lie. We could use the original approach to find these with commands like fsolve(sin(x)+x*cos(x)=0, x,1..2.5), but it's just as easy to use the function $f(x)$ now that it has been defined. Thus we use the following commands to find the five eigenvalues:

```
fsolve( f(x)= 0, x, 1..2.5);
fsolve( f(x)= 0, x, 4..6);
fsolve( f(x)= 0, x, 7..9);
fsolve( f(x)= 0, x, 10.5..12);
```

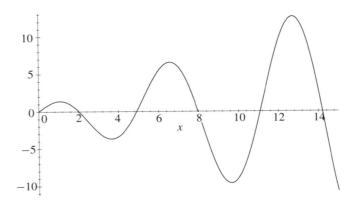

Figure 5.6.    Computer plot of $f(x) = \sin x + x \cos x$.

```
fsolve( f(x)= 0, x, 14..15);
```

This produces the values listed in Table 3.1 back in Chapter 3, where we first considered this problem.

## 5.3.    Fourier Analysis

As a purely instructional tool, computer plots of truncated Fourier series for well-known functions can be of great value. While the work in the first section with computer plots was absolutely essential in studying a given heat flow, the work we do here is merely to obtain a better understanding of the sense in which a Fourier series actually represents a given function. For the sake of simplicity, we consider only Fourier sine series. The representation of a function $\phi$ defined on $[0, 1]$ is

$$\phi(x) = \sum_{n=1}^{\infty} A_n \sin n\pi x, \tag{5.2}$$

for $x \in [0, 1]$, and $A_n = 2 \int_0^1 \phi(x) \sin n\pi x \, dx$, the $n$th Fourier sine coefficient of $\phi$. So far, throughout all of our work we have taken equation (5.2) for granted. What we have been assuming is that (1) the series on the right side of the equation actually converges, and (2) it converges to the number $\phi(x)$. As we will see (and have alluded to before), these two assumptions may not always be true. The questions raised here began with a heated controversy between Euler and D'Alembert in the eighteenth century over the possibility of so representing an arbitrary function by a series of sine waves (at that time, mathematicians still had not devised the definition of convergence of a series and also were not in common agreement about the classes of functions they were dealing with). There are precise answers to these questions and others that arose, which you can find in many advanced texts on Fourier analysis (also called *harmonic analysis* because of the connection between the sine waves $\sin n\pi x$ and the musical scale). We intend to explore these questions here in an informal way. For this we will compare the graph of $\phi$ with that of the function obtained by truncating the Fourier series after $k$ terms:

$$\phi(x, k) \equiv \sum_{n=1}^{k} A_n \sin n\pi x.$$

**Example 5.3**   Suppose we take

$$\phi(x) = 1 - x^2,$$

for $x \in [0, 1]$. Using Table D.1 in Appendix D, we find that the Fourier coefficients are given by the expression

$$A_n = 2\left[\frac{1}{n\pi} + \frac{2[1 - (-1)^n]}{n^3\pi^3}\right].$$

Some *Maple* code to define the truncated Fourier series and plot this for several values of $k$ is

```
A := n-> 2*(1/(n*Pi) + 2*(1-(-1)^n)/(n^3*Pi^3));
p := proc(x,k) s:=0; for n from 1 to k do
        s:= s + A(n)*sin(n*Pi*x) od; end;
plot({1-x^2, p(x,20)},x=0..1);
plot({1-x^2, p(x,50)},x=0..1);
e:=(x,k)->abs(1-x^{2}-p(x,k));
plot(e(x,20),x=0..1);plot(e(x,50),x=0..1);
```

The resulting plots are shown in Figures 5.7–5.8. Also shown in these figures are the plots of the *errors*, i.e., the plots of the functions $e(x, k) = |\phi(x) - \phi(x, k)|$ for $k = 20, 50$. Each type of plot displays qualitatively how well the partial Fourier series approximates the given function. We see that the values of $e(x, 20)$ are all less than $0.1$, except near $x = 0$. More precisely $e(x, 20) < 0.1$ for all $x \in [0.078, 1]$. Using 50 terms, we see that $e(x, 50) < 0.05$ for all $x \in [0.031, 1]$. In general (with the exception of $x = 0$), the errors will approach zero as the number of terms increases: $\lim_{k\to\infty} e(x, k) = 0$. The exceptional point $x = 0$ is due to a discontinuity in the *odd, periodic extension* of $\phi$. This function, which

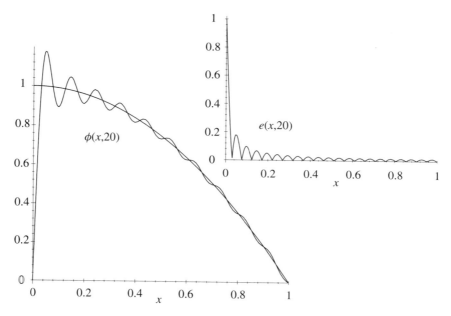

Figure 5.7.    Plot of $\phi(x, 20)$, which is the function given by 20 terms in the Fourier series for $\phi(x) = 1 - x^2$. Also shown in the upper right is the plot of $e(x, 20) = |\phi(x) - \phi(x, 20)|$, which displays the errors in the 20 term approximation.

we denote by $\phi_o$, is defined by first extending $\phi$ to an odd function on $[-1, 1]$,

$$\phi_o(x) = \begin{cases} -\phi(-x) & \text{if } -1 \leq x < 0, \\ \phi(x) & \text{if } 0 \leq x \leq 1, \end{cases}$$

and then extending to a periodic function (of period 2) on all of $\mathbb{R}$. It is this function $\phi_o$ that the Fourier sine series is trying to represent:

$$\phi_o(x) = \sum_{n=1}^{\infty} A_n \sin n\pi x, \tag{5.3}$$

for $-\infty < x < \infty$. Figure 5.9 shows this for 50 terms of the series, i.e., a plot of $\phi(x, 50)$ on the interval $[-2, 2]$. In this example, $\phi_o$ is piecewise continuous and infinitely differentiable on the open intervals between its points of discontinuity ($x = 0, \pm 2, \pm 4, \ldots$). The theory (see [Pin 91, pages 67–75], [Str 92, pages 124–128] for elementary accounts) tells us that the series representation (5.3) is valid; i.e., $\lim_{k \to \infty} \phi(x, k) = \phi_o(x)$ at all points $x$ other than the points of discontinuity, and further, at a point $x$ of discontinuity, the Fourier series converges to the

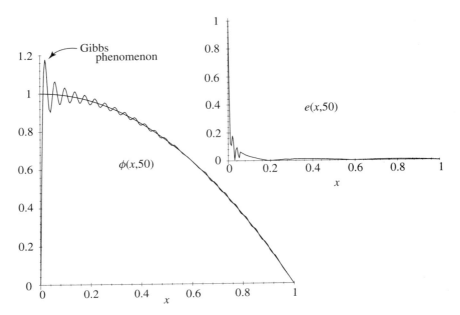

Figure 5.8.   Plot of $\phi(x, 50)$, which is the function given by 50 terms in the Fourier series for $\phi(x) = 1 - x^2$. Also shown, in the upper right, is the plot of $e(x, 50) = |\phi(x) - \phi(x, 50)|$, which displays the errors in the 50 term approximation.

average value in the "jump" at the discontinuity:

$$\sum_{n=1}^{\infty} A_n \sin n\pi x = \frac{1}{2}\left(\phi_o(x+) + \phi_o(x-)\right).$$

Here $\phi_o(x\pm) = \lim_{s \to x\pm} \phi_o(s)$ are the one-sided limits at $x$. For this example, the average value in the jump is 0, and this is exhibited in Figure 5.9 (each term $A_n \sin n\pi x$ in the series is 0 at the points of of discontinuity). The discontinuities also cause the *Gibbs phenomenon*: The approximations $\phi(\cdot, k)$, near a point of discontinuity will have errors larger than almost 9% of the magnitude of the jump, and this persists no matter how large $k$ is (see [ON 91, pages 1038-1039]). In this example, the magnitude of the jump is 2, and so the errors $e(x, k)$ are larger than approximately 0.18 for $x$ near a point of discontinuity. This is exhibited clearly in Figures 5.7– 5.8.

Another result from the theory is that (under the conditions above) the series not only converges *pointwise* to the value of the function at points of continuity ($\lim_{k \to \infty} \phi(x, k) = \phi_o(x)$ for every $x$ at which $\phi_o$ is continuous), but moreover, on each closed interval not containing a point of discontinuity, the convergence

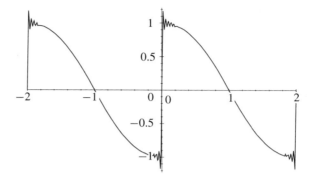

Figure 5.9.    Plot of $\phi(x, 50)$ on the interval $[-2, 2]$. This is the approximation to the odd, periodic extension $\phi_o(x)$ by 50 terms in the Fourier series.

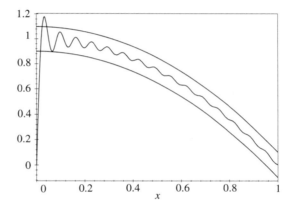

Figure 5.10.    An illustration of uniform convergence of the Fourier series on the interval $[0.04, 1]$. Here $\varepsilon = 0.1$, and the plots are the graphs of $\phi(x, 20)$, $\phi(x) + \varepsilon$, and $\phi(x) - \varepsilon$.

of the series is *uniform*. Uniform convergence is a stronger type of convergence than pointwise convergence, and its meaning is as follows. Suppose $[a, b]$ is an interval not containing a point of discontinuity. Then given $\varepsilon > 0$, there is a $K$ such that for all $k \geq K$, the graph of $\phi(\cdot, k)$ lies between the graphs of $\phi_o + \varepsilon$ and $\phi_o - \varepsilon$ on the interval $[a, b]$. Figure 5.10 illustrates this concept for this example. The interval considered is $[0.04, 1]$ and in the figure, $\varepsilon = 0.1$. By experimentation (doing plots for various values of $k$), we find that the first value $K$ for which the graph of $\phi(\cdot, K)$ lies between the graphs of $\phi \pm \varepsilon$ on $[0.04, 1]$ is $K = 30$.

## Exercises 5.3

The first group of exercises deals with Fourier analysis per se, unconnected with any heat problem. Expand each function $\phi$ in the type of Fourier series indicated.

$$\phi(x) = \sum_{n=0}^{\infty} A_n X_n(x).$$

Write the code for defining the approximation function

$$\phi(x, k) = \sum_{n=0}^{k} A_n X_n(x),$$

and plot the graphs of $\phi(x, k)$ for various values of $k$. Choose small values of $k$ where the approximations are not good as well as large values of $k$ where the approximations are better. Identify any points of discontinuity in the even $\phi_e$ or odd $\phi_o$ periodic extensions of $\phi$. Note: The even periodic extension is used when representing $\phi$ by a cosine series. Study the uniform convergence of the series on appropriately chosen subintervals $[a, b]$ of $[0, 1]$, by selecting several values of $\varepsilon$ and determining the first value of $k$ for which the graph of $\phi(\cdot, k)$ lies completely between the graphs of $\phi \pm \varepsilon$ on the interval $[a, b]$. Check that the Gibbs phenomenon is always about 9% of the magnitude of the jump at discontinuities.

1. $x = \sum_{n=1}^{\infty} A_n \sin n\pi x$.
2. $x = A_0 + \sum_{n=1}^{\infty} A_n \cos n\pi x$.
3. $x^2 = A_0 + \sum_{n=1}^{\infty} A_n \cos n\pi x$.
4. $x - x^2 = \sum_{n=1}^{\infty} A_n \sin n\pi x$.
5. $x - x^2 = A_0 + \sum_{n=1}^{\infty} A_n \cos n\pi x$.
6. $x = \sum_{n=1}^{\infty} A_n \sin \frac{2n-1}{2} \pi x$.
7. $1 = \sum_{n=1}^{\infty} A_n \cos \lambda_n x$ (where the $\lambda_n$'s are the solutions of $\cot \lambda = \lambda$).
8. $3x^2 - 5x + 4 = A_0 + \sum_{n=1}^{\infty} A_n \cos n\pi x$.
9. $10xe^{-4x} = \sum_{n=1}^{\infty} A_n \sin n\pi x$.
10. $x^{\sin 4\pi x} = \sum_{n=1}^{\infty} A_n \sin n\pi x$. Note: here and in the next problem you will have to use *Maple*'s numerical integration capability, since the the Fourier coeffficients

$$A_n = 2 \int_0^1 x^{\sin 4\pi x} \sin n\pi x \, dx$$

are not computable by hand. Some *Maple* code for this is

```
p:=x=>x^(sin(4*Pi*x)); A:=array(1..200);
for n to 200 do
  A[n]:=evalf(2*int(p(x)*sin(n*Pi*x),x=0..1) od:
```

*Maple*'s basic integration command is int, and it will attempt numerous routines to evaluate the integral in closed form or as a true value. If this is not possible (as it is here), it returns the integral unevaluated. To force *Maple* to use its numerical integration schemes, use evalf in conjunction with int, as we have done in the above code. This can be inefficient, since all the routines for exactly calculating the integral are tried first before *Maple* gives up and invokes the numerical schemes. If you wish, you can combine evalf with the inert form Int of the integration command to, theoretically, achieve a faster evaluation (although it's not noticeable here).

11. $x^{-1} \sin 4\pi x = \sum_{n=1}^{\infty} A_n \sin n\pi x$. If you are familiar with the cosine integral Ci$(x)$, you can compute the Fourier coefficients by hand. Otherwise, use numerical integration as in the last problem.

12. $\phi(x) = \sum_{n=1}^{\infty} A_n \sin n\pi x$, where

$$\phi(x) = \begin{cases} x & \text{for } 0 \leq x \leq 1/2, \\ x - 1 & \text{for } 1/2 \leq x \leq 1. \end{cases} \tag{5.4}$$

13. $\phi(x) = \sum_{n=1}^{\infty} A_n \sin n\pi x$, where

$$\phi(x) = \begin{cases} 1 & \text{for } 0 \leq x \leq 1/2, \\ -1 & \text{for } 1/2 \leq x \leq 1. \end{cases} \tag{5.5}$$

14. $\phi(x) = \sum_{n=1}^{\infty} A_n \sin n\pi x$, where

$$\phi(x) = \begin{cases} 0 & \text{for } 0 \leq x \leq 1/2, \\ x - 1/2 & \text{for } 1/2 \leq x \leq 1. \end{cases} \tag{5.6}$$

15. $\phi(x) = \sum_{n=1}^{\infty} A_n \sin n\pi x$, where

$$\phi(x) = \begin{cases} \sin 2\pi x & \text{for } 0 \leq x \leq 1/2, \\ 0 & \text{for } 1/2 \leq x \leq 1. \end{cases} \tag{5.7}$$

16. $\phi(x) = \sum_{n=1}^{\infty} A_n \sin n\pi x$, where $\phi(x) = H_{a,p,b}(x)$, and $H_{a,p,b}$ is the *triangular pulse function*

$$H_{a,p,b}(x) = \begin{cases} 0 & \text{for } 0 \leq x < a, \\ (x-a)/(p-a) & \text{for } a \leq x < p, \\ (x-b)/(p-b) & \text{for } p \leq x < b, \\ 0 & \text{for } b \leq x \leq 1. \end{cases} \tag{5.8}$$

Write a general code for this, but do the plots for $H_{1/4,3/8,1/2}$.

The next group of problems deals with finding Sturm–Liouville eigenvalues $-\lambda_n^2$ numerically (actually, all we need are the $\lambda_n$'s themselves). In each problem:

(a) Numerically compute a *sufficient* number of the $\lambda_n$'s. The actual number of these needed depends on the problem and on the function $f$ being approximated. You might want to store these values in an array for use in part (b).

(b) For the function $f(x) = x$ compute the Fourier series approximation

$$f(x, k) = \sum_{n=1}^{k} A_n X_n(x)$$

to $f(x)$ in terms of the eigenfunctions $X_n$. Then explore the goodness of these approximations by plotting $f(x, k)$ for several values of $k$ (small and large).

17.

$$X'(0) = 0,$$

$$X'(1) + X(1) = 0.$$

18.

$$-X'(0) + X(0) = 0,$$

$$X(1) = 0.$$

19.

$$-X'(0) + X(0) = 0,$$

$$X'(1) + X(1) = 0.$$

In the next group of exercises:

(a) Find the series representation for the solution $u(x, t)$ of the IBVP.

(b) Write the computer code defining the function $u(x, t, k)$, which is the sum of the first $k$ terms of the series (also add in the $S(x, t)$ and the 0th eigenfunction term if these are present in the solution).

(c) By experimenting with a few trial plots of $u(x, t, k)$, determine a "good" value to use for $k$ and several good intermediate times $0 < t < \infty$, that plotted together give a picture of the heat flow.

(d) Plot $u(x, t, k)$ for the values of $k$ and times $t$ you chose in part (c) (all plots in the same picture, please!), and label the different curves with the times $t$.

(e) Plot the evolution of the temperature over time at selected points along the rod. If there is a steady state, try to determine how long it takes for the temperature distribution to essentially reach this state (for all practical purposes). Do the selected points along the rod all appear to reach their steady-state values at about the same time? Why or why not?

(f) Based on the above parts of the analysis, summarize what your plots display and make comments about diffusion of heat through the rod and how the BCs, ICs, and internal heat source (if any) influence the diffusion.

*Note*: Most of these IBVPs you have already solved in previous exercises (marked by a box: $\boxed{\leadsto}$ ), so just use those answers here. Otherwise you need to work the problem from scratch. *Also*: For convenience, we take $\alpha = 1$ in these problems.

20. $\boxed{\leadsto}$ (*Insulated boundaries*) (See 2(a) in Exercises 4.1)

$$u_t = u_{xx} \qquad\qquad \text{for } 0 < x < 1, t > 0,$$

$$u_x(0, t) = 0 \qquad\qquad \text{for } t > 0,$$

$$u_x(1, t) = 0 \qquad\qquad \text{for } t > 0,$$

$$u(x, 0) = a(x - x^2) \qquad \text{for } 0 < x < 1.$$

Here $a$ is a parameter. Include this in your code, but do the problem for one choice of $a$ (say $a = 1$ or 2). How does the choice of $a$ influence the heat flow?

21. $\boxed{\leadsto}$ (*Fixed temperature and insulated boundaries*) (See 2(b) in Exercises 4.1)

$$u_t = u_{xx} \qquad \text{for } 0 < x < 1, t > 0,$$

$$u(0, t) = 0 \qquad \text{for } t > 0,$$

$$u_x(1, t) = 0 \qquad \text{for } t > 0,$$

$$u(x, 0) = x \qquad \text{for } 0 < x < 1.$$

22. $\boxed{\leadsto}$ (*Insulated and specified flux*) (See 2(c) in Exercises 4.1)

$$u_t = u_{xx} \qquad\qquad \text{for } 0 < x < 1, t > 0,$$

$$u_x(0, t) = 0 \qquad\qquad \text{for } t > 0,$$

$$u_x(1, t) + u(1, t) = 0 \qquad \text{for } t > 0,$$

$$u(x, 0) = 1 \qquad\qquad \text{for } 0 < x < 1.$$

23. ◻ (*Fixed temperature at the boundaries*) (See 2(f) in Exercises 4.1)

$$u_t = \alpha^2 u_{xx} \qquad\qquad \text{for } 0 < x < 1, t > 0,$$

$$u(0, t) = 0 \qquad\qquad \text{for } t > 0,$$

$$u(1, t) = 0 \qquad\qquad \text{for } t > 0,$$

$$u(x, 0) = H_{a,p,b}(x) \qquad \text{for } 0 < x < 1.$$

Here $H_{a,p,b}$ is the triangular pulse function. Do the analysis for the symmetrical case $(a, p, b) = (1/4, 1/2, 3/4)$ and the asymmetrical case $(a, p, b) = (1/8, 3/8, 5/8)$. Compare and contrast.

24. ◻ (See 1(a) in Exercises 4.2)

$$u_t = \alpha^2 u_{xx} \qquad \text{for } 0 < x < 1, t > 0,$$

$$u(0, t) = 5 \qquad \text{for } t > 0,$$

$$u(1, t) = 2 \qquad \text{for } t > 0,$$

$$u(x, 0) = 0 \qquad \text{for } 0 < x < 1.$$

25. ◻ (See 1(d) in Exercises 4.2)

$$u_t = \alpha^2 u_{xx} + x(1 + t)^{-2} \qquad \text{for } 0 < x < 1, t > 0,$$

$$u(0, t) = 0 \qquad\qquad\qquad\qquad \text{for } t > 0,$$

$$u(1, t) = t(1 + t)^{-1} \qquad\qquad\quad \text{for } t > 0,$$

$$u(x, 0) = \phi(x) \qquad\qquad\qquad\quad \text{for } 0 < x < 1.$$

Do the analysis for $\phi(x) = x - x^2$ and for $\phi(x) = 0$.

26. ◻ (See 2(a) in Exercises 4.2)

$$u_t = \alpha^2 u_{xx} - 6\alpha^2 \qquad \text{for } 0 < x < 1, t > 0,$$

$$u_x(0, t) = -2 \qquad\qquad \text{for } t > 0,$$

$$u_x(1, t) = 4 \qquad\qquad\; \text{for } t > 0,$$

$$u(x, 0) = 2x \qquad\qquad\; \text{for } 0 < x < 1.$$

27. ⊡ (See 1 in Exercises 4.3)

$$u_t = \alpha^2 u_{xx} + 1 \qquad \text{for } 0 < x < 1, t > 0,$$

$$u(0, t) = 0 \qquad \text{for } t > 0,$$

$$u(1, t) = 0 \qquad \text{for } t > 0,$$

$$u(x, 0) = 1 \qquad \text{for } 0 < x < 1.$$

28.

$$u_t = \alpha^2 u_{xx} + xe^{-t} \qquad \text{for } 0 < x < 1, t > 0,$$

$$u(0, t) = 0 \qquad \text{for } t > 0,$$

$$u(1, t) = 0 \qquad \text{for } t > 0,$$

$$u(x, 0) = 1 \qquad \text{for } 0 < x < 1.$$

29.

$$u_t = \alpha^2 u_{xx} + x - x^2 \qquad \text{for } 0 < x < 1, t > 0,$$

$$u_x(0, t) = 0 \qquad \text{for } t > 0,$$

$$u_x(1, t) = 0 \qquad \text{for } t > 0,$$

$$u(x, 0) = 1 \qquad \text{for } 0 < x < 1.$$

30. ⊡ (See 10 in Exercises 4.3)

$$u_t = \alpha^2 u_{xx} + 1 \qquad \text{for } 0 < x < 1, t > 0,$$

$$u(0, t) = 0 \qquad \text{for } t > 0,$$

$$u(1, t) = 1 \qquad \text{for } t > 0,$$

$$u(x, 0) = x \qquad \text{for } 0 < x < 1.$$

31. ⊡ (See 11 in Exercises 4.3)

$$u_t = \alpha^2 u_{xx} \qquad \text{for } 0 < x < 1, t > 0,$$

$$u(0, t) = 0 \qquad \text{for } t > 0,$$

$$u(1, t) = \sin t \qquad \text{for } t > 0,$$

$$u(x, 0) = 1 \qquad \text{for } 0 < x < 1.$$

In this problem make sure you look at early times, before the heat flow settles into a sinusoidal pattern. Determine approximately how long this takes.

The next exercise deals with certain integral formulas that come up in calculating Fourier coefficients:

32. For $a > 0$ and $p = 0, 1, 2, \ldots$, define polynomials $h_p(x, a)$ by

$$h_p(x, a) = \sum_{k=0}^{[p/2]} \frac{(-1)^k p!\, x^{p-2k}}{(p - 2k)!\, a^{2k+1}}.$$

By convention, $h_0(x, a) \equiv \frac{1}{a}$, and $[p/2]$ is the greatest integer $\leq p/2$. Show that the following integral formulas hold:

(a)

$$\int x^p \sin ax \, dx = -h_p(x, a) \cos ax + \frac{p}{a} h_{p-1}(x, a) \sin ax.$$

(b)

$$\int x^p \cos ax \, dx = h_p(x, a) \sin ax + \frac{p}{a} h_{p-1}(x, a) \cos ax.$$

# 6 Two-Dimensional Heat Flow

All of the analysis we used in the preceeding chapters to solve the 1-dimensional heat flow problem carries over to the solution of similar problems in two dimensions. Thus we will use the technique of separation of variables to get the expression of the general solution in terms of a series involving Sturm–Liouville eigenfunctions and eigenvalues, and then look at the resulting Fourier analysis that arises from particularizing the general solution so as to satisfy the IC. All the ingredients are basically the same, except that now there is two times as much work. However, what we wish to discuss here is more than just a routine extension of the 1-dimensional theory. As we shall see, the geometry has an unusual and interesting effect on the solutions in higher dimensions, and the technique of transforming from non-homogeneous to homogeneous BCs is perhaps an unexpected extension of the 1-D case.

## 6.1. The 2-D Heat Equation

The two-dimensional heat flow problem (like the 1-D problem) is a special case (idealization) of the 3-D heat flow problem discussed in the general derivation of the heat equation. We would like again to describe the circumstances whereby the heat flow is forced to be in two dimensions.

Thus, suppose the solid $\Omega$ is a thin plate (see Figure 6.1) sandwiched between a two-dimensional region $R$ in the $x$-$y$ plane and an identical region $R'$ centered above $R$ at a distance $d$ from the $x$-$y$ plane. The boundary $\partial\Omega$ of the solid $\Omega$ then consists of the two flat surfaces $R$ and $R'$, together with the vertical curved surface $\Gamma$, as shown. To ensure that the heat flow is two-dimensional, we make the following assumptions:

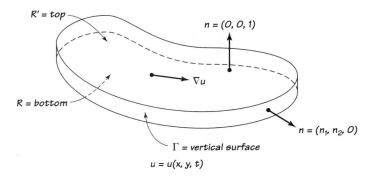

Figure 6.1.    The region $\Omega$: a thin plate.

(1) The top and bottom surfaces $R$ and $R'$ are insulated.

(2) On the vertical surface $\Gamma$, the BCs do not depend on the $z$ coordinate.

(3) The initial temperature distribution $\phi$ does not depend on $z$; i.e., $\phi = \phi(x, y)$ is a function of $x$ and $y$ only.

(4) The internal heat source density does not depend on $z$; that is, $F = F(x, y, t)$.

Under these assumptions, we look for a solution of the 3-D IBVP that is a function of $x$, $y$, and $t$ only: $u = u(x, y, t)$. In this case, one has $\nabla u = (u_x, u_y, 0)$, so that the heat flux vector field $v = -\nabla u$ is parallel to the $x$-$y$ plane at all points $(x, y, z)$ in $\Omega$. Because of this, one sees that $\nabla u \cdot n = 0$ at all points on the bottom and top boundary surfaces $R$ and $R'$. This is so because the outward-directed normal is $n = (0, 0, -1)$ at all points on $R$ and is $n = (0, 0, 1)$ at all points on $R'$. Thus two of the boundary conditions, insulated top and bottom, are automatically satisfied. To express the BCs on the remaining part, the vertical surface $\Gamma$, let $\partial R$ denote the curve that bounds the region $R$, that is, $\partial R$ = the boundary of $R$ (see Figure 6.2). Then the vertical surface $\Gamma$ is obtained by translating $\partial R$ in the $z$ direction. Because of the geometry involved, the outward-directed normal to the vertical surface is parallel to the $x$-$y$ plane and does not depend on $z$; namely, $n = (n_1(x, y), n_2(x, y), 0)$ at all points $(x, y, z)$ on $\Gamma$. Thus the remaining BC reduces to

$$\kappa(x, y) \left[ u_x(x, y, t) n_1(x, y) + u_y(x, y, t) n_2(x, y) \right] + h(x, y) u(x, y)$$

$$= g(x, y, t),$$

for all points $(x, y)$ on $\partial R$. The initial condition is

$$u(x, y, 0) = \phi(x, y), \tag{6.1}$$

for all points in $R$. Finally, the (non-homogeneous) heat equation, under the above assumptions, reduces to

$$u_t = \alpha^2(u_{xx} + u_{yy}) + F. \qquad (6.2)$$

Thus we can dispense with the 3-dimensional region $\Omega$ and focus attention on the 2-dimensional planar region $R$ with boundary curve $\partial R$, and now let $n$ denote the outward directed normal to this curve at each point (see Figure 6.2).

Thus the two-dimensional heat IBVP is to determine the function $u = u(x, y, t)$ that satifies

$$u_t = \alpha^2 \nabla^2 u + F \qquad \text{in } R, \text{ for } t > 0, \qquad (6.3)$$

$$\kappa \nabla u \cdot n + hu = g \qquad \text{on } \partial R \text{ for } t > 0, \qquad (6.4)$$

$$u = \phi \qquad \text{in } R, \text{ for } t = 0. \qquad (6.5)$$

The boundary conditions given in equation (6.5) generally will vary from point to point along the boundary $\partial R$, and in many examples the boundary will consist of several pieces, $\partial R = C_1 \cup C_2 \cup \cdots \cup C_r$, with perhaps different specifications on each piece, thus giving $r$ separate boundary conditions. Usually, $\kappa$ and $h$ will be constant functions on each piece of the boundary.

As was mentioned above, the analytic solution of the 2-D IBVP, especially the success of the separation of variables technique, depends on the geometry of the region $R$. In this chapter we limit the discussion to rectangular regions $R$ and defer the more complicated geometries to later chapters.

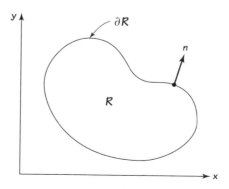

Figure 6.2.    The region $R$ and its boundary curve $\partial R$.

## 6.2.   Rectangular Regions $R$

A general rectangular region $R$ is

$$R = \{ (x, y) \,|\, 0 \le x \le L,\, 0 \le y \le M \}.$$

For simplicity, we discuss only the case where $R$ is the *unit square*:

$$L = 1 = M.$$

The boundary consists of four pieces: two horizontal sides and two vertical sides. See Figure 6.3. Note that the outward-directed normals $n$ to the boundary curves (which here are just straight line segments) are $n = (-1, 0)$ and $n = (1, 0)$ along the left and right vertcal sides, and $n = (0, -1)$ and $n = (0, 1)$ along the bottom and top horizontal sides respectively. Thus the general BCs in equation(6.5) naturally split up into the following four conditions:

$$-\kappa_1 u_x(0, y, t) + h_1 u(0, y, t) = g_1(y, t), \tag{6.6}$$

$$\kappa_2 u_x(1, y, t) + h_2 u(1, y, t) = g_2(y, t), \tag{6.7}$$

$$-\kappa_3 u_y(x, 0, t) + h_3 u(x, 0, t) = g_3(x, t), \tag{6.8}$$

$$\kappa_4 u_y(x, 1, t) + h_4 u(x, 1, t) = g_4(x, t), \tag{6.9}$$

for all $0 \le x \le 1$ and $0 \le y \le 1$. These BCs are rather tedious to write down and confusing to look at, and so when specifying the BCs in the ensuing examples, we will just draw a picture and label the sides of the square with the BCs, as in Figure 6.3.

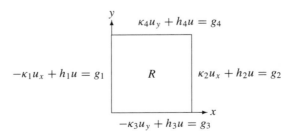

Figure 6.3.   The unit square $R$ and its boundary $\partial R$ (the sides).

## 6.3.    The Homogeneous Case

We begin first with the homogeneous case: a homogeneous heat equation and the BCs (6.6)–(6.9) homogeneous: $g_i = 0, i = 1, 2, 3, 4$. As might be expected, the separation of variables technique in a 2-D IBVP leads to a pair of Sturm–Liouville problems, and thus gives the solution in terms of a double summation series involving products of the eigenfunctions from each of the two SL problems. Here is how this works out. We look for a solution of the IBVP of the form

$$u(x, y, t) = X(x)Y(y)T(t).$$

Substituting this in the heat equation $u_t = \alpha^2(u_{xx} + u_{yy})$ gives

$$T'(t)X(x)Y(y) = \alpha^2 \left[ T(t)X''(x)Y(y) + T(t)X(x)Y''(y) \right].$$

Separating the parts of this equation that depend on $t$ from those that depend on $x, y$, one arrives at

$$\frac{T'(t)}{\alpha^2 T(t)} = \frac{X''(x)}{X(x)} + \frac{Y''(y)}{Y(y)}.$$

Since the left-hand side of this equation depends only on $t$, while the right-hand side depends only on $x, y$, it must be the case that each side is constant; i.e., for some constant $k$,

$$\frac{T'(t)}{\alpha^2 T(t)} = k, \tag{6.10}$$

$$\frac{X''(x)}{X(x)} + \frac{Y''(y)}{Y(y)} = k. \tag{6.11}$$

Note that in order for equation (6.11) to hold, each of the ratios there must be constant; i.e., $X''(x)/X(x) = k_1$ and $Y''(y)/Y(y) = k_2$, for some constants $k_1$ and $k_2$. Then $k = k_1 + k_2$. Thus we get out of all of this three differential equations:

$$X''(x) = k_1 X(x), \tag{6.12}$$

$$Y''(y) = k_2 Y(y), \tag{6.13}$$

$$T'(t) = k\alpha^2 T(t). \tag{6.14}$$

Next, if we substitute $u(x, y, t) = X(x)Y(y)T(t)$ into the homogeneous BCs for $u$ (equations (6.6)–(6.9), with $g_i = 0, i = 1, 2, 3, 4$), we get

$$\left[-\kappa_1 X'(0) + h_1 X(0)\right] Y(y)T(t), = 0 \tag{6.15}$$

$$\left[\kappa_2 X'(1) + h_2 X(1)\right] Y(y)T(t), = 0 \tag{6.16}$$

$$\left[-\kappa_3 Y'(0) + h_3 Y(0)\right] X(x)T(t), = 0 \tag{6.17}$$

$$\left[\kappa_4 Y'(1) + h_4 Y(1)\right] X(x)T(t). = 0 \tag{6.18}$$

To satisfy these conditions, it is necessary that the expression in the brackets be zero in each of the above four equations. Relabeling the constants, we arrive at the standard Sturm–Liouville BCs on $X$ and $Y$. Thus we get the following two SL problems:

$$X''(x) = k_1 X(x), \tag{6.19}$$

$$-\kappa_1 X'(0) + h_1 X(0) = 0, \tag{6.20}$$

$$\kappa_2 X'(1) + h_2 X(1) = 0. \tag{6.21}$$

and

$$Y''(y) = k_2 Y(y), \tag{6.22}$$

$$-\kappa_3 Y'(0) + h_3 Y(0) = 0, \tag{6.23}$$

$$\kappa_4 Y'(1) + h_4 Y(1) = 0. \tag{6.24}$$

As we have seen in the previous chapters, the solutions of these SL problems involve eigenvalues and eigenfunctions

$$\{\lambda_n\}_{n=0}^{\infty} \quad \text{and} \quad \{X_n\}_{n=0}^{\infty}$$

for the first SL problem and

$$\{\mu_m\}_{m=0}^{\infty} \quad \text{and} \quad \{Y_m\}_{m=0}^{\infty}$$

for the second problem. As in the one-dimensional case, one or the other of the above SL problems may not have 0 as an eigenvalue. *By convention* we take $\lambda_0 = 0$ and $\mu_0 = 0$, and also by convention we take $X_0 = 0$ if 0 is not an eigenvalue of the SL problem for $X$, and we take $Y_0 = 0$ if 0 is not an eigenvalue of the SL problem for $Y$. Now, the separation constant is $k = k_1 + k_2$, and each of $k_1$, $k_2$ can assume infinitely many values: $k_1 = -\lambda_n^2$, $k_2 = -\mu_m^2$, for $n$, $m = 0, 1, 2, \ldots$. Then too, $k = -(\lambda_n^2 + \mu_m^2)$ is a doubly indexed infinitude of values.

Now, going back to the equation for $T$ (equation (6.14)) and inserting one of the above values for $k$, we get $T'(t) = -(\lambda_n^2 + \mu_m^2)\alpha^2 T(t)$. This, of course, has solution $T(t) = \exp(-(\lambda_n^2 + \mu_m^2)\alpha^2 t)$ (up to a constant multiple).

If we put all the above results together, we get a double summation series

$$u(x, y, t) = \sum_{n=0}^{\infty}\sum_{m=0}^{\infty} A_{nm} e^{-(\lambda_n^2 + \mu_m^2)\alpha^2 t} X_n(x) Y_m(y), \qquad (6.25)$$

which, at least formally, satisfies the homogeneous PDE + BCs. This can be thought of as the general solution of the homogeneous problem, since it involves infinitely many arbitrary constants: $\{A_{nm}\}_{n,m=0}^{\infty}$. These constants are the (generalized) Fourier cofficients for the initial temperature function $\phi(x, y)$. Namely, if we take $t = 0$ in equation (6.25), all the exponential factors become 1, and satisfying the IC amounts to choosing the $A_{nm}$'s so that

$$\sum_{n=0}^{\infty}\sum_{m=0}^{\infty} A_{nm} X_n(x) Y_m(y) = \phi(x, y). \qquad (6.26)$$

The possibility of representing a given function $\phi(x, y)$ of two variables in terms of such a series involving products of Sturm–Liouville eigenfunctions is analogous to the Fourier analysis we discussed for the one-dimensional problem. In fact, using the results from the 1-D case, it is not hard to prove the *orthogonality relations*

$$\int_0^1 \int_0^1 X_n(x) Y_m(y) X_r(x) Y_s(y)\, dx dy = 0, \qquad (6.27)$$

for all pairs of integers $(n, m)$, $(r, s)$, with $(n, m) \neq (r, s)$. Using these orthogonality relations, one can show (heuristically at least) that if $\phi$ is represented by the series in equation (6.26), then necessarily the constants $\{A_{nm}\}_{n,m=0}^{\infty}$ are given by the integral formula

$$A_{nm} = (L_n M_m)^{-1} \int_0^1 \int_0^1 \phi(x, y) X_n(x) Y_m(y)\, dx dy. \qquad (6.28)$$

Here $L_n \equiv \int_0^1 X_n^2(x) dx$ and $M_m \equiv \int_0^1 Y_m^2(y) dy$ are constants that must be computed once the eigenfunctions are determined for any particular problem. Thus, having computed the $A_{nm}$'s from formula (6.28), we have completely solved the IBVP, at least formally, with the series representation in equation (6.25).

**Example 6.1**   Consider the homogeneous IBVP (fixed temperature of 0 on the boundary)

$$u_t = u_{xx} + u_{yy} \qquad\qquad \text{in } R, \text{ for } t > 0, \qquad (6.29)$$

$$u = 0 \qquad\qquad \text{on } \partial R, \text{ for } t > 0, \qquad (6.30)$$

$$u(x, y, 0) = 16(x - x^2)(y - y^2) \qquad \text{for } x, y \text{ in } R. \qquad (6.31)$$

Here the corresponding Sturm–Liouville problems have BCs $X(0) = 0 = X(1)$ and $Y(0) = 0 = Y(1)$. Thus $X_n(x) = \sin n\pi x$ and $Y_m(y) = \sin m\pi y$, for $n, m = 1, 2, 3, \ldots$, and so the general solution is

$$u(x, y, t) = \sum_{n=1}^{\infty} \sum_{m=1}^{\infty} A_{nm} e^{-(n^2+m^2)\pi^2 t} \sin n\pi x \sin m\pi y. \qquad (6.32)$$

To satify the IC, we need to compute the Fourier coefficients in the representation

$$\sum_{n=1}^{\infty} \sum_{m=1}^{\infty} A_{nm} \sin n\pi x \sin m\pi y = 16(x - x^2)(y - y^2).$$

Note that for the eigenfunctions here, the constants $L_n$ and $M_m$ are both equal to $1/2$ for all $n$ and $m$. Thus, using properties of double integrals and some of our previous computations of Fourier coefficients in the 1-D case, we get

$$A_{nm} = 4 \int_0^1 \int_0^1 16(x - x^2)(y - y^2) \sin n\pi x \sin m\pi y \, dx dy$$

$$= 16 \left[ 2 \int_0^1 (x - x^2) \sin n\pi x \, dx \right] \left[ 2 \int_0^1 (y - y^2) \sin n\pi y \, dy \right]$$

$$= \frac{256[1 - (-1)^n][(1 - (-1)^m]}{n^3 m^3 \pi^6}.$$

This, of course, gives that $A_{nm} = 0$ unless both $n$ and $m$ are odd. Inserting these coefficients back into the general solution gives the particular solution of the IBVP

$$u(x, y, t) = \qquad\qquad\qquad\qquad\qquad\qquad\qquad\qquad (6.33)$$

$$\sum_{k=1}^{\infty} \sum_{p=1}^{\infty} \frac{1024 e^{-[(2k-1)^2+(2p-1)^2]\pi^2 t}}{(2k-1)^3(2p-1)^3 \pi^6} \sin(2k-1)\pi x \, \sin(2p-1)\pi y.$$

The steady-state solution in this problem is the identically 0 temperature distribution, and we see that the above solution tends to this steady state in the limit: $\lim_{t \to \infty} u(x, y, t) = 0$. The graph of the initial temperature distribution $z = \phi(x, y)$ is easy to sketch by hand, since each section of this surface by a plane $x = a$ (with $0 < a < 1$) perpendicular to the $x$-axis is a parabola $z = k(y - y^2)$ (where $k = a - a^2$). Similarly, the sections by planes perpendicular to the $y$-axis are parabolas. The graph of the steady state $S(x, y) = 0$ is the $x$-$y$ plane. For $0 < t < \infty$ the graph of the intermediate temperature distribution $u(x, y, t)$ is a surface that passes through the edges of the square $R$ and has a shape that is intermediate between the graph of $\phi$ and the flat graph of $S$. This not too hard to visualize and sketch roughly. The exact computer sketches of this are shown in Figure 6.4. The information displayed in the figure is as expected. The heat flows from the hot region at the center of the plate toward the boundaries, which are held at zero degrees, and this process continues until the plate cools down to zero degees everywhere. Because of the symmetry in the initial temperature distribution $\phi$, the heat flow is perfectly symmetric as well. Figure 6.5 displays the evolution over time of the temperature at four points in the plate: $(1/8, 1/2)$, $(1/4, 1/2)$, $(3/8, 1/2)$, and $(1/2, 1/2)$. The information displayed indicates that it takes about 0.2 seconds for these four points to cool to zero degrees. One would infer from this that 0.2 seconds is the time it takes for the entire plate to reach its steady state.

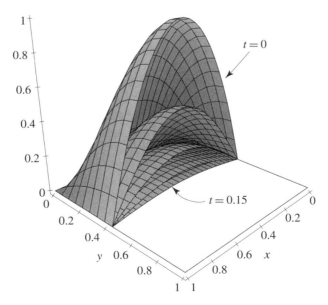

Figure 6.4.   Evolution of the temperature distribution in the plate in Example 6.1 from initial state $\phi$ to final state $S = 0$ (surfaces are cut for viewing).

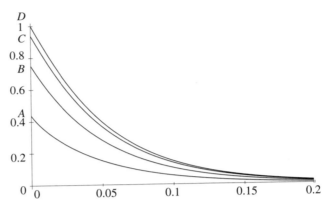

Figure 6.5.   Graphs showing the evolution over time of the temperature at four points in the plate: $A = (1/8, 1/2)$, $B = (1/4, 1/2)$, $C = (3/8, 1/2)$, and $D = (1/2, 1/2)$ in Example 6.1.

This example was selected because it is simple and displays all the main features in representing the temperature distribution $u$ in terms of a series. From a physical point of view, the related example where the BCs are the same but the initial temperature distribution is $\phi(x, y) = \sin \pi x \sin \pi y$ is qualitatively the same, but mathematically it is considerably simpler. This is because $\phi$ is the product of eigenfunctions for the problem, and so the series solution for $u$ reduces to just one term:

$$u(x, y, t) = e^{-2\pi^2 t} \sin \pi x \, \sin \pi y.$$

(Verify this !) The form of this readily displays how each intermediate temperature distribution is similar to the initial temperature distribution and decays to to the steady state of zero degrees throughout the plate.

**Example 6.2**   In this example let us change the BCs to ones where all the sides of the square are insulated. Thus consider

$$u_t = u_{xx} + u_{yy} \qquad\qquad \text{in } R, \text{ for } t > 0, \qquad (6.34)$$

$$\nabla u \cdot n = 0 \qquad\qquad \text{on } \partial R, \text{ for } t > 0, \qquad (6.35)$$

$$u(x, y, 0) = 18(x - x^2)(y - y^2) \qquad \text{for } x, y \text{ in } R. \qquad (6.36)$$

Now, the BCs in the coresponding Sturm–Liouville problem are $X'(0) = 0 = X'(1)$ and $Y'(0) = 0 = Y'(1)$, so that 0 is an eigenvalue for both problems. Using our previous work in the 1-D case gives the general solution in this example:

$$u(x, y, t) = \sum_{n=0}^{\infty} \sum_{m=0}^{\infty} A_{nm} e^{-(n^2+m^2)\pi^2 t} \cos n\pi x \cos m\pi y. \qquad (6.37)$$

To satisfy the IC, we proceed pretty much as before, except now the $A_{nm}$'s are the Fourier coefficients of $(x-x^2)(y-y^2)$ in the double *cosine* series expansion, so the integral formulas are slightly different when one of $n, m$ is zero. This is so because $L_0 = 1 = M_0$, while $L_n = 1/2 = M_m$ for all other values of $n$ and $m$. Thus

$$A_{00} = 18 \int_0^1 \int_0^1 (x - x^2)(y - y^2) \, dx dy$$

$$= 18 \int_0^1 (x - x^2) dx \int_0^1 (y - y^2) dy$$

$$= 18 \cdot \frac{1}{6} \cdot \frac{1}{6} = \frac{1}{2}.$$

Next, for one of $n, m$ zero and the other nonzero, say $m = 0, n \neq 0$, we get

$$A_{n0} = 2 \int_0^1 \int_0^1 18(x - x^2)(y - y^2) \cos n\pi x \, dx dy$$

$$= 18 \cdot 2 \int_0^1 (x - x^2) \cos n\pi x \, dx \int_0^1 (y - y^2) dy$$

$$= 18 \cdot \frac{2((-1)^{n+1} - 1)}{n^2 \pi^2} \cdot \frac{1}{6}.$$

As can be seen, these coefficients are zero unless $n$ is even, in which case $A_{2k,0} = -3/k^2\pi^2$. Because of the symmetry in the problem, we see that similarly, $A_{0m}$ is zero unless $m$ is even, in which case $A_{0,2p} = -3/p^2\pi^2$. Finally, for $n \neq 0$ and $m \neq 0$, we find

$$A_{nm} = 4 \int_0^1 \int_0^1 18(x - x^2)(y - y^2) \cos n\pi x \cos m\pi y \, dx dy$$

$$= 18 \left[ 2 \int_0^1 (x - x^2) \cos n\pi x \, dx \right] \left[ 2 \int_0^1 (y - y^2) \cos m\pi y \, dy \right]$$

$$= \frac{18 \cdot 4[(-1)^{n+1} - 1][(-1)^{m+1} - 1]}{n^2 m^2 \pi^4}.$$

These coefficients, then, are zero unless both $n$ and $m$ are even. Thus if we insert these values for the $A_{nm}$'s back into the general solution above and separate the series, we get the particular solution of the IBVP in this example:

$$u(x, y, t) = \frac{1}{2} - \frac{3}{\pi^2} \sum_{k=1}^{\infty} \frac{1}{k^2} e^{-4k^2\pi^2 t} \cos 2k\pi x$$

$$- \frac{3}{\pi^2} \sum_{p=1}^{\infty} \frac{1}{p^2} e^{-4p^2\pi^2 t} \cos 2p\pi y \tag{6.38}$$

$$+ \frac{18}{\pi^4} \sum_{k=1}^{\infty} \sum_{p=1}^{\infty} \frac{e^{-4(k^2+p^2)\pi^2 t}}{k^2 p^2} \cos 2k\pi x \cos 2p\pi y.$$

Well, this is certainly a very complicated expression for the solution. For the most part, we will not separate out the various subseries in the total series as we have done in equation (6.38), but rather just give the form of the general solution and the separate computation of the coefficients $A_{nm}$ (see the comments at the end of this section). However, the explicity displayed form in equation (6.38) helps exhibit that the solution tends to the steady-state temperature distribution $\frac{1}{2}$:

$$\lim_{t \to \infty} u(x, y, t) = \frac{1}{2}.$$

In this example we know that the steady states are constant functions $w(x, y) \equiv T$, and since the sides of $R$ are insulated, we know from physical principles that the heat will eventually be distributed uniformly throughout $R$. The eventual uniform temperature $T = 1/2$ comes from the Fourier coefficient $A_{00} = \int_0^1 \int_0^1 \phi(x, y) \, dx dy$, which is the average value of $\phi$ on $R$. See Figure 6.6. Naturally, other ICs, that is, other choices of $\phi$, will give different final temperatures $T = A_{00}$.

This qualitative understanding of the evolution of $u(x, y, t)$ with time may be all that is needed in some circumstances. If, however, we need more exact information about the temperature, then the series representation (6.38) (together with a computer) is the most useful expression of $u(x, y, t)$. To illustrate, we give the following *Maple* code to define the approximate solution $u(x, y, t, K, P)$, which is by definition what we get by truncating the infinite sums in (6.38) and just suming as $k = 1, \ldots, K$, and $p = 1, \ldots, P$. For convenience in the programming, we introduce the function $g(k, t) = e^{-4k^2\pi^2 t}/k^2$. The *Maple* code we need is

```
g := (k,t) -> exp(-4*k^2*Pi^2*t)/(k^2);
u := proc(x,y,t,K,P) s:=0;
```

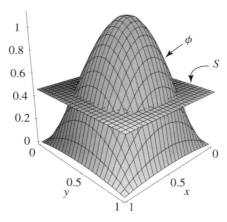

Figure 6.6.   Graph of the IC and the steady state for Example 6.2.

```
for k from 1 to K do
for p from 1 to P do
s:= s- 3*g(k,t)*cos(2*k*Pi*x)/(Pi^2*P)
- 3*g(p,t)*cos(2*p*Pi*y)/(Pi^2*K)
+ 18*g(k,t)*g(p,t)*cos(2*k*Pi*x)*cos(2*p*Pi*y)/Pi^4
od; od;
s := s + 1/2; end;
```

Before plotting the temperature distributions at various times, it is often advisable to investigate how the temperatures at selected points in the plate vary over time. This will give some information about how rapidly the various temperatures approach their steady-state values. In this example, the heat flows from the hot region at the center toward the boundaries, which, being insulated, heat up. Thus the temperature drops toward 1/2 at the center, rises toward 1/2 at the boundary, and does something in between. To see exactly what, we choose points $A = (0, 1/4)$, $B = (1/20, 1/4)$, $C = (2/20, 1/4)$, $D = (3/20, 1/4)$, $E = (4/20, 1/4)$ in the plate and plot the time evolution of temperatures at these points. This is shown in Figure 6.7. The information displayed in the figure indicates that while the temperature rises continuously toward 1/2 degree at the boundary point $A = (0, 1/4)$ and at the point $B = (1/20, 1/4)$, the temperatures at the other three points, which are more interior to the plate, at first drop and then rise continuously toward the steady-state value. This is due to the fact that these points initially lose heat as it flows away toward the boundary and then experience heat gain as the heat from the center of the plate reaches them.

Figure 6.7 also shows that it takes approximately 0.1 seconds for the plate to reach its steady-state distribution. Using this information, we can generate plots

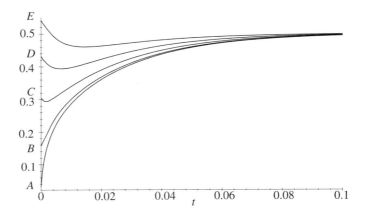

Figure 6.7.   The time evolution of temperatures at the points $A = (0, 1/4)$, $B = (1/20, 1/4)$, $C = (2/20, 1/4)$, $D = (3/20, 1/4)$, $E = (4/20, 1/4)$ in the plate for Example 6.2.

of the (approximate) temperature distributions at several intermediate times, before the temperature becomes uniformly $1/2$ in the plate. These plots are shown in Figures 6.8–6.9. Each of these was generated by *Maple* commands like

```
with(plots):
p1 := plot3d(u(x,y,.05,10,10),x=0..1,y=0..1):
p2 := spacecurve({[s,0,0],[s,1,0],[0,s,0],[1,s,0]},s=0..1):
display3d({p1,p2});
```

Note the use of a colon ":" to terminate the first three commands. This prevents the output from being displayed on the screen. The spacecurve command

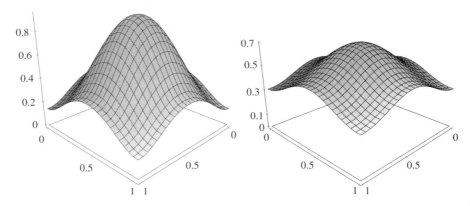

Figure 6.8.   Plots for Example 6.2 showing the temperatures in the plates at times $t = .01$ and $t = .03$.

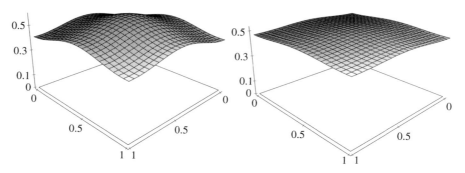

Figure 6.9.    Plots for Example 6.2, showing the temperatures in the plate at times $t = 0.05$ and $t = 0.08$.

draws the boundary of the plate for use in the display. You might also want to use *Maple*'s `animate3d` to create a moving picture of the evolution of the temperature distribution toward its steady state. This can be interesting and informative, but from a scientific standpoint, we are usually interested in doing just the reverse, namely, freezing the motion in some complex dynamical system, such as heat flow or fluid flow, so that it can be more carefully analyzed.

One further remark on the use of the computer here and in the exercises. In coding the series solution to the IBVP in Example 6.2, you may prefer to use directly the results we found for the Fourier coefficients and store them in an array as follows:

```
f:=n->((-1)^(n+1)-1)/(n^2*Pi^2);
A := array(0..100,0..100);
for n to 100 do
  A[n,0]:= evalf(6*f(n));
  A[0,n] := A[n,0] od;
for n to 100 do for m to 100 do
  A[n,m] := evalf(72*f(n)*f(m));
  A[m,n] := A[n,m] od od;
  A[0,0] := 0.5;
```

Thus, even though these coefficients are 0 for odd values of $n$, $m$, you may wish to ignore this and use the above coding, which is conceptually simpler. The series solution (or at least its approximation) may now be coded as follows:

```
u := proc(x,y,t,N) s:= 0;
      for n from 0 to N do for m from 0 to N do
      s := s + A[n,m]*exp(-a[n,m]*t)*cos(n*Pi*x)*cos(m*Pi*y)
      od; od; end;
```

Here we are assuming that the values of $a_{nm} = (n^2+m^2)\pi^2$ have been previously stored in an array a. The use of the arrays is more efficient, since then $A_{nm}$ and $a_{nm}$ do not have to be computed each time the procedure is called with new values of $x$ and $y$. For example, when plotting $u(x, y, 0.05, 40)$, the procedure is called 625 times (with the default grid size of $25 \times 25$). On the other hand, having all those zeros in the array A is not very efficient. With a fast computer, the efficiencies and inefficiencies in programming are hardly noticeable.

## Exercises 6.3

1. Expand the following functions in a double Fourier sine series

$$\phi(x, y) = \sum_{n,m=1}^{\infty} A_{nm} \sin n\pi x \, \sin m\pi y.$$

You may use any of your previous calculations for single Fourier sine series.

(a) $\phi(x, y) = xy$.

(b) $\phi(x, y) = x^2 y^2 - y + 1$.

(c) $\phi(x, y) = 1$.

(d) $\phi(x, y) = 5 \sin 3\pi x \, \sin 2\pi y - 2 \sin 4\pi x \, \sin \pi y$.

(e)

$$\phi(x, y) = \begin{cases} \sin 2\pi x \sin 2\pi y & \text{if } 0 \le x, y \le 1/2, \\ 0 & \text{otherwise.} \end{cases}$$

(f) $\phi(x, y) = H_{0,1/4,1/2}(x) H_{0,1/4,1/2}(y)$, where $H_{a,p,b}$ is the triangular pulse function introduced earlier.

(g) $\phi(x, y) = 100xye^{-4(x+y)}$.

(h) $\phi(x, y) = x^{\sin 4\pi x} y^{\sin 4\pi y}$.

2. Expand the following functions in a double Fourier cosine series

$$\phi(x, y) = \sum_{n,m=0}^{\infty} A_{nm} \cos n\pi x \, \cos m\pi y.$$

You may use any of your previous calculations for single Fourier cosine series.

(a) $\phi(x, y) = xy$.

(b) $\phi(x, y) = x^2 y^2 - y + 1$.

(c) $\phi(x, y) = 1$.

(d) $\phi(x, y) = 5 \cos 3\pi x \cos 2\pi y - 2 \cos 4\pi x \cos \pi y$.

(e)

$$\phi(x, y) = \begin{cases} \sin 2\pi x \sin 2\pi y & \text{if } 0 \le x, y \le 1/2, \\ 0 & \text{otherwise.} \end{cases}$$

(f) $\phi(x, y) = 2H_{0,1/4,1/2}(x)H_{0,1/4,1/2}(y)$, where $H_{a,p,b}$ is the triangular pulse function introduced earlier.

3. For each of the parts you did in Exercise 1, draw computer plots of $\phi$ and its truncated Fourier series

$$\phi(x, y, N) = \sum_{n,m=1}^{N} A_{nm} \sin \pi x \sin m\pi y.$$

Use a range of values for $N$, from small values, where the approximation is not so good, to large values, where it is. For clarity, you should plot these surfaces in separate pictures. Also, plot the graph of $\phi$ for comparison. Note: the graph of $\phi(x, y, N)$, in the Fourier sine series case, will be a surface that passes through the boundary of the square $R$ (why?) even though the graph of $\phi(x, y)$ does not. Thus for the graphs of the former, you might want to cut away, for readability, the parts of the surface near the boundary. For example, just plot $\phi(x, y, N)$ for $x$ and $y$ in the interval [0.05, 0.95], or whatever is appropriate. Also include the boundary of the plate in the plot.

4. Solve the two-dimensional heat problem

$$u_t = \alpha^2 \nabla^2 u \qquad \text{in } R, \text{ for } t > 0,$$

$$u = \phi \qquad \text{in } R, \text{ for } t = 0,$$

where $R$ is the unit square, and the initial condition and boundary conditions are given below. Also describe the steady-state, eventual temperature distribution and evolution of the temperature in the plate. Attempt some hand-drawn sketches of the temperature evolution.

(a) Initial condition: $\phi(x, y) = 1$, and BCs

$$u_y = 0$$

$u = 0$      $R$      $u = 0$

$$u_y = 0$$

How and why is this related to a corresponding 1-D heat flow problem? Now let $\phi(x, y) = 2H_{0,1/4,1/2}(x)H_{0,1/4,1/2}(y)$, and rework this problem. How and why does this change things?

(b) Initial condition: $\phi(x, y) = 2H_{0,1/4,1/2}(x)H_{0,1/4,1/2}(y)$, and BCs:

$$u = 0$$

$$u = 0 \quad \boxed{\quad R \quad} \quad u = 0$$

$$u_y = 0$$

(c) Boundary conditions

$$u_y = 0$$

$$u_x = 0 \quad \boxed{\quad R \quad} \quad u = 0$$

$$u_y = 0$$

and initial condition:

(i) $\phi(x, y) = xy$.

(ii) $\phi(x, y) = f(x)f(y)$, where

$$f(x) = \begin{cases} \sin 2\pi x & \text{if } 0 \le x \le 1/2, \\ 0 & \text{if } 1/2 < x \le 1. \end{cases}$$

(iii) $\phi(x, y) = H_{0,1/4,1/2}(x)H_{0,1/4,1/2}(y)$.

(d) Boundary conditions

$$u_y = 0$$

$$u_x = 0 \quad \boxed{\quad R \quad} \quad u = 0$$

$$u = 0$$

and initial condition:

(i) $\phi(x, y) = H_{0,1/4,1/2}(x) H_{0,1/4,1/2}(y)$.

(ii) $\phi(x, y) = H_{0,1/4,1/2}(x) y$.

5. For each of the parts you did in Exercise 4, use a computer to (a) plot the time evolution of the temperature at several appropriately selected points in the plate; (b) plot the initial temperature distribution and several intermediate temperature distributions, at selected times, for the plate as a whole and also for a line in the plate like $y = 1/4$. Make sure you use enough terms in the series to get an accurate representation of these distributions. Analyze and comment on these graphical studies and combine them with the commentary and work on Exercise 4.

6. Suppose that in the 2-D homogeneous heat IBVP for the square, the BCs are either (i) fixed temperature 0 on the boundary, or (ii) totally insulated on the boundary. Show that if the initial temperature function $\phi$ is symmetric, $\phi(x, y) = \phi(y, x)$, for all $x$ and $y$, then the temperature function $u$ at each fixed time $t > 0$ is also symmetric. Interpret this result geometrically. How would you generalize this result to other types of BCs ?

7. Examples 6.1 and 6.2 in the text have symmetrical initial temperature distributions $\phi$, and this makes the heat flow fairly simple and predictable. As an in-depth and longer project, study the asymmetrical case, by taking $\phi(x, y) = H_{a,p,b}(x) H_{a',p',b'}(y)$ and exploring the heat flow for various choices of $a, p, b, a', p', b'$ that make $\phi$ asymmetric.

8. Show how the orthogonality relations (6.27) follow from the 1-D case. Use this, along with heuristic arguments like those in equations (4.10)–(4.12) for the 1-D case, to derive the integral formula (6.28) for the Fourier coefficients.

9. (*General rectangular plates*) The text discusses the series solution of the homogeneous IBVP for a square plate. Discuss the changes that occur for the general rectangular plate $R = [0, L] \times [0, M]$.

## 6.4.   The Semi-Homogeneous Problem

We now consider the solution of the general semi-homogeneous heat problem, which, of course, includes the homogeneous heat problem as a special case. Here the boundary conditions are still homogeneous, but there is heat being added/extracted within the plate $R$ as dictated by the heat source/sink density $F$. Just as in one dimension, we shall see (in the next section) that any two-dimensional heat problem can be transformed into a semi-homogeneous problem, and thus solving semi-homogeneous problems is an essential step in the process of solving any heat problem.

The 2-D semi-homogeneous problem for the unit square $R$ is

$$u_t - \alpha^2 \nabla^2 u = F \qquad \text{in } R, \text{ for } t > 0,$$

$$-\kappa_1 u_x(0, y, t) + h_1 u(0, y, t) = 0 \qquad 0 \le y \le 1,$$

$$\kappa_2 u_x(1, y, t) + h_2 u(1, y, t) = 0 \qquad 0 \le y \le 1,$$

$$-\kappa_3 u_y(x, 0, t) + h_3 u(x, 0, t) = 0 \qquad 0 \le x \le 1,$$

$$\kappa_4 u_y(x, 1, t) + h_4 u(x, 1, t) = 0 \qquad 0 \le x \le 1,$$

$$u = \phi \qquad \text{in } R, \text{ for } t = 0,$$

and the solution of this problem is entirely similar to the 1-D semi-homogeneous problem. As in the 1-D case, we look for a solution that has the form

$$u(x, y, t) = \sum_{n=0}^{\infty} \sum_{m=0}^{\infty} T_{nm}(t) X_n(x) Y_m(y), \qquad (6.39)$$

where the $X_n$'s and $Y_m$'s are the Sturm–Liouville eigenfunctions for the two respective Sturm–Liouville problems

$$X'' = -\lambda^2 X, \qquad (6.40)$$

$$-\kappa_1 X'(0) + h_1 X(0) = 0, \qquad (6.41)$$

$$\kappa_2 X'(1) + h_2 X(1) = 0, \qquad (6.42)$$

and

$$Y'' = -\mu^2 Y, \qquad (6.43)$$

$$-\kappa_3 Y'(0) + h_3 Y(0) = 0, \qquad (6.44)$$

$$\kappa_4 Y'(1) + h_4 Y(1) = 0. \qquad (6.45)$$

Thus the $X_n$'s and $Y_m$'s are determined immediately, and because they satisfy the above Sturm–Liouville BCs, it is easy to see that the series expression (6.39) for $u$ satisfies the heat IBVP BCs for any choice of the functions $T_{nm}(t)$. These $T_{nm}$'s must be specifically chosen so that $u$ satisfies the non-homogeneous PDE and the IC. If we substitute the series expression (6.39) into the PDE, this will give us ODEs that the $T_{nm}$'s must satisfy. To see this, note that since $X_n'' = -\lambda_n^2 X_n$

and $Y_m'' = -\mu_m^2 Y_m$, we find that

$$u_t - \alpha^2(u_{xx} + u_{yy}) = \sum_{n=0}^{\infty}\sum_{m=0}^{\infty}\left[T_{nm}' X_n Y_m - \alpha^2(T_{nm} X_n'' Y_m + T_{nm} X_n Y_m'')\right]$$

$$= \sum_{n=0}^{\infty}\sum_{m=0}^{\infty}\left[T_{nm}' + \alpha^2(\lambda_n^2 + \mu_m^2)T_{nm}\right]X_n Y_m.$$

Thus to satisfy the PDE $u_t - \alpha^2(u_{xx} + u_{yy}) = F$, we need to require the $T_{nm}$'s to be such that

$$\sum_{n=0}^{\infty}\sum_{m=0}^{\infty}\left[T_{nm}' + \alpha^2(\lambda_n^2 + \mu_m^2)T_{nm}\right]X_n Y_m = F. \tag{6.46}$$

To satisfy the IC on $u$, the $T_{nm}$'s must also be such that

$$\sum_{n=0}^{\infty}\sum_{m=0}^{\infty}T_{nm}(0)X_n(x)Y_m(y) = \phi(x, y). \tag{6.47}$$

Thus expanding $F$ and $\phi$ in their generalized Fourier series with respect to the eigenfunctions

$$F(x, y, t) = \sum_{n=0}^{\infty}\sum_{m=0}^{\infty}F_{nm}(t)X_n(x)Y_m(y), \tag{6.48}$$

$$\phi(x, y) = \sum_{n=0}^{\infty}\sum_{m=0}^{\infty}A_{nm}X_n(x)Y_m(y) \tag{6.49}$$

and equating the coefficients of like terms in the series on each side of (6.46) and (6.47) gives the explicit conditions on the $T_{nm}$'s

$$T_{nm}' + a_{nm}T_{nm} = F_{nm}, \tag{6.50}$$

$$T_{nm}(0) = A_{nm}, \tag{6.51}$$

where for convenience we have introduced the notation

$$a_{nm} \equiv \alpha^2(\lambda_n^2 + \mu_m^2). \tag{6.52}$$

The initial value problem (6.50)–(6.51) completely determines $T_{nm}$, since $F_{nm}$ and $A_{nm}$ are known, being computed from the integral formulas

$$F_{nm}(t) = (L_n M_m)^{-1} \int_0^1 \int_0^1 F(x, y, t) X_n(x) Y_m(y) dx dy, \qquad (6.53)$$

$$A_{nm} = (L_n M_m)^{-1} \int_0^1 \int_0^1 \phi(x, y) X_n(x) Y_m(y) dx dy. \qquad (6.54)$$

Using techniques from the study of elementary differential equations, it is easy to derive the following integral formula for the $T_{nm}$'s:

**Time-Dependent Factors**

$$T_{nm}(t) = \left[ A_{nm} + \int_0^t F_{nm}(s) e^{a_{nm} s} \, ds \right] e^{-a_{nm} t}. \qquad (6.55)$$

This is the same as in the 1-D case, but now we have two subscripts, $n$ and $m$. Also note that

$$T_{00}(t) = A_{00} + \int_0^t F_{00}(s) ds, \qquad (6.56)$$

and this factor is pertinent only when all four sides are insulated, since by convention, $X_0 = 1$ and $Y_0 = 1$ in this case, and in every other case either $X_0$ or $Y_0$ is zero.

**Example 6.3**    Consider the heat problem where the temperature is fixed at 0 at all points along the boundary and there is an internal source of heat with density

$$F(x, y) = \begin{cases} 30 \sin 2\pi x \, \sin 2\pi y & \text{if } (x, y) \in [0, 1/2] \times [0, 1/2], \\ 0 & \text{otherwise.} \end{cases} \qquad (6.57)$$

Also assume that the initial temperature distribution is identically zero: $\phi = 0$ in $R$.

With these boundary conditions ($u = 0$, on $\partial R$), the solution has the form

$$u(x, y, t) = \sum_{n=1}^{\infty} \sum_{m=1}^{\infty} T_{nm}(t) \sin n\pi x \sin m\pi y,$$

and all we have to do is compute the time-dependent amplitudes $T_{nm}$. For this we first calculate the Fourier sine coefficients $F_{nm}$ of $F$. Note these are constants (i.e., do not depend on $t$), and since $F$ is a product $F(x, y) = 30f(x)f(y)$, we first just compute

$$f_n = 2 \int_0^{1/2} \sin 2\pi x \sin n\pi x \, dx$$

$$= \frac{\sin[(n-2)\pi/2]}{(n-2)\pi} - \frac{\sin[(n+2)\pi/2]}{(n+2)\pi},$$

provided that $n \neq 2$, and for $n = 2$ we get

$$f_2 = 1/2.$$

Note that for $n$ even (and larger than 2), $f_n = 0$, while for $n$ odd, say $n = 2k - 1$

$$f_{2k-1} = \frac{4(-1)^k}{(2k-3)(2k+1)\pi}.$$

Using these calculations, we find

$$F_{nm} = 4 \int_0^{1/2} \int_0^{1/2} 30(\sin 2\pi x \sin 2\pi y) \sin n\pi x \sin m\pi y \, dxdy$$

$$= 30 \left[ 2 \int_0^1 \sin 2\pi x \sin n\pi x \, dx \right] \left[ 2 \int_0^1 \sin 2\pi y \sin m\pi y \, dy \right]$$

$$= 30 f_n f_m.$$

Next, it is clear that the Fourier coefficients of the initial temperature distribution $\phi \equiv 0$ are $A_{nm} = 0$ for every $n, m$. Thus we find for the $T_{nm}$'s,

$$T_{nm}(t) = \left[ 0 + F_{nm} \int_0^t e^{a_{nm}s} ds \right] e^{-a_{nm}t}$$

$$= \frac{F_{nm}}{a_{nm}} [1 - e^{-a_{nm}t}].$$

Here $a_{nm} = (n^2 + m^2)\pi^2\alpha^2$. Thus the complete solution of this semi-homogeneous problem is

$$u(x, y, t) = \sum_{n=1}^{\infty} \sum_{m=1}^{\infty} \frac{F_{nm}}{a_{nm}} [1 - e^{-a_{nm}t}] \sin n\pi x \sin m\pi y. \tag{6.58}$$

It is easy to see from this that the temperature distribution in the plate tends to a steady-state temperature distribution $S$ given by

$$S(x, y) = \lim_{t \to \infty} u(x, y, t)$$

$$= \sum_{n=1}^{\infty} \sum_{m=1}^{\infty} \frac{F_{nm}}{a_{nm}} \sin n\pi x \sin m\pi y. \qquad (6.59)$$

In order to analyze graphically the flow of heat in the plate, we do the following. From the nature of the problem, we know that the plate is initially at 0 degrees and then begins to heat up (due to the internal heat source $F$), while the boundary is held at fixed temperature 0. As time increases, the temperature distribution in the plate approaches the steady state $S$. Theoretically, this takes infinitely long, but practically, it takes only a finite amount of time before $u(x, y, t)$ and $S(x, y)$ are indiscernibly different. To get an estimate of how long this takes, we plot the variation of temperature $u(1/4, 1/4, t)$ at the point $(1/4, 1/4)$ about which the internal heat source is centered. This is shown in Figure 6.10, along with a plot of the internal heat source $F$. From the way the point $(1/4, 1/4)$ heats up, it would appear that the temperature distribution will reach its steady state by approximately $t = 0.4$. Notice also that the temperature changes very rapidly between $t = 0$ and $t = 0.1$, and after $t = 0.1$, the changes are slow and small as the temperature approaches its steady-state value of approximately 0.89 degrees. Based on this, we select some intermediate times for plotting. The next two figures, Figures 6.11–6.12, show the temperatures in the plate for times $t = 0.005, 0.01, 0.04, 0.1$. A view of the steady-state temperature distribution

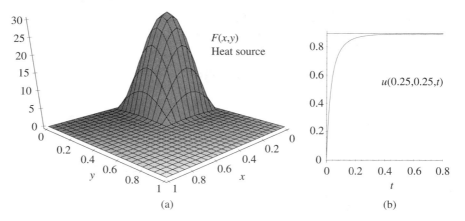

(a)                                  (b)

Figure 6.10.   For Example 6.3, (a) plot of the internal heat source $F$, (b) the variation of temperature $u(1/4, 1/4, t)$ at the point $(1/4, 1/4)$ about which the internal heat source is centered.

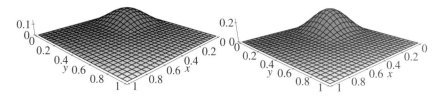

Figure 6.11.    The temperatures in the plate at times $t = 0.005$ and $t = 0.01$ for Example 6.3.

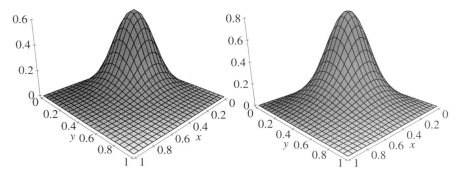

Figure 6.12.    The temperatures in the plate at times $t = 0.04$ and $t = 0.1$ for Example 6.3.

$S$ in the plate is shown in Figure 6.13. You should compare this with the graph of the heat source $F$ to see how the shape of $F$ influences the shape of $S$. Also shown in the same figure is a plot of the *isotherms* for $S$. Each curve is a level curve for the graph of $S$ and thus is a curve in which all points are at the same temperature.

## Exercises 6.4

For each of the following semi-homogeneous problems, find the series solution of the IBVP and determine the eventual temperature distribution and steady states (if any). Use a computer to thoroughly analyze the evolution of the temperature in the plate. In particular, do the following: (a) plot the eventual temperature distribution (if any) and (in a separate plot) the isotherms of this distribution; (b) plot the time evolution of the temperature at selected points in the plate and (where appropriate) try to empirically determine the time at which the temperature distribution in the plate is approximately the same as the eventual temperature distribution; (c) plot the temperature distributions along lines $x = a$ or $y = b$ in the plate at appropriately selected times $t$ (choose lines that you feel best display the heat flow); (d) plot the temperature distributions in the plate at appropriately selected times $t$; (e) comment on and discuss the nature of your results. Note: In some of these problems the heat distribution changes very rapidly initially, so make sure you do not overlook some interesting behavior near $t = 0$. Also: It's a good idea to check your computation of the

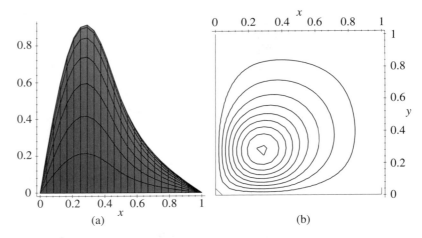

Figure 6.13.   (a) View of the steady-state temperature distribution in the plate for Example 6.3;
(b) a plot of the isotherms for this distribution

Fourier coefficients $A_{nm}$ and $F_{nm}$ by plotting the graphs of the series approximations to $\phi$ and $F$.

1.  Internal heat source $F(x, y) = 100H_{0,1/4,1/2}(x)y$, initial condition $\phi(x, y) = 4(x - x^2)$, and BCs

$$u_y = 0$$

$$u = 0 \qquad R \qquad u = 0$$

$$u = 0$$

2.  Internal heat source $F(x, y) = 100H_{0,1/4,1/2}(x)H_{0,1/4,1/2}(y)$, initial condition $\phi(x, y) = 0$, and BCs

$$u_y = 0$$

$$u = 0 \qquad R \qquad u = 0$$

$$u_y = 0$$

3. Internal heat source $F(x, y) = 10xy - 5/2$, initial condition $\phi(x, y) = 0$, and BCs

$$u_y = 0$$

$$u_x = 0 \quad \boxed{\quad R \quad} \quad u_x = 0$$

$$u_y = 0$$

*Note:* The heat source $F$ is actually a heat sink at points in the plate where it is negative (heat is being extracted there). Does this explain why there can be a (finite) eventual temperature distribution $u_\infty$, even though the boundary is totally insulated? Suppose we change $F$ to $F(x, y) = 10xy$. What difference does this make?

4. Internal heat source $F(x, y) = 4e^{-t/4}H_{0,1/4,1/2}(x)H_{0,1/4,1/2}(y)$, initial condition $\phi(x, y) = 0$, and BCs

$$u_y = 0$$

$$u_x = 0 \quad \boxed{\quad R \quad} \quad u_x = 0$$

$$u_y = 0$$

5. (*Steady states*) Suppose in the general 2-D semi-homogeneous problem (for the square) that the internal heat source does not depend on the time: $F = F(x, y)$.

   (a) Determine all the steady states. Unlike the corresponding exercise for 1-D steady states (Chapter 4), you will have to use series here. Show in particular that there is a unique steady state unless all four sides of the square are insulated. When all four sides of the square are insulated, show that there are either no steady states or infinitely many.

   (b) Under what condition does there exist an eventual temperature distribution

$$u_\infty(x, y) = \lim_{t \to \infty} u(x, y, t).$$

   When this condition is met, show that $u_\infty$ coincides with one of the steady states.

6. With the same assumption as in the last exercise, study the effect of the diffusivity constant $p = \alpha^2$ on the heat flow. Namely, for a given problem, consider the effect of varying $p$ and holding everything else fixed. For the cases that have steady states,

let $S_p$ denote the steady state for the problem with a given value of $p$. Show that

$$S_p(x, y) = \frac{1}{p} \cdot S_1(x, y),$$

and comment on why this seems physically reasonable; i.e., small diffusivity values result in large steady-state temperature distributions. Also discuss how $p$ affects the approximate time it takes for the plate to reach its steady-state temperature distribution. Use Example 6.3 to illustrate the ideas in your discussion. Namely, let $u(x, y, t, N, p)$ be the approximate solution to the problem (using double sums up to $N$) for a given value of $p$. Study the behavior at the point $(1/4, 1/4)$, which is the hottest point in the steady-state temperature distribution for any value of $p$ (verify this). For this example, find (approximately) the value of the diffusivity $p$ so that the temperature in the plate never exceeds 1 degree and so that the hottest spot reaches 1 degree in temperature after 1 second has elapsed.

## 6.5.  Transforming to Homogeneous BCs

In a 2-D heat problem with non-homogeneous BCs, the technique for transforming to homogeneous BCs is similar to the 1-D case, but with some interesting features that while present in the 1-D case, were not entirely conspicuous. The technique does not depend on the shape of the region $R$, and so we return to the case of a general region before resuming the discussion of the rectangular region.

Thus we consider the general IBVP

$$u_t = \alpha^2 \nabla^2 u + F \qquad \text{in } R, \text{ for } t > 0 \qquad (6.60)$$

$$\kappa \nabla u \cdot n + hu = g \qquad \text{on } \partial R \text{ for } t > 0, \qquad (6.61)$$

$$u = \phi \qquad \text{in } R, \text{ for } t = 0, \qquad (6.62)$$

with the function $g = g(x, y, t)$ not identically zero at points $(x, y)$ along the boundary $\partial R$.

As in the 1-D case, we look for a solution $u$ of (6.60)–(6.62), which has the form

$$u = S + U,$$

where $S = S(x, y, t)$ is chosen to satisfy the non-homogeneous BCs (6.61). That is, assume we can find an $S$ such that

$$\kappa \nabla S \cdot n + hS = g$$

on the boundary $\partial R$, for $t > 0$. Then since $\nabla u = \nabla S + \nabla U$, we see that (6.61) reduces to $\kappa \nabla U \cdot n + hU = 0$, which are homogeneous BCs on $U$.

Next, since $u = S + U$, we have

$$u_t = S_t + U_t,$$

$$\nabla^2 u = \nabla^2 S + \nabla^2 U,$$

and thus substituting $u = S + U$ into the PDE gives (after rearranging slightly)

$$U_t = \alpha^2 \nabla^2 U + (F + \alpha^2 \nabla^2 S - S_t).$$

This is a non-homogeneous heat equation for $U$ and involves the function $S$ which we have chosen. The expression in the parentheses in this equation is the transformed heat source/sink function. To simplify this, we try to choose an $S$ that not only satisfies the non-homogeneous BCs as above, but also satisfies

$$\nabla^2 S = 0.$$

Thus, *assuming* that we can determine an $S$ that satisfies these two conditions, we see that the original IBVP for $u$ gets transformed into the new IBVP with homogeneous BCs for $U$

$$U_t = \alpha^2 \nabla^2 U + F - S_t \qquad \text{in } R, \text{ for } t > 0, \qquad (6.63)$$

$$\kappa \nabla U \cdot n + hU = 0 \qquad \text{on } \partial R, \text{ for } t > 0, \qquad (6.64)$$

$$U = \phi - S \qquad \text{in } R, \text{ for } t = 0. \qquad (6.65)$$

Note that this is entirely analogous to the 1-D case, except there we did not explicitly mention that $S$, in addition to satisfying the non-homogeneous BCs, was also to satisfy $\nabla^2 S = 0$. In the 1-D case this last equation is simply $S_{xx} = 0$, which has the general solution $S(x,t) = A(t)x + B(t)$, with $A, B$ arbitrary functions of $t$.

Also note that even though the IBVP (6.63)–(6.65) has homogeneous BCs, in general the PDE is still non-homogeneous. Since the transformed problem for $U$ is semi-homogeneous, we know how to solve it (at least when $R$ is the unit square). Thus to find the solution $u = S + U$ of the original problem, we need to develop a method for finding $S$. To simplify the discussion, we will do this only for the

**Special Case:**  Assume that the original non-homogeneous BCs do not depend on $t$, i.e., that the $g$ in (6.61) does not depend on $t$.

Thus we can assume that $S$ does not depend on $t$ either, so that $S_t = 0$. Now the problem is to find an $S = S(x, y)$ that satisfies

$$\nabla^2 S = 0 \qquad \text{in } R, \tag{6.66}$$

$$\kappa \nabla S \cdot n + hS = g \qquad \text{on } \partial R. \tag{6.67}$$

This is a new type of problem, known as a *boundary value problem*, or just BVP for brevity. The PDE (6.66) is *Laplace's equation*, and it occurs in many other physical situations. Solutions of Laplace's equation are called *harmonic* functions. The problem then is to find a solution of Laplace's equation that satisfies the given boundary conditions. There are two cases of this that have special names:

### The Dirichlet Problem

Find a solution $S$ of Laplace's equation that has specified values on the boundary:

$$\nabla^2 S = 0 \qquad \text{in } R, \tag{6.68}$$

$$S = g \qquad \text{on } \partial R. \tag{6.69}$$

### The Neumann Problem

Find a solution $S$ of Laplace's equation that has specified normal derivative $\partial S / \partial n \equiv \nabla S \cdot n$ on the boundary:

$$\nabla^2 S = 0 \qquad \text{in } R, \tag{6.70}$$

$$\nabla S \cdot n = g \qquad \text{on } \partial R. \tag{6.71}$$

In equation (6.71) we require that the given function $g$ have the property that its line integral around the boundary $\partial R$ be zero. Otherwise, there is an inconsistency in the Neumann problem. See the exercises in the next chapter.

The general BVP (6.66)–(6.67), with $\kappa \neq 0$ and $h \neq 0$, is sometimes called the *Robin problem*. To solve these types of BVPs for a general region $R$ can become quite involved. In the next chapter we will see how for rectangular regions the separation of variables technique enables us to obtain series solutions of these problems. The theoretical solution of BVPs for general regions $\Omega$ in $\mathbb{R}^3$ is discussed in Chapter 13. The solution of BVPs is important in many other situations (like electrostatics and steady fluid flow) besides in connection with solving heat IBVPs.

Thus the complete series solution of the 2-D heat IBVP for the square will, in general, require the series solution of the corresponding BVP for $S$ (next chapter). However, there are special cases of the BVP for $S$ that can be solved without series. The next example and some of the ensuing exercises illustrate this.

**Example 6.4**    Consider the IBVP with homogeneous heat equation $u_t = \nabla^2 u$, with Dirichlet boundary conditions

$$u = x$$

$$u = 0 \qquad R \qquad u = y$$

$$u = 0$$

and initial condition $u(x, y, 0) = 1$, for $(x, y) \in R$. Transforming to homogeneous BCs gives $u = S + U$ with $S$ a solution of Laplace's equation that also satisfies the above BCs. It's easy to see that:

$$S(x, y) = xy$$

is the required solution of this BVP. The exercises in the next chapter will show that for special BCs like the ones here, closed-form polynomial solutions of the

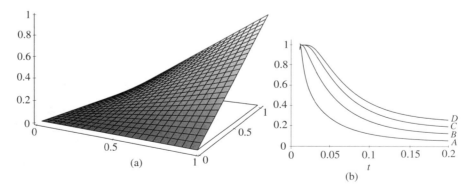

Figure 6.14.    For Example 6.4, plots of (a) the steady state $S(x, y) = xy$ (b) the temperature evolution in the plate at the points $A = (1/8, 1/2)$, $B = (1/4, 1/2)$, $C = (3/8, 1/2)$, $D = (1/2, 1/2)$.

BVP are available and can be found systematically. That is how we got the above $S$. You should verify that it does satisfy the BVP being considered here.

With $S$ now known, the transformed problem for $U$ is $U_t = \nabla^2 U$, homogeneous BCs $U = 0$ on $\partial R$, and initial condition $U(x, y, 0) = 1 - xy$ in $R$. This is easily solved, and then the solution of the original problem is

$$u(x, y, t) = xy + \sum_{n=1}^{\infty} \sum_{m=1}^{\infty} A_{nm} e^{-a_{nm}t} \sin n\pi x \, \sin m\pi y, \tag{6.72}$$

where $a_{nm} = (n^2 + m^2)\pi^2$ and

$$A_{nm} = \frac{1 + (-1)^{n+1}}{n\pi} \cdot \frac{1 + (-1)^{m+1}}{m\pi} - \frac{2[1 - (-1)^n]}{n\pi} \cdot \frac{2[1 - (-1)^m]}{m\pi}.$$

The latter are the double Fourier sine coefficients for $1 - xy$ and are readily written down from previous calculations (see the tables in Appendix D).

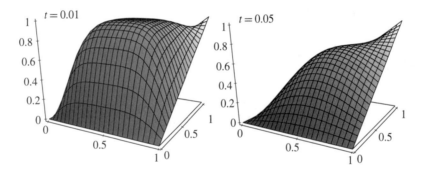

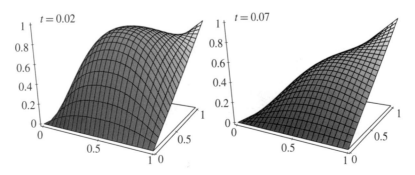

Figure 6.15.   The temperature distributions in the plate in Example 6.4 at the times $t = 0.01, 0.02, 0.05, 0.07$.

We see from equation (6.72) that $S(x, y) = xy$ is the eventual temperature distribution in the plate and is also the unique steady state for the IBVP (see Figure 6.14). This need not always be the case. In general, the $S$ used to transform to homogeneous BCs will *not* be a steady state (as, e.g., in the case of an non-homogeneous heat equation).

The graph in Figure 6.14(b) shows that it takes about 2/10 of a second for all the temperatures in the plate to reach their steady-state values. Figure 6.15 shows the temperature distributions in the plate at the indicated times.

**Exercises 6.5**

1. Solve the following IBVPs with homogeneous PDE $u_t = \nabla^2 u$ and the indicated BCs and ICs. In each case find the series solution and steady states and do a computer analysis of the heat flow as you did in the previous exercise set.
   *Note*: You should be able to transform each of these to homogeneous BCs with a polynomial of degree $\leq 1$, i.e., of the form

$$S(x, y) = Dx + Ey + F.$$

(a) Initial condition $\phi(x, y) = 0$ and BCs

$$u = 1$$

$$u = 1 \quad\boxed{\quad R \quad}\quad u = 1$$

$$u = 1$$

(b) Initial condition $\phi(x, y) = 1$ and BCs

$$u = 1$$

$$u = y \quad\boxed{\quad R \quad}\quad u = y$$

$$u = 0$$

(c) Initial condition $\phi(x, y) = 1$ and BCs

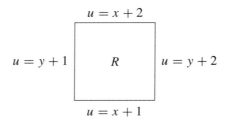

$$u = x + 2$$

$u = y + 1$    $R$    $u = y + 2$

$$u = x + 1$$

(d) Initial condition $\phi(x, y) = 2$ and BCs

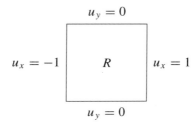

$$u_y = 0$$

$u_x = -1$    $R$    $u_x = 1$

$$u_y = 0$$

2. (*Poisson's equation*) The nonhomogeneous version of Laplace's equation is known as *Poisson's equation*. It has the form $\nabla^2 S = f$, where $f$ is a given function. This PDE occurs in electrostatics and many other areas of mathematical physics. The corresponding BVP is

$$\nabla^2 S = f \qquad \text{in } R, \tag{6.73}$$

$$\kappa \nabla S \cdot n + hS = g \qquad \text{on } \partial R \tag{6.74}$$

and is a generalization of the BVP (6.66)–(6.67). This exercise discusses how to solve this BVP when $R$ is the unit square.

(a) Show how to solve the *semi-homogeneous* BVP

$$\nabla^2 S = f \qquad \text{in } R, \tag{6.75}$$

$$\kappa \nabla S \cdot n + hS = 0 \qquad \text{on } \partial R \tag{6.76}$$

by means of a series. *Hint:* consider $S$ as a steady state for a corresponding heat problem and use equations (6.39) and (6.50) with $T_{nm}$ a constant (not a function of $t$).

(b) Show that if $S^1$ is a solution of the semi-homogeneous BVP (6.75)–(6.76) and if $S^2$ is a solution of the BVP (6.66)–(6.67), then $S = S^1 + S^2$ is a solution of the Poisson BVP (6.73)–(6.74). *Note:* As mentioned, finding $S^2$ by series methods is discussed in the next chapter.

(c) Here is an alternative to the approach in parts (a) and (b). Show that if $S^2$ is a (suitably differentiable) function on $R$ that satisfies the BC: $\kappa \nabla S^2 \cdot n + h S^2 = g$, then letting

$$S = S^2 + S^1$$

transforms the BVP (6.73)–(6.74) into a semi-homogeneous BVP for $S^1$. This latter BVP can be solved as in part (a). The fact that such a function $S^2$ exists is a theoretical result, and in practice it is difficult to determine (you can always impose the additional restriction $\nabla^2 S^2 = 0$ and use the series method from the next chapter).

# 7 Boundary Value Problems

In the last chapter, we saw how the general boundary value problem (BVP)

$$\nabla^2 S = 0 \qquad \text{in } R,$$

$$\kappa \nabla S \cdot n + hS = g \qquad \text{on } \partial R$$

for a planar region $R$ occurs in the study of the heat equation. In this chapter we discuss how to find series solutions of such BVPs. For the most part we confine ourselves just to Dirichlet- and Neumann-type boundary conditions, and in all cases the region $R$ is the unit square.

## 7.1. The Dirichlet Problem for the Unit Square

The Dirichlet problem occurs in the study of many physical phenomena. As we have seen, it arises in transforming a heat flow problem with nonhomogeneous BCs to one with homogeneous BCs. In this circumstance it is often the case that the solution $S$ of the Dirichlet problem is the steady state temperature distribution for the heat problem. Specifically, for the unit square $R$, consider the IBVP

$$u_t = \alpha^2 \nabla^2 u,$$

$$u(0, y, t) = g_1(y),$$

$$u(1, y, t) = g_2(y),$$

$$u(x, 0, t) = g_3(x),$$

$$u(x, 1, t) = g_4(x),$$

$$u(x, y, 0) = \phi(x, y),$$

z
$g_1(y)$
$g_4(x)$
y
$g_2(y)$
$g_3(x)$
x

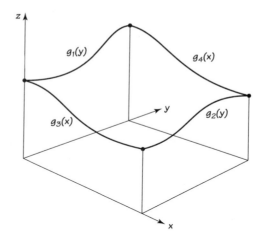

Figure 7.1. Graph of the fixed-temperature BCs.

where $g_1, g_2$ are given functions of $y$, and $g_3, g_4$ are given functions of $x$. These functions specify the fixed temperatures on the four sides of the square $R$. Figure 7.1 graphically depicts these BCs by plotting the graphs of $g_1, g_2, g_3, g_4$ over the various pieces of the boundary.

As was done previously, we will specify these BCs in the abbreviated form shown in Figure 7.2.

To solve this type of IBVP, we transform $u = S + U$ as in the last chapter and get that $S$ must satisfy the Dirichlet problem

$$\nabla^2 S = 0 \qquad \text{in } R,$$

$$S = g \qquad \text{on } \partial R,$$

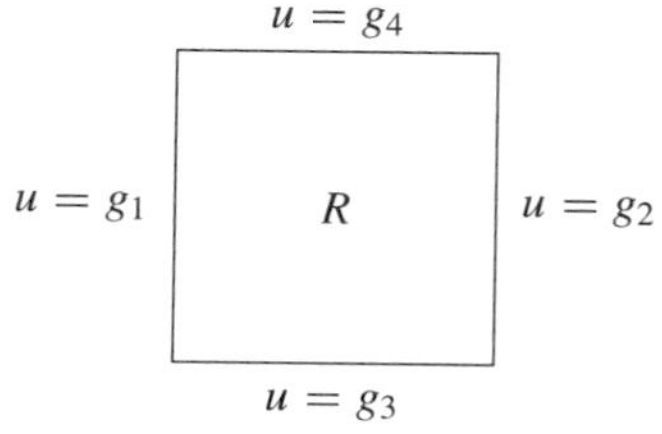

$u = g_4$
$u = g_1$
$R$
$u = g_2$
$u = g_3$

Figure 7.2. Specification of the BCs in a square plate $R$ with its boundary held at fixed temperatures $g_1, g_2, g_3, g_4$ as shown.

while $U$ must satisfy

$$U_t = \nabla^2 U \qquad \text{in } R \tag{7.1}$$

$$U = 0 \qquad \text{on } \partial R, \tag{7.2}$$

$$U = \phi - S \qquad \text{in } R, \text{ at } t = 0. \tag{7.3}$$

We know how to solve this heat problem. It has general solution

$$U(x, y, t) = \sum_{n=1}^{\infty} \sum_{m=1}^{\infty} A_{nm} e^{-(n^2+m^2)\pi^2\alpha^2 t} \sin n\pi x \sin m\pi y, \tag{7.4}$$

where the $A_{nm}$'s are the Fourier coefficients of $\phi - S$. Thus to get the complete solution

$$u(x, y, t) = S(x, y) + U(x, y, t),$$

we need a method for solving the Dirichlet problem for $S$. We explain the method by looking at a number of examples. It is important to note that for the type of heat problem we are considering here, the $S$ represents the steady-state temperature distribution that the temperature in the plate approaches as $t \to \infty$.

**Example 7.1**   Suppose the Dirichlet problem for $S$ has the BCs shown in Figure 7.3. To find solutions of Laplace's equation $\nabla^2 S = 0$, we use separation of variables; i.e., look for a solution of the form $S(x, y) = X(x)Y(y)$. Substituting in the PDE and rearranging gives

$$\frac{X''(x)}{X(x)} = -\frac{Y''(y)}{Y(y)} = k.$$

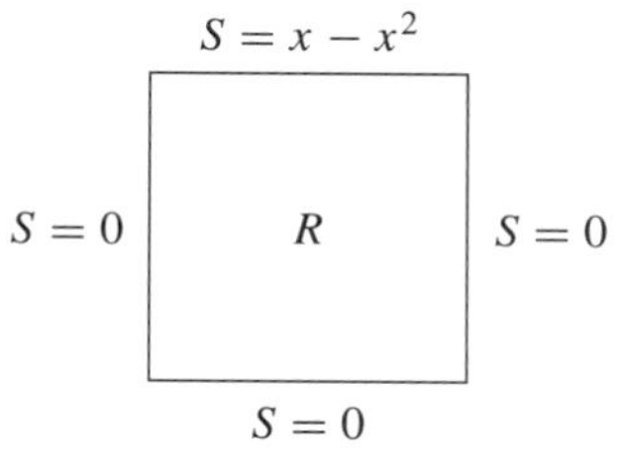

Figure 7.3.   BCs for Example 7.1

Here the reasoning is that these two ratios must be constant, say equal to $k$. Thus we get the two DE's:

$$X'' = kX,$$

$$Y'' = -kY.$$

Next, substituting $S = XY$ in the BCs and eliminating unnecessary factors, we get

$$X(0) = 0,$$

$$X(1) = 0,$$

$$Y(0) = 0.$$

The other boundary condition on $S$, which does not lead to a BC on $Y$, is $X(x)Y(1) = x - x^2$. The strategy is to worry about this last BC on $S$ after we find a general solution of $\nabla^2 S = 0$ that satisfies the other three BCs. The Sturm–Liouville problem for $X$ gives us $X_n(x) = \sin n\pi x$ and $k = -n^2\pi^2$. Now the DE for $Y$ becomes

$$Y'' = n^2\pi^2 Y,$$

which has general solution $Y(y) = A \sinh n\pi y + B \cosh n\pi y$. If we substitute this in the third BC above, we get $Y(0) = B = 0$. Therefore, $Y = Y_n = A_n \sinh n\pi y$. Using the superposition principle, the above reasoning shows that

$$S(x, y) = \sum_{n=1}^{\infty} A_n \sin n\pi x \, \sinh n\pi y \qquad (7.5)$$

satisfies Laplace's equation and three of the boundary conditions on $S$. To satisfy the last BC on $S$, we substitute the above series in the BC to get

$$\sum_{n=1}^{\infty} A_n \sinh n\pi \, \sin n\pi x = x - x^2. \qquad (7.6)$$

Thus we see that $A_n \sinh n\pi$ is the $n^{\text{th}}$ Fourier sine coefficient of $x - x^2$.

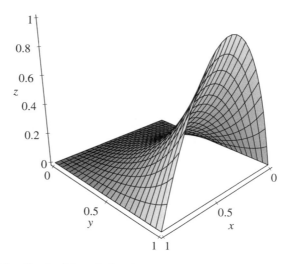

Figure 7.4.   Graph of the solution $S$ to the Dirichlet problem in Example 7.1

Consequently,

$$A_n = \frac{4(1 - (-1)^n)}{n^3 \pi^3 \sinh n\pi}.$$

This completely determimes the solution $S$ of the Dirichlet problem. The graph of $S$ is shown in Figure 7.4. Notice that the surface passes through the given boundary curves (just as it is required to do).

**Example 7.2**   Consider solving $\nabla^2 S = 0$ with the BCs shown in Figure 7.5. This Dirichlet problem is clearly related to the one in the last example, and we can proceed to solve it in either of two ways. The first way is to go through the separation of variables argument as before, getting the same Sturm–Liouville

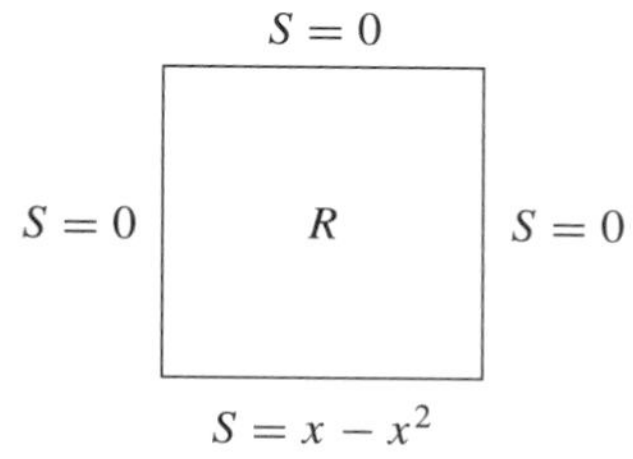

$$S = 0$$

$$S = 0 \qquad R \qquad S = 0$$

$$S = x - x^2$$

Figure 7.5.   Dirichlet BCs for Example 7.2

problem for $X$

$$X'' = kX,$$

$$X(0) = 0,$$

$$X(1) = 0,$$

and the partial problem for $Y$,

$$Y'' = -kY,$$

$$Y(1) = 0.$$

So again we get $X_n(x) = \sin n\pi x$, and $k = -n^2\pi^2$, $n = 1, 2, 3, \ldots$. Now the DE for $Y$ is

$$Y'' = n^2\pi^2 Y.$$

To find solutions of the this that satisfy $Y(1) = 0$, we exploit the fact that the hyperbolic sine function vanishes at zero. Thus a little guesswork leads us to taking

$$Y(y) = \sinh n\pi(1 - y)$$

for the functions satisfying the DE plus the single BC $Y(1) = 0$. Putting everything together gives a series

$$S(x, y) = \sum_{n=1}^{\infty} A_n \sin n\pi x \, \sinh n\pi(1 - y), \qquad (7.7)$$

which satisfies Laplace's equations and the three homogeneous BCs. In the series the $A_n$'s are arbitrary and are to be chosen such that $S$ satisfies the last BC $S(x, 0) = x - x^2$. Thus we see that the $A'_n s$ are exactly the same as in the last example. This completes the solution of the problem.

The similarity of the solution here with that in the last problem suggests the second way of doing this problem. Namely, let $S^0$ denote the solution obtained in the last example. It satisfies Laplace's equation and the BCs there. Now suppose we define a new function $S$ on the square $R$ by

$$S(x, y) = S^0(x, 1 - y),$$

for $(x, y) \in R$. Then (using the chain rule) it is easy to verify that $S$ satisfies Laplace's equation. Furthermore, it is straightforward to check that $S$ satisfies the BCs in this problem. Using the series for $S^0$ and the above definition of $S$, we get the series expression for $S$. In essence, this method of solution of the problem amounts to flipping the square $R$ over (and of course, using the work we already did on the other problem).

**Example 7.3**   Consider solving $\nabla^2 S = 0$ with the BCs shown in Figure 7.6. With the experience gained in the previous examples, we should be able to guess that the series solution for this problem should have the form

$$S(x, y) = \sum_{n=1}^{\infty} A_n \sinh n\pi (1 - x) \sin n\pi y. \qquad (7.8)$$

Clearly, each of the terms in the series satisfies Laplace's equation (check it out) and the BCs $S(1, y) = 0$, $S(x, 0) = 0$, $S(x, 1) = 0$. To satisfy the last BC $S(0, y) = 1$ amounts to chosing the $A_n$'s in the series to satisfy

$$\sum_{n=1}^{\infty} (A_n \sinh n\pi) \sin n\pi y = 1;$$

i.e., the factors $A_n \sinh n\pi$ are the Fourier sine coefficients of the constant 1 function. Thus

$$A_n = \frac{2[1 - (-1)^n]}{n\pi \sinh n\pi},$$

and this completes the solution of the problem. The graph of the solution $S$ is shown in Figure 7.7.

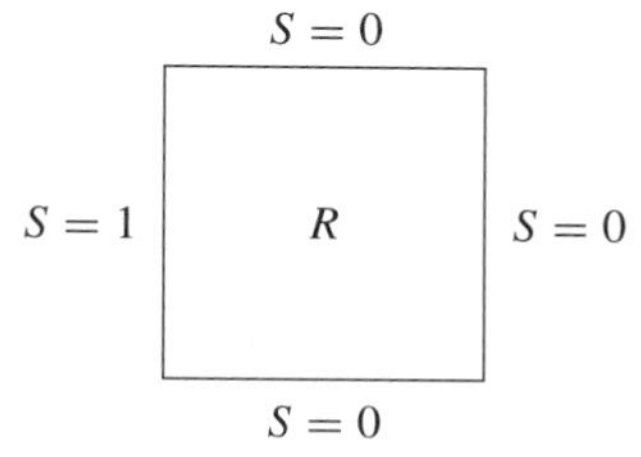

Figure 7.6.   Dirichlet BCs for Example 7.3.

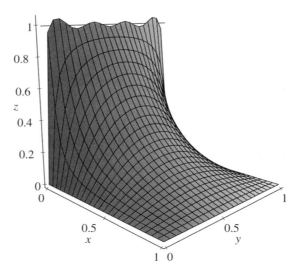

Figure 7.7.   Plot of the solution $S$ for the BVP in Example 7.3.

**Example 7.4**   In this last example we consider the geometric interpretation of the solution $S$. For this, consider the Dirichlet problem with the BCs shown in Figure 7.8. This arises from a heat problem where the temperature on three sides of the square is held fixed at 0 degrees and along the fourth side is fixed at temperatures that vary linearly from 0 to 1 degree. If we heat the plate up initially with any temperature distribution and keep the boundary temperatures fixed at their prescribed values thereafter, then the temperature $u$ has the form

$$u = S + U,$$

with $U$ being transient $\lim_{t\to\infty} U(x, y, t) = 0$, and $S$ being the steady state, or eventual temperature distribution. The graph of $S$ is a surface passing through the given boundary curves on $\partial R$. There are many surfaces that do this, but the one we want must also satisfy Laplace's equation $\nabla^2 S = 0$. You may be able to take a guess as to what this surface looks like, but to get a precise picture, we first determine the series representation for $S$ and then plot it with a computer. A short calculation gives

$$S(x, y) = \sum_{n=1}^{\infty} \frac{2(-1)^{n+1}}{n\pi \sinh n\pi} \sin n\pi x \sinh n\pi y$$

as the series for $S$. For the computer plot we look at the approximation to $S$ given

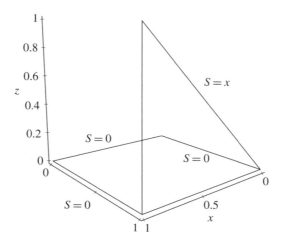

Figure 7.8.    Dirichlet BCs for Example 7.4

by the sum of the first $k$ terms of the series:

$$S(x, y, k) = \sum_{n=1}^{k} \frac{2(-1)^{n+1}}{n\pi \sinh n\pi} \sin n\pi x \sinh n\pi y.$$

For a small number of terms, say $k = 5$, a computer plot gives the graph of $S(x, y, 5)$ in Figure 7.9. This clearly is not a good approximation to $S$, since along the top boundary of $R$ the graph of $S(x, 1) = x$ is a straight line. The graph of $S(x, 1, 5)$ is an approximation to this straight line by the sum of 5 sine waves, which is far too few to get the right shape. Note also that there is a discontinuity in the fixed boundary temperatures at the vertex $(1, 1)$ of $R$, and Figure 7.9 shows this Gibbs-type phenomenon for the surface $S$. A fairly accurate approximation to $S$ can be obtained by using 100 terms of the series. The graph of $S(x, y, 100)$ is shown in Figure 7.10. This surface represents the eventual temperature distribution throughout the plate.

## 7.1.1.   Solution of the Dirichlet Problem for the Square

The above examples are all rather special types of Dirichlet problems in that the BCs on three sides of the square are zero. However, we can always reduce the solution of the general Dirichlet problem to solving four Dirichlet problems of this special type. This reduction exploits the fact that Laplace's equation is linear: The sum of any number of solutions is also a solution. Thus we take the general Dirichlet problem with $g_i$, $i = 1, 2, 3, 4$, specifying the boundary values

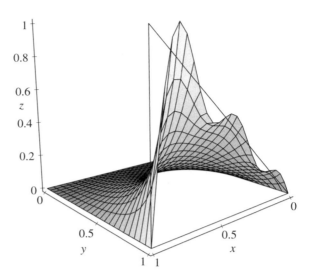

Figure 7.9.   Plot of the approximation $S(x, y, 5)$ to $S(x, y)$ in Example 7.4.

of $S$, and we split $S$ up into a sum of four functions,

$$S = S^1 + S^2 + S^3 + S^4,$$

with the functions $S^i, i = 1, 2, 3, 4$, satisfying the special types of BCs with zero values on three sides of the square. This is shown in Figure 7.11. Now if we solve each of these four Dirichlet problems, expressing $S^i, i = 1, 2, 3, 4$, as a series, the solution of the general Dirichlet problem results as the series obtained by combining these four special series. Based upon our experience with the examples above, it is not hard to see what the solution is.

**Series solution of the Dirichlet problem for the square**

$$S(x, y) = \sum_{n=1}^{\infty} \left[ A_n^2 \sinh n\pi x + A_n^1 \sinh n\pi (1 - x) \right] \sin n\pi y$$

$$+ \sum_{n=1}^{\infty} \left[ A_n^4 \sinh n\pi y + A_n^3 \sinh n\pi (1 - y) \right] \sin n\pi x.$$

The coefficients $A_n^i$ are given by

$$A_n^i = \frac{2}{\sinh n\pi} \int_0^1 g_i(u) \sin n\pi u \, du, \tag{7.9}$$

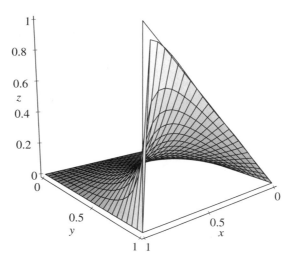

Figure 7.10.   Plot of the approximation $S(x, y, 100)$ to $S(x, y)$ in Example 7.4.

for $i = 1, 2, 3, 4$.

## 7.1.2.   Heat Flow

The solution $S$ of any Dirichlet problem can be viewed as a steady-state solution of a corresponding heat problem. As such, the graph of $S$ displays the temperature $S(x, y)$ at the point $(x, y)$ in $R$. Even though the distribution $S$ of temperatures is constant in time, you should realize that there is a flow of heat going on within the plate $R$. To see this, consider Example 7.4 above, where three sides of $R$ are held at temperature 0 and the fourth side is held at temperatures that vary in a linear fashion along the side. To maintain these temperatures on the fourth side, heat must be supplied, and this heat will flow out through the plate toward the colder sides, where in order to maintain the zero degrees on the colder sides, the

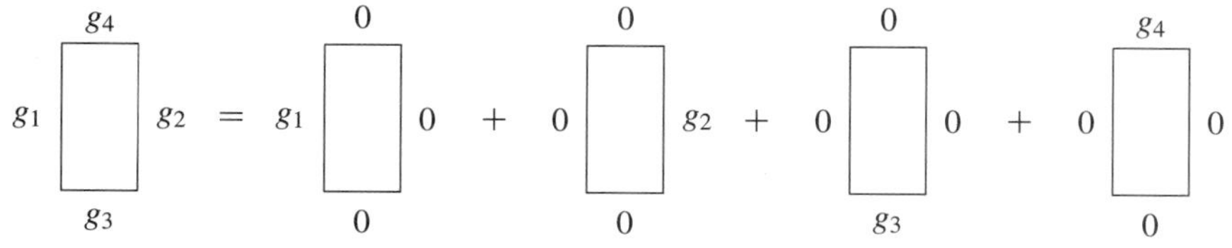

Figure 7.11.   Splitting the Dirichlet problem into four parts

heat must continually be extracted. To obtain a detailed picture of this heat flow, we plot the flow (see Chapter 1) generated by the heat flux vector field,

$$v = -\nabla S = (-S_x, -S_y).$$

Note that since $S$ is given analytically by a series

$$S(x, y) = \sum_{n=1}^{\infty} \frac{2(-1)^{n+1}}{n\pi \sinh n\pi} \sin n\pi x \sinh n\pi y,$$

each component of the flux vector $v$ is given by a series

$$v_1 = -S_x(x, y) = -\sum_{n=1}^{\infty} \frac{2(-1)^{n+1}}{\sinh n\pi} \cos n\pi x \sinh n\pi y,$$

$$v_2 = -S_y(x, y) = -\sum_{n=1}^{\infty} \frac{2(-1)^{n+1}}{\sinh n\pi} \sin n\pi x \cosh n\pi y.$$

To use any computer software package for plotting the flow, we must, of course, replace the above series expressions for the components of $v$ by their approximations by the sums of the first $k$ terms in the series (*Maple* has the facility to sum from 1 to $\infty$, but I do not think you want to use it for the task at hand). Also note that the investigation we did above in Example 7.4 revealed that $k = 50$ terms gave a good approximation to $S$. To shorten the execution time, it's a good idea to make the coefficients into an array structure and compute them just once before doing the plot. Some *Maple* code for doing all of this is as follows:

```
A:=array(1..100):
for n from 1 to 100
   do
     A[n] := evalf( 2*(-1)^(n+1)/sinh(n*Pi))
   od:
Sx := proc(x,y,k) s:=0;
      for n from 1 to k do
      s := s + evalf(A[n]*cos(n*Pi*x)*sinh(n*Pi*y))
      od; end;
Sy := proc(x,y,k) s:=0;
      for n from 1 to k do
      s := s + evalf(A[n]*sin(n*Pi*x)*cosh(n*Pi*y))
      od; end;
with(DEtools):
```

```
system:=D(x)(t)=-Sx(x,y,k),D(y)(t)=-Sy(x,y,k);
init := [0,.2,1],[0,.3,1],[0,.4,1],[0,.5,1],[0,.6,1],
        [0,.65,1],[0,.7,1],[0,.8,1]:
DEplot([system],[x,y],t=0..2,{init},x=0..1,y=0..1,
        stepsize=.05);
```

The plot produced by *Maple* is shown in Figure 7.12. There are several things to note here:

(1) *Maple* requires that the system of differential equations contain no symbols or parameters other than $x$ and $y$ when the DEplot command is invoked. In our case, because both $S_x$ and $S_y$ involve $\pi$ in their summations, it was necessary to use evalf in their definitions to force *Maple* to evaluate $\pi$ as a floating-point number.

(2) In the DEplot command, [x,y] tells *Maple* what the independent variables in the vector field are, and t=0..2 indicates that the flow lines $\phi_t(x_0, y_0)$ are to be plotted for each initial condition $(x_0, y_0)$, as $t$ ranges from 0 to 2.

(3) The initial conditions are given in the list called init and each has the form $[0, x_0, y_0]$, which indicates you want the integral curve that passes through $(x_0, y_0)$ at time $t = 0$. Thus the integral curve $t \mapsto \phi_t(x_0, y_0)$, for $t \in [0, 2]$, gets plotted. Note the way we chose the initial conditions so that $(x_0, y_0)$ is a point on the top side of the square $R$. This was done because we know that the heat flow will be away from this side toward one of the other three sides. Figure 7.12 shows this clearly and also shows that at time $t = 2$, some of

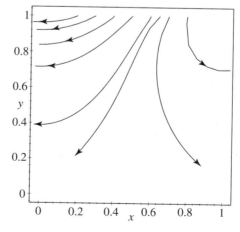

Figure 7.12.    The heat flow lines for the temperature distribution in Example 7.4.

the flow lines have not yet reached the sides toward which they are flowing. This is so because each flow line is traced out at a different speed. *Maple* presently does not have a feature in its DE package that allows us to see the flow being traced out dynamically in time (see the CD-ROM for a simple procedure called DDEplot, which will draw the flow lines dynamically).

(4) *Mountain Slope Analogy:* A good way to view the heat flow lines in Figure 7.12 and the speeds at which heat flows along them is as follows. Think of the surface plot of $S$ as a mountain slope. Figure 7.13 shows this plot again, but now with contour lines (and a different angle of view). To help your visualization, Figure 7.13 also shows the usual 2-D contour map of the same mountain slope. Now view yourself as a skier on this mountain slope. Starting at a point on the higher side and skiing so as to maximize your speed, you will make a track on the slope whose projection is precisely one of the heat flow lines shown in Figure 7.12. This is so because (Chapter 1) the gradient of $S$ at any point is perpendicular to the level curve (i.e., contour) through that point. The speed at which you ski down the slope corresponds to the speed at which heat flows along the heat flow line. Near the summit your speed of descent will be breathtaking indeed (experts only)! Also recall that in traditional heat flow terminology, the contour lines in Figure 7.12 are the isotherms (lines of equal temperature) of the heat distribution, and to maintain this distribution, heat flows from hot to cold along lines that intersect each isotherm orthogonally.

(5) Because the flow lines are traversed at different speeds, some will reach the boundary of $R$ more quickly than others. Bear in mind that the flow line

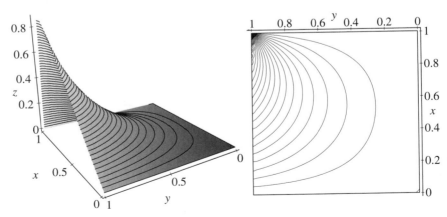

Figure 7.13.    The temperature distribution $S$, viewed as a mountain slope with contours drawn in.

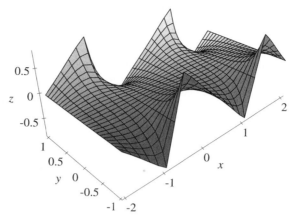

Figure 7.14.    The graph of $S$ over the square $[-2, 2] \times [-1, 1]$. This gives an extended view of the mountain slope that is the graph of $S$.

is computed all the way up to time $t = 2$, even though we have specified for *Maple* to show us only the part of it that lies in $R$. *Beware:* You cannot necessarily get around this by choosing a large ending time, say $t = 60$. In this particular example, some of the flow lines after leaving the square $R$ may encounter numerical difficulty. This is so since the series representation for $S$ is not necessarily valid outside $R$. Because of the $\sin n\pi x$ terms in the series, the mountain slope extends periodically in the $x$ direction as far as you can see. Figure 7.14 shows partly what the terrain looks like out there. However, in the $y$ direction, the mountain terrain is undefined for $y$ outside the interval $[-1, 1]$ and ski trails that enter these regions are dangerous indeed. In Release 3 of *Maple*, you can include the option: `limitrange=true`, in the `DEplot` and this will stop the computation when the integral curve leaves the viewing rectangle. Release 4 has this option as a default but it does not always work (see material on the CD-ROM).

## Exercises 7.1

The following problems deal with the Dirichlet problem

$$\nabla^2 S = 0 \qquad \text{in } R, \text{ for } t > 0,$$

$$S = g \qquad \text{in } R, \text{ for } t = 0,$$

where $R$ is the unit square and the boundary conditions are as indicated.

1. Find the series representation for $S$ if the BCs are:

(a)

$$S = 0$$

| | | |
|---|---|---|
| $S = \sin \pi y$ | $R$ | $S = y^2$ |

$$S = 0$$

(b)

$$S = x$$

| | | |
|---|---|---|
| $S = 0$ | $R$ | $S = y^2$ |

$$S = 0$$

(c)

$$S = 1$$

| | | |
|---|---|---|
| $S = y$ | $R$ | $S = y^2$ |

$$S = 0$$

(d)

$$S = 2x$$

| | | |
|---|---|---|
| $S = 1 - y^2$ | $R$ | $S = 1 + y^2$ |

$$S = 1$$

(e) In this problem use the idea in Example 7.2 to express your answer in the form
$S(x, y) = S^0(x, y) + S^0(x, 1 - y)$. Thus only the series for $S^0$ is needed.

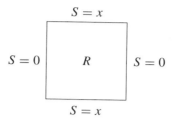

$$S = x$$
$$S = 0 \quad\quad R \quad\quad S = 0$$
$$S = x$$

2. Determine the conditions on the coefficients $A$ through $F$ in order for the second-degree polynomial

$$S(x, y) = Ax^2 + Bxy + Cy^2 + Dx + Ey + F$$

to be a solution of Laplace's equation.

   Find polynomial solutions of the Dirichlet problem with the following BCs. If you wish, you can compare these with the corresponding series solutions. The series solutions will be quite complicated, so the polynomial solution is always preferable. *However*, polynomial solutions are possible only when the boundary conditions are rather special.

(a)

$$S = x$$
$$S = 0 \quad\quad R \quad\quad S = 1$$
$$S = x$$

(b)

$$S = x - x^2$$
$$S = y^2 - y \quad\quad R \quad\quad S = y^2 - y$$
$$S = x - x^2$$

(c)

$$S = x^2 - 1$$

$$S = -y^2 + 3y - 3 \qquad R \qquad S = -y^2 + 5y - 4$$

$$S = x^2 - 2x - 3$$

3. In the following problems find the solution $S$ and use a computer to (i) plot the graph of $S$ (or at least its approximation $S(x, y, N)$ for suitable values of $N$); (ii) plot some of the isotherms for the plate; (iii) plot some of the heat flow lines in the plate;, and (iv) use the mountain slope analogy to discuss the nature of the heat flow and distribution of temperatures in the plate.

(a)

$$S = \sin \pi x$$

$$S = \sin \pi y \qquad R \qquad S = 0$$

$$S = \sin \pi x$$

(b)  Same as Exercise 1(e) above.

4. Suppose $g : [0, 1] \to \mathbb{R}$ is a given function, and consider the Dirichlet problem with BCs

$$S = g$$

$$S = 0 \qquad R \qquad S = 0$$

$$S = 0$$

Let $S^0$ denote the solution to this problem, and show how the solutions of the following problems can be expressed in terms of $S^0$: (i) $S = g$ on one of the sides and $S = 0$ on the remaining sides; (ii) $S = g$ on two of the sides and $S = 0$ on the remaining sides; (iii) $S = g$ on all four sides.

5. Let $g(x) = 1_{[a,b]}(x)$ be the step function for the interval $[a, b]$. Use the results from Exercise 4 to study the solutions of the following Dirichlet problems:

(a)

$S = g(x)$

$S = 0$     $R$     $S = g(y)$

$S = 0$

(b)

$S = g(x)$

$S = 0$     $R$     $S = 0$

$S = g(x)$

## 7.2.  The Neumann Problem for the Unit Square

We have previously discussed how the Neumann problem arises in connection
with a heat IBVP for an arbitrary planar region $R$. If we specialize that discussion
to the case when $R$ is the unit square, then the Neumann problem can be solved
by series, much like what we did for the Dirichlet problem. There are, however,
important and interesting differences. The details are as follows.

The heat problem with Neumann BCs that we are interested in consists of the
heat equation

$$u_t = \alpha^2 \nabla^2 u \qquad \text{in } R, \text{ for } t > 0$$

(for simplicity we just consider the case of a homogeneous PDE), the usual initial
condition

$$u(x, y, 0) = \phi(x, y),$$

for $(x, y)$ in $R$, and the boundary conditions shown in Figure 7.15.

*Note:*   We have combined the conductivity constants $\kappa_i, i = 1, .., 4$, into the
given functions $g_i$. Also recall that $-\nabla u \cdot n$ is the heat flux in the direction of

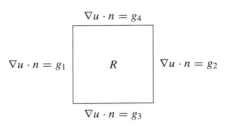

Figure 7.15.    Neumann BCs for the unit square.

the outward-directed normal $n$. Thus writing the Neumann BCs as $-\nabla u \cdot n = -g_i$, $i = 1, 2, 3, 4$, shows that when $g_i$ is negative, there is heat flux outward across the $i$th side (heat loss to the interior), and when $g_i$ is positive, the heat flux is inward (heat gain).

Recall that we can be more specific than this about the exact form for these BCs if we take into account what the outward directed unit normal $n$ is on the four parts of the boundary. For example, on the left-hand vertical side, $n = (-1, 0)$ and consequently, $\nabla u \cdot n = -u_x$. Thus the Neumann BCs for the unit square are more specifically given by Figure 7.16.

Transforming to homogeneous BCs amounts to looking for a solution of the non-homogeneous problem of the form: $u = S + U$, where $S$ is chosen to satisfy Laplace's equation $\nabla^2 S = 0$ and the non-homogeneous BCs for $u$ (Figure 7.16). This is the Neumann problem for $S$. Note that we are assuming that the boundary data $g_i, i = 1, \ldots, 4$, do not depend on $t$, and so $S = S(x, y)$ will not depend on $t$ either.

If we can find a solution to the Neumann problem, then the other part of the heat problem is to determine $U$ as a solution of the transformed IBVP

$$U_t = \nabla^2 U \qquad \text{in } R, \qquad (7.10)$$

$$\nabla U \cdot n = 0 \qquad \text{on } \partial R, \qquad (7.11)$$

$$U = \phi - S \qquad \text{in } R, \text{ at } t = 0. \qquad (7.12)$$

This is just the standard insulated BC heat problem (with shifted initial temperature distribution $\phi - S$). As we have seen, the solution of this problem is

$$U(x, y, t) = A_{00} + \sum_{(n,m)\neq 0} A_{nm} e^{-(n^2+m^2)\pi^2\alpha^2 t} \cos n\pi x \cos m\pi y, \qquad (7.13)$$

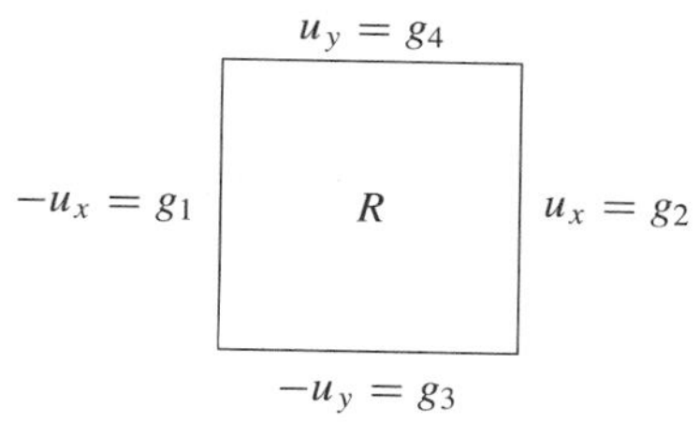

Figure 7.16.    Simplified Neumann BCs for the unit square.

where the $A_{nm}$'s are the Fourier coefficients of $\phi - S$.

This then completes the solution of the original non-homogeneous problem, the solution being given by

$$u(x, y, t) = S(x, y) + A_{00} + \sum_{(n,m)\neq 0} A_{nm} e^{-(n^2+m^2)\pi^2\alpha^2 t} \cos n\pi x \cos m\pi y.$$

From this we see that the steady-state (eventual) temperature distribution is

$$\lim_{t\to\infty} u(x, y, t) = S(x, y) + A_{00},$$

and that the series part of $u$ represents the transient temperature distribution.

Recall that the solution of the Neumann problem is *not* unique, and thus since $u = S + U$ depends on the solution $S$ of the Neumann problem (and $U$ also depends on $S$ through the Fourier coefficients of $\phi - S$), it appears that $u$ will be different for different choices of $S$. The exercises show that this is not the case. They also show that the steady state and the transient state do not depend on the $S$ chosen.

We will discuss various particular examples of Neumann problems here and leave the solution of the general Neumann problem as an exercise. One aspect of the Neumann problem that makes it different from a Dirichlet problem is that the given boundary functions $g_i$, $i = 1, \ldots, 4$, must satisfy a certain condition in order for the Neumann problem to have any solution at all. This condition on the $g_i$'s is the following

**Solvability Condition**

$$\int_0^1 g_1(y)dy + \int_0^1 g_2(y)dy + \int_0^1 g_3(x)dx + \int_0^1 g_4(x)dx = 0.$$

The necessity of this condition will be illustrated in the following examples. Further discussion can be found in the exercises.

**Example 7.5**   Consider the Neumann problem with the boundary conditions shown in Figure 7.17. Note first that the solvability condition is satisfied:

$$\int_0^1 \left( y - \frac{1}{2} \right) dy = \left( \frac{1}{2} y^2 - \frac{1}{2} y \right) \Big|_0^1 = 0.$$

As with the Dirichlet problems we discussed, the separation of variables technique can be used to get elementary solutions of Laplace's equation and three of the BCs (the homogeneous ones). Thus we assume that $S$ has the form $S(x, y) = X(x)Y(y)$, substitute in Laplace's equation, and rearrange to get

$$\frac{X''(x)}{X(x)} = -\frac{Y''(y)}{Y(y)} = k,$$

for some constant $k$. This leads to the pair of DEs

$$X'' = kX,$$

$$Y'' = -kY.$$

In addition, the three homogeneous BCs on $S$ lead to the following BCs on $X, Y$:

$$X'(0) = 0,$$

$$Y'(0) = 0,$$

$$Y'(1) = 0.$$

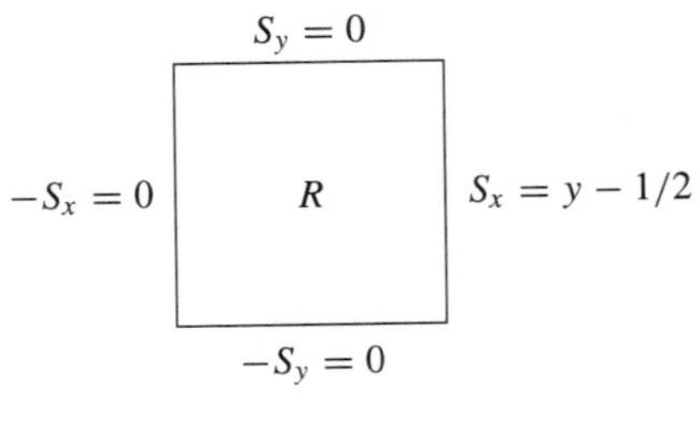

Figure 7.17.   Neumann BCs for Example 7.5.

We thus have a complete Sturm–Liouville problem for $Y$, with eigenfunctions

$$Y_n(y) = \cos n\pi y \qquad n = 0, 1, 2, \ldots.$$

The eigenvalues are $-k = -n^2\pi^2$, and using these values for $k$ in the partial Sturm–Liouville problem for $X$ gives $X'' = n^2\pi^2 X$, and $X'(0) = 0$. Thus we take $X$ to be a hyperbolic cosine function:

$$X_n(x) = \cosh n\pi x.$$

We conclude that an elementary solution of Laplace's equation that also satisfies three of the Neumann BCs is

$$\cosh n\pi x \, \cos n\pi y, \qquad n = 0, 1, 2, \ldots.$$

Note that for $n = 0$, the above function is just the constant 1 function. With a little bit of experience you should be able to write down immediately from the boundary conditions the elementary solutions (as above) *without* going through the separation of variables technique. We will do this from now on. In essence, we will solve all our Neumann problems by using elementary solutions that are a product of cosine and hyperbolic cosine functions.

Now, in order to satisfy the remaining boundary condition, we combine the above elementary solutions in a series,

$$S(x, y) = A_0 + \sum_{n=1}^{\infty} A_n \cosh n\pi x \, \cos n\pi y, \qquad (7.14)$$

which gives, by the superposition principle, a function that (formally) satisfies Laplace's equation and three of the BCs. The $A_n$'s, which are arbitrary constants, are chosen such that this series representation satisfies the last BC $S_x(1, y) = y - 1/2$. Taking the partial derivative of the above expression gives

$$S_x(x, y) = \sum_{n=1}^{\infty} n\pi A_n \sinh n\pi x \, \cos n\pi y,$$

and thus the boundary conditon is

$$\sum_{n=1}^{\infty} (n\pi A_n \sinh n\pi) \cos n\pi y = y - 1/2. \qquad (7.15)$$

Thus the numbers $n\pi \sinh n\pi A_n$ are the Fourier cosine coefficients for the function $y - 1/2$. Note that the constant term is absent in the above cosine series, but this is as it should be, since the constant term in the Fourier cosine series for $y - 1/2$ is

$$\int_0^1 \left(y - \frac{1}{2}\right) dy = 0.$$

*This is why we needed the solvability condition for the Neumann problem.*

For $n > 0$ we equate the $n$th coefficient in the series on the left side of equation (7.15) with the $n$th Fourier cosine series of $y - 1/2$ to get

$$n\pi \sinh n\pi A_n = 2 \int_0^1 (y - \frac{1}{2}) \cos n\pi y \, dy$$

$$= 2 \int_0^1 y \cos n\pi y \, dy$$

$$= \frac{2[(-1)^n - 1]}{n^2 \pi^2}.$$

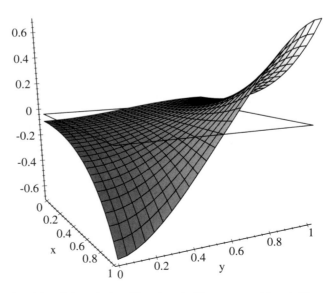

Figure 7.18.   Plot of the solution $S(x, y)$ to the Neumann problem in Example 7.5.

Solving this for $A_n$ gives what we need. In summary, a solution of the Neumann problem in this example is given by the series

$$S(x, y) = A_0 + \sum_{n=1}^{\infty} \frac{2[(-1)^n - 1]}{n^3 \pi^3 \sinh n\pi} \cosh n\pi x \, \cos n\pi y. \tag{7.16}$$

Notice that we said that this is *a* solution and not *the* solution, since as we have mentioned several times, there are infinitely many solutions of any Neumann problem, and they all differ by a constant. We see this in the above series solution: The coefficient $A_0$ is not determined by the boundary conditions. We are free to choose any value for it. For the choice $A_0 = 0$, the graph of $S(x, y)$ is shown in Figure 7.18. Notice that the curves shown on the surface have the required tangent line slopes at the boundary of $R$, namely 0 on three of the boundaries and varying as $y - 1/2$ on the fourth boundary. In addition, the surface represents the steady-state temperature distribution for the corresponding heat problem with these Neumann BCs. As mentioned above, the actual steady state is $A_0 + S(x, y)$ for some constant $A_0$. The graph of this is the same as in Figure 7.18, except translated along the $z$-axis by the amount $A_0$. The interpretation of the BCs is that three of the sides are insulated (no heat flux across them), while the vertical, right-hand side has a specified outward heat flux that varies as $-\nabla S \cdot n = -(y - 1/2)$. Thus heat flows outward across the bottom half of the boundary (where $1/2 - y > 0$) and inward across the top half (where $1/2 - y < 0$).

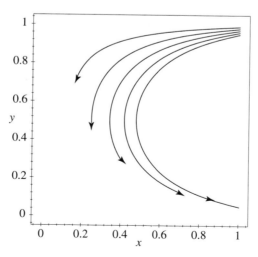

Figure 7.19.    Heat flow lines for the steady-state temperature distribution in Example 7.5.

Some of the heat flow lines are shown in Figure 7.19. Based on the mountain slope analogy, our choice of initial points consisted of high points on the slope: $(1, 0.99)$, $(1, 0.98)$, $(1, 0.97)$, $(1, 0.96)$, $(1, 0.95)$. Again, skiing down, following the path of maximum descent, will trace out the heat flow lines shown. Because three of the sides are insulated, you will never end up at a point there.

**Example 7.6**    Consider the Neumann problem with the boundary conditions as indicated in Figure 7.20 It is easy to see that the solvability condition is satisfied:

$$\int_0^1 \left( -\frac{3}{2} y^2 + \frac{1}{2} \right) dy = 0.$$

Based on our experience from the last example, we see that a solution here has the form

$$S(x, y) = A_0 + \sum_{n=1}^{\infty} A_n \cosh n\pi (1 - x) \cos n\pi y. \qquad (7.17)$$

You should verify that this satisfies Laplace's equation and the three homogeneous BCs in this example. As in the last example, the $A_n$'s are chosen such that the last BC, $-S_x(0, y) = -\frac{3}{2} y^2 + \frac{1}{2}$, is satisfied. Thus we need

$$-\sum_{n=1}^{\infty} A_n ((-n\pi) \sinh n\pi) \cos n\pi y = -\frac{3}{2} y^2 + \frac{1}{2}.$$

Equating the coefficients on the left side with the corresponding Fourier cosine

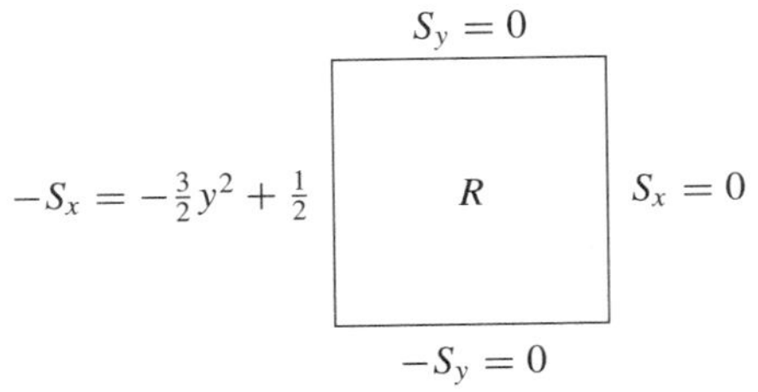

Figure 7.20.    Neumann BCs for Example 7.6

coefficients of $-\frac{3}{2}y^2 + \frac{1}{2}$ and solving for $A_n$ gives

$$
\begin{aligned}
A_n &= \frac{2}{n\pi \sinh n\pi} \int_0^1 \left( -\frac{3}{2}y^2 + \frac{1}{2} \right) \cos n\pi y \, dy \\
&= \frac{-3}{n\pi \sinh n\pi} \int_0^1 y^2 \cos n\pi y \, dy \\
&= \frac{6(-1)^{n+1}}{n^3 \pi^3 \sinh n\pi}.
\end{aligned}
$$

This is for $n > 0$. As before, $A_0$ is not determined by the BCs and is completely arbitrary.

The preceeding two examples were special in the sense that the boundary conditions on three sides of the square were homogeneous. We will now use this to solve a more general case (the most general case is left for the exercises).

**Example 7.7**    Consider the Neumann problem with BCs as shown in Figure 7.21. We first check that the solvability condition is satisfied for these boundary conditions:

$$
\int_0^1 -\frac{3}{2}y^2 dy + \int_0^1 y \, dy = -\frac{1}{2} + \frac{1}{2} = 0.
$$

Based on our experience with the Dirichlet problem, we might try to decompose this Neumann problem into two other Neumann problems, each of which has homogeneous BCs on three sides of the square. This is shown in Figure 7.22.

However, this decomposition gives us two Neumann problems, neither of which is solvable, since neither one satisfies the solvability condition. This difficulty is circumvented, however, if we decompose the original problem into

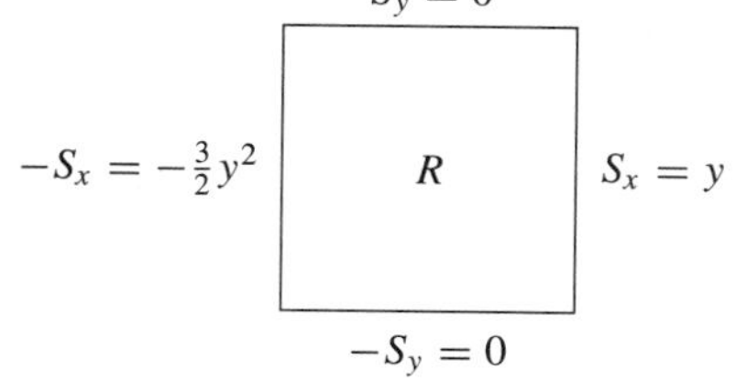

Figure 7.21.    Neumann BCs for Example 7.7.

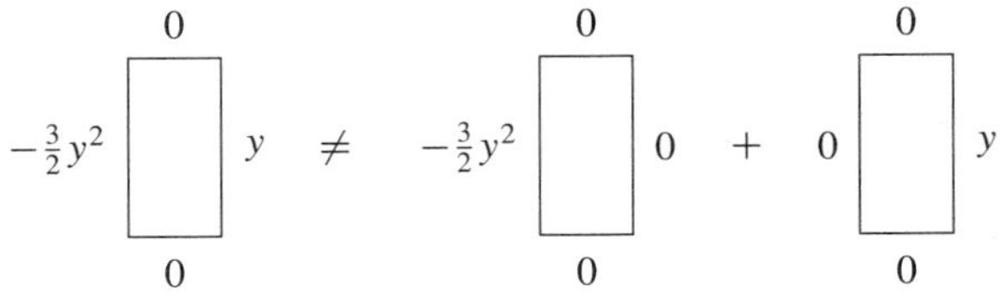

Figure 7.22.   An incorrect way to decompose the Neumann BCs in Example 7.7.

the sum of *three* Neumann problems, as shown in Figure 7.23. It is easy to check that now each of the three separate Neumann problems satisfies the the solvability condition. Also, if we denote the solutions of these problems by $S^1$, $S^2$, $S^3$, then

$$S \equiv S^1 + S^2 + S^3$$

gives a solution of the original Neumann problem. This is so, since the sum of any number of solutions of Laplace's equation is also a solution, and the evaluation of $S^1 + S^2 + S^3$ at points along the boundary gives the desired values (check this out !).

Now, we already know how to solve the second and third problems for $S^2$ and $S^3$; this we did in the examples above. Thus to finish the problem, we need to develop a technique for solving the first problem for $S^1$. By the way, you should notice that the constants $-1/2$ and $1/2$ are the values of the integrals of $-3y^2/2$ and $y$ from 0 to 1. To find $S^1$, we try a second-order polynomial in $x$, $y$ of the form

$$S^1(x, y) = Ax^2 + Bxy - Ay^2 + Dx + Ey + F.$$

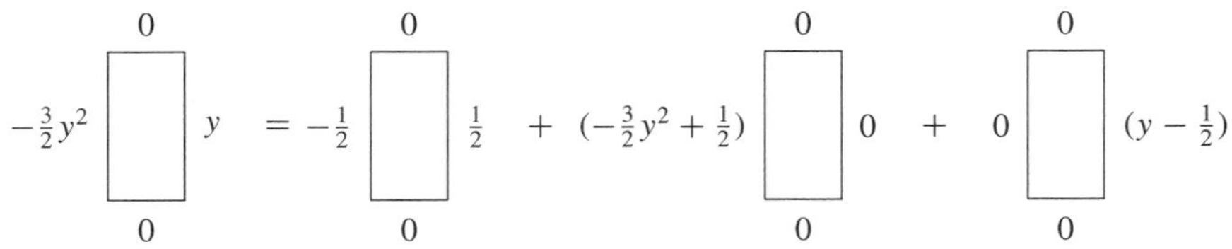

Figure 7.23.   The correct way to decompose the Neumann BCs in Example 7.7.

This automatically satisfies Laplace's equation for any choice of the constants. We thus try to determine the constants by substituting $S^1$ into the BCs. First we compute

$$S^1_x = 2Ax + By + D,$$

$$S^1_y = Bx - 2Ay + E.$$

Thus the BCs on the left and right sides of the square are

$$-S^1_x(0, y) = -(By + D) = -1/2,$$

$$S^1_x(1, y) = 2A + By + D = 1/2.$$

This gives

$$A = 0, B = 0, D = 1/2.$$

The BCs on the top and bottom of the square lead to the equations

$$-S^1_y(x, 0) = -(Bx + E) = 0,$$

$$S^1_y(x, 1) = Bx - 2A + E = 0.$$

We already know that $A = 0$, $B = 0$, and so from the above equations we get additionally that $E = 0$. In summary, we have determined that

$$S^1(x, y) = \frac{1}{2}x + F,$$

where $F$ is an arbitrary constant. Note that the special nature of the BCs here makes the quadratic terms $Ax^2 + Bxy - Ay^2$ drop out, and you might have omitted them from the start. However, for more general BCs the quadratic terms will be needed.

Summarizing: The solution of the Neumann problem in this example is

$$S(x, y) = A_0 + \frac{1}{2}x + \sum_{n=1}^{\infty} \frac{2[(-1)^n - 1]}{n^3\pi^3 \sinh n\pi} \cosh n\pi x \, \cos n\pi y$$

$$+ \sum_{n=1}^{\infty} \frac{6(-1)^{n+1}}{n^3\pi^3 \sinh n\pi} \cosh n\pi (1 - x) \, \cos n\pi y.$$

Note that we have combined the three arbitrary constants from the separate solutions of the three Neumann problems into one arbitrary constant $A_0$ in the above equation.

**Exercises 7.2**

1. Solve the Neumann problem where the flux is constant on all four boundaries:

$$S_y = -(a + b + c)$$

$$-S_x = a \qquad R \qquad S_x = c$$

$$-S_y = b$$

Do this by looking for a solution that is a second-degree polynomial:

$$S(x, y) = Ax^2 + Bxy - Ay^2 + Dx + Ey + F.$$

*Note*: The isotherms (level curves) of $S$ are conic sections. Determine which types of conic sections can occur.

2. Solve the following Neumann problems and use a computer to (i) plot your solutions as a surface over the unit square (include the unit square in the plot), and (ii) plot the flow for the heat flux vector field $v = -\nabla S$. Comment on the plots and how they reflect the nature of the steady-state heat distribution $S$.

(a)

$$S_y = 0$$

$$-S_x = -1 \qquad R \qquad S_x = 0$$

$$-S_y = 1$$

(b)

$$S_y = 1$$

$$-S_x = 0 \qquad R \qquad S_x = -1$$

$$-S_y = 0$$

(c)

$$S_y = 0$$

$$-S_x = 2 \qquad R \qquad S_x = 1$$

$$-S_y = -3$$

(d)

$$S_y = -2$$

$$-S_x = 2 \qquad R \qquad S_x = 1$$

$$-S_y = -1$$

3. Solve the following Neumann problems. Generally, you will need a series for part of your answer.

(a)

$$S_y = 0$$

$$-S_x = 0 \qquad R \qquad S_x = y^2 - y + \tfrac{1}{6}$$

$$-S_y = 0$$

(b)

$$S_y = 0$$

$$-S_x = 0 \qquad R \qquad S_x = 0$$

$$-S_y = x^2 - x + \tfrac{1}{6}$$

(c)

$$S_y = 0$$

$$-S_x = 2 - 3y^2 \quad\boxed{R}\quad S_x = 0$$

$$-S_y = -2x$$

(d)

$$S_y = 2 - 3x^2$$

$$-S_x = 0 \quad\boxed{R}\quad S_x = 0$$

$$-S_y = -2x$$

(e)

$$S_y = 1 - x$$

$$-S_x = -y^2 \quad\boxed{R}\quad S_x = \tfrac{5}{6} + y$$

$$-S_y = -x$$

4. Solve the general Neumann problem for the square by decomposing it into 5 separate Neumann problems.

5. Use a computer to plot the solutions of the Neumann problems in Examples 7.6 and 7.7. Include the boundary square in your plots for reference. Also plot the flow of the heat flux vector field and interpret the results.

6. Use a computer to plot the solutions of the Neumann problems in Exercises 3(d) and 3(e). Interpret your results in terms of heat flow.

7. (*The solvability condition*) Consider the 3-D Neumann problem for a solid region $\Omega$ in $\mathbb{R}^3$:

$$\nabla^2 S = 0 \qquad \text{in } \Omega,$$

$$\nabla S \cdot n = g \qquad \text{on } \partial\Omega.$$

Here $g$ is a given continuous function on the boundary of $\Omega$.

(a) Show that if the Neumann problem has a solution $S$, then necessarily, the given boundary function $g$ has the property

$$\int_{\partial\Omega} g \, dA = 0. \qquad (7.18)$$

Thus if we want to have any chance of solving a Neumann problem, we must assume that the given boundary function $g$ has this property.

(b) The solvability condition for the 2-D Neumann problem on a planar region $R$ can be derived in various ways. One way is to expand $R$ into a 3-D region $\Omega$ as follows. Let the region $\Omega$ have the shape shown in Figure 7.24. The top and bottom of $\Omega$ are identical planar regions $R$, $R'$. Next we extend the boundary data $g$ in the 2-D Neumann problem to boundary data $\hat{g}$ for a 3-D Neumann problem on $\Omega$. Suppose $g : \partial R \times (0, \infty) \to \mathbb{R}$ is a given time-dependent function on the boundary $\partial R$ of the planar region $R$. Extend $g$ to a (time-dependent) function $\hat{g}$ on the boundary of $\Omega$ as follows:

$$\hat{g}(x, y, z, t) = g(x, y, t) \qquad \text{for } (x, y) \in \partial R, 0 < z < h,$$

$$\hat{g}(x, y, 0, t) = 0 \qquad \text{for } (x, y) \in R,$$

$$\hat{g}(x, y, h, t) = 0 \qquad \text{for } (x, y) \in R.$$

Suppose that the boundary of $R$, that is, the curve $\partial R$, is parametrized by a map $\alpha : I \to \mathbb{R}^3$, with

$$\alpha(s) = \Big(\alpha_1(s), \alpha_2(s), 0\Big),$$

for $s \in I$. Using this, construct suitable parametrizations of the three surfaces that constitute the boundary $\partial\Omega$ of $\Omega$, and then show that

$$\int_{\partial\Omega} \hat{g} \, dA = 0$$

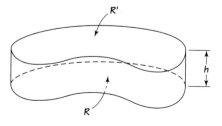

Figure 7.24.   A "cylindrical" region $\Omega$ of height $h$ over the planar region $R$.

if and only if

$$\int_{\partial R} g \, dL = 0.$$

Also, interpret the 3-D boundary conditions on $\Omega$ and why, for $h$ small, they seem to be a natural extension of the 2-D boundary conditions on $R$.

(c) Each scalar field $S$ on $R$ extends to a scalar field $\widehat{S}$ on $\Omega$ defined by $\widehat{S}(x, y, z) = S(x, y)$. Show that $\widehat{S}$ satifies the 3-D Neumann problem on $\Omega$ if and only if $S$ satisfies the 2-D Neumann problem on $R$. Hence get
**The 2-D Solvability Condition:**

$$\int_{\partial R} g \, dL = 0.$$

(d) An alternative, perhaps quicker, derivation of the solvability condition is as follows. Use Green's theorem to show that

$$\int_{\partial R} \nabla S \cdot n \, dL = \int_R \nabla^2 S \, dA,$$

and use this to get the solvability condition. Here $n$ is the outward directed normal to the boundary curve (which is denoted by $\nu$ in Appendix A).

(e) Consider the special case when $R$ is the unit square and $g$ is defined on the boundary in terms of the four maps $g_1, g_2, g_3, g_4$, as shown in Figure 7.16. With $R$ parametrized by the identity map $\alpha(x, y) = (x, y)$ and $\gamma_1, \gamma_2, \gamma_3, \gamma_4$ being the standard parametrization of the boundary $\partial R$ (see Appendix A), show that

$$\int_{\partial R} g \, dL = \int_0^1 g_3(x, t) \, dx + \int_0^1 g_4(x, t) \, dx \int_0^1 g_2(y, t) \, dy + \int_0^1 g_1(y, t) \, dy.$$

Thus in considering the Neumann problem for $R$ the unit square, we must always impose the condition

$$\int_0^1 g_3(x, t) dx + \int_0^1 g_4(x, t) dx \int_0^1 g_2(y, t) dy + \int_0^1 g_1(y, t) dy = 0$$

for consistency. Show that this condition is satisfied in Exercise 1(d) above. Also, discuss how the series solution method breaks down if we change the function $g_2$ there to $g_2(y) = y - y^2$.

## 7.3.   Mixed BVPs

The following two examples illustrate how to solve BVPs with boundary conditions that are *neither* of the Dirichlet type nor the Neumann type, but rather a mixture of these.

**Example 7.8**   Consider solving Laplace's equation

$$\nabla^2 S = 0$$

on the unit square with the boundary conditions shown in Figure 7.25. Here two opposite sides are insulated and the other two held at the fixed temperatures shown. Separating variables $S(x, y) = X(x)Y(y)$ leads to a complete Sturm–Liouville problem for $X$:

$$X'' = kX,$$

$$X'(0) = 0,$$

$$X'(1) = 0.$$

This yields eigenvalues $k = -n^2\pi^2$ and eigenfunctions $X_n = \cos n\pi x$, for $n = 0, 1, 2, \ldots$. *Note* that 0 is an eigenvalue and $X_0 = 1$ is the corresponding eigenfunction. The partial Sturm–Liouville problem for $Y$ is

$$Y'' = n^2\pi^2 Y,$$

$$Y(0) = 0,$$

and so we choose $Y_n = \sinh n\pi y$, for $n = 1, 2, \ldots$, and of course for $n = 0$ we choose $Y(y) = y$. Thus a series solution of Laplace's equation that satisfies the

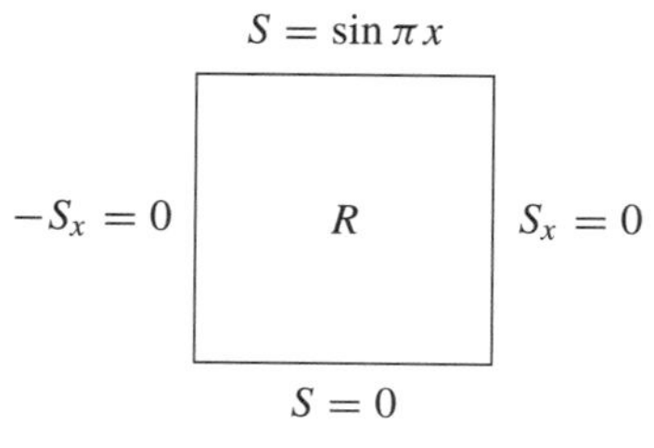

Figure 7.25.   Mixed BCs for Example 7.8.

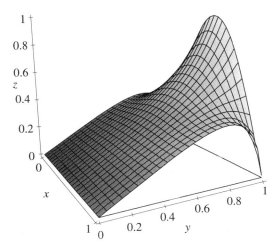

Figure 7.26.   Plot of the solution $S$ to the mixed BVP in Example 7.8.

three homogeneous BCs is

$$S(x, y) = A_0 y + \sum_{n=1}^{\infty} A_n \cos n\pi x \, \sinh n\pi y. \qquad (7.19)$$

To satisfy the last BC, we must choose the $A_n$'s such that

$$S(x, 1) = A_0 + \sum_{n=1}^{\infty} (A_n \sinh n\pi) \cos n\pi x \stackrel{\text{need}}{=} \sin \pi x.$$

Consequently,

$$A_0 = \int_0^1 \sin \pi x \, dx = 2/\pi,$$

and

$$(\sinh n\pi) A_n = 2 \int_0^1 \sin \pi x \, \cos n\pi x \, dx$$

$$= \int_0^1 [\sin(n+1)\pi x - \sin(n-1)\pi x] dx$$

$$= \begin{cases} 0 & \text{if } n = 2k - 1, \\ -4/(\pi(4k^2 - 1)) & \text{if } n = 2k. \end{cases}$$

Thus the solution of this mixed boundary value problem is

$$S(x, y) = \frac{2}{\pi} y - \frac{4}{\pi} \sum_{k=1}^{\infty} \frac{\cos 2k\pi x \, \sinh 2k\pi y}{(4k^2 - 1) \sinh 2k\pi}. \tag{7.20}$$

The graph of this (approximated by 20 terms of the series) is shown in Figure 7.26. Notice that the tangent lines to the surface are horizontal at points along the insulated sides of the boundary. The graph represents a steady-state distribution of temperature in a plate subject to the given BCs.

**Example 7.9**    This example is the same as the last one, except that we change the BC along the top of the square to $S(x, 1) = 1 + \cos 2\pi x$. The general series solution (7.19) in the last example works here also, and we have left to determine the $A_n$'s so that the new BC along the top of the square holds. As we have seen many times before, the special form of the function $1 + \cos 2\pi x$ in the BC eliminates the need to use the integral formulas to compute the $A_n$'s. Comparing with the terms in the series, we see that $A_0 = 1$ and $A_{-2} = \sinh 2\pi$. All the other $A_n$'s are zero. Thus we find that the solution in this example is simply

$$S(x, y) = y + \frac{\cos 2\pi x \, \sinh 2\pi y}{\sinh 2\pi}.$$

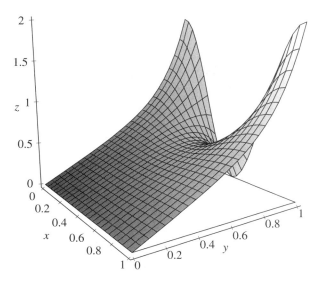

Figure 7.27.    Plot of the solution $S$ to the mixed BVP in Example 7.9.

The graph of this solution is shown in Figure 7.27.

## Exercises 7.3

1. Solve the following mixed boundary value problems and use a computer to plot the solution.

(a)

$$S_y + S = x$$

$$S = 0 \quad \boxed{\phantom{XX} R \phantom{XX}} \quad S = 0$$

$$S = 0$$

(b)

$$S_y + S = x$$

$$S = 0 \quad \boxed{\phantom{XX} R \phantom{XX}} \quad S = 0$$

$$-S_y + S = 0$$

(c) In this problem $K_{a,b}$ is the parabolic pulse function, $m \geq 0$ is a constant, and the BCs are

$$S_y + S = 0$$

$$S_x = 0 \quad \boxed{\phantom{XX} R \phantom{XX}} \quad S_x = 0$$

$$-S_y + S = mK_{a,b}(x)$$

For the computer plots, investigate how the parameters $a$, $b$, $m$ affect the solution by using several appropriate sets of values. Use $m = 0$ in one of these sets.

2. Describe the series solution for a general mixed (or Robin) problem with BCs

$$\kappa_4 S_y + h_4 S = g_4(x)$$

$$-\kappa_1 S_x + h_1 S = 0 \qquad \boxed{R} \qquad \kappa_2 S_x + h_2 S = 0$$

$$-\kappa_3 S_y + h_3 S = 0$$

Use this to describe the series solution of the general Robin problem by decomposing it into 4 problems, each of which has homogeneous BCs on three sides of the boundary (like the one above).

3. Solve the following heat IBVPs and use a computer to analyze the heat flow. Examine, explain, and interpret your results using all the techniques previously discussed. In each case assume that the heat equation is homogeneous: $u_t = \nabla^2 u$.

(a) Initial temperature $\phi(x, y) = 2K_{0,1/2}(x)K_{0,1/2}(y)$ and boundary conditions

$$u = 1 + \cos 2\pi x$$

$$-u_x = 0 \qquad \boxed{R} \qquad u_x = 0$$

$$u = 0$$

(b) Initial temperature $\phi(x, y) = 2K_{1/4,1/2}(x)K_{1/4,1/2}(y)$ and boundary conditions

$$u_y = 0$$

$$-u_x = 0 \qquad \boxed{R} \qquad u_x = y - 1/2$$

$$-u_y = 0$$

*Hint*: See Example 7.5 and note that in transforming to homogeneous BCs $u = S + U$, the $S$ will be given by a series

$$S(x, y) = \sum_{n=1}^{\infty} B_n \cosh n\pi x \, \cos n\pi y$$

(and not in closed form). To find the coefficients in the IC for the transformed problem, you may wish first to expand each $\cosh n\pi x$ in terms of its Fourier cosine series:

$$\cosh n\pi x = \sum_{m=0}^{\infty} C_{nm} \cos m\pi x.$$

# 8  3-D Heat Flow

The analytic techniques and methods we used to solve heat flow problems in one and two dimensions extend in a straightforward manner to the general, and more realistic, case of heat flow in 3-D bodies or regions in space. Thus for $\Omega$ a region in $\mathbb{R}^3$, the general heat IBVP is: Find a function: $u = u(x, y, z, t)$, defined on $\Omega$ and its boundary $\partial\Omega$, that satisfies

$$u_t = \alpha^2 \nabla^2 u + F \qquad \text{in } \Omega, \text{ for } t > 0,$$

$$\kappa \nabla u \cdot n + hu = g \qquad \text{on } \partial\Omega \text{ for } t > 0,$$

$$u = \phi \qquad \text{in } \Omega, \text{ for } t = 0.$$

In keeping with our developments in one and two dimensions, we limit our discussion in this chapter to the case where

$$\Omega = \text{the unit cube.}$$

Later (Chapters 12 and 13) we will learn how to handle other, more interesting, geometries. The case for $\Omega = $ the unit cube should be easy for you to do, now that you have studied the analogous problems for the unit interval and the unit square. Indeed, this development could be a good exercise, but for the sake of completeness and understanding, we briefly outline this here.

In a sense, this chapter could be viewed as a summary and review of all the techniques that were used in solving 1-D and 2-D heat problems. Indeed, these are just special cases of the 3-D problem where the data forces the solution to depend on only one or two of the spatial variables.

## 8.1.  The Heat Equation for the Unit Cube

The unit cube $\Omega$ has boundary $\partial\Omega$ consisting of six faces, and in keeping with the customary usage (viewing $\Omega$ as a box), we will refer to these faces as the top, bottom, and four sides (front, back, left, and right sides). For the geometry of the cube, the BC

$$\kappa \nabla u \cdot n + hu = g \qquad \text{on } \partial\Omega \text{ for } t > 0$$

actually consists of six different BCs (one for each face of the cube), which we write as

$$-\kappa_1 u_x(0, y, z, t) + h_1 u(0, y, z, t) = g_1(y, z, t),$$

$$\kappa_2 u_x(1, y, z, t) + h_2 u(1, y, z, t) = g_2(y, z, t),$$

$$-\kappa_3 u_y(x, 0, z, t) + h_3 u(x, 0, z, t) = g_3(x, z, t),$$

$$\kappa_4 u_y(x, 1, z, t) + h_4 u(x, 1, z, t) = g_4(x, z, t),$$

$$-\kappa_5 u_z(x, y, 1, t) + h_5 u(x, y, 0, t) = g_5(x, y, t),$$

$$\kappa_6 u_z(x, y, 1, t) + h_6 u(x, y, 1, t) = g_6(x, y, t),$$

for all $0 \le x \le 1$, $0 \le y \le 1$, $0 \le z \le 1$, and $t > 0$. The labeling $g_i$, $i = 1, \ldots, 6$, of the six boundary functions extends the notation used earlier for the 2-D problem. For convenience, we will represent the cube by its perspective projection on the $x$-$y$ plane as shown in Figure 8.1 and label the boundaries as indicated.

## 8.2.  The Semi-Homogeneous Problem

The semi-homogeneous heat problem has homogeneous BCs $g_i = 0, i = 1, \ldots, 6$, and the internal heat source/sink function $F$ in the heat equation is (possibly) not zero. Based on our work in one and two dimensions, you can easily see that the solution of this problem is described as follows.

**Theorem 8.1 (Semi-Homogeneous Problem)**    *The general solution u of the semi-homogeneous heat equation and homogeneous BCs for the unit cube has a formal series representation given by*

$$u(x, y, z, t) = \sum_{n=0}^{\infty}\sum_{m=0}^{\infty}\sum_{\ell=0}^{\infty} T_{nm\ell}(t) X_n(x) Y_m(y) Z_\ell(z). \qquad (8.1)$$

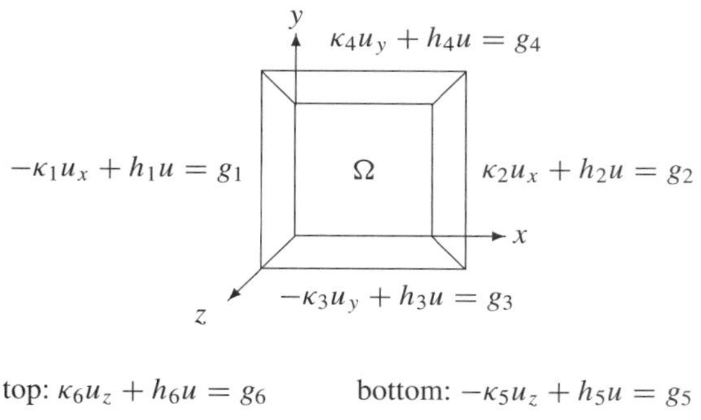

Figure 8.1.  Specification of the BCs on the faces of the unit cube $\Omega$. The cube, or box, is represented by its projection on the $x$-$y$ plane. The view is looking down on the top of the box. The boundary functions $g_1$, $g_2$ refer to the left and right sides of the box, and $g_3$, $g_4$ refer to the front and back of the box. The functions $g_5$, $g_6$ are the boundary functions for the bottom and top of the box, respectively.

*The functions appearing in this series are:*

**(I) The Sturm–Liouville Eigenfunctions**    $X_n$, $Y_m$, $Z_\ell$, *which satisfy*

$$X_n'' = -\lambda_n^2 X_n,$$

$$Y_m'' = -\mu_m^2 Y_m,$$

$$Z_\ell'' = -\nu_\ell^2 Z_\ell,$$

*and*

$$-\kappa_1 X_n'(0) + h_1 X_n(0) = 0,$$

$$\kappa_2 X_n'(1) + h_2 X_n(1) = 0,$$

$$-\kappa_3 Y_m'(0) + h_3 Y_m(0) = 0,$$

$$\kappa_4 Y_m'(1) + h_4 Y_m(1) = 0,$$

$$-\kappa_5 Z_\ell'(0) + h_5 Z_\ell(0) = 0,$$

$$\kappa_6 Z_\ell'(1) + h_6 Z_\ell(1) = 0,$$

*with corresponding Sturm–Liouville eigenvalues* $\lambda_n$, $\mu_m$, *and* $\nu_\ell$.

**(II) The Time-Dependent Factors**   $T_{nm\ell}$, *given by*

$$T_{nm\ell}(t) = \left[ A_{nm\ell} + \int_0^t F_{nm\ell}(s)e^{a_{nm\ell}s}\,ds \right] e^{-a_{nm\ell}t}. \tag{8.2}$$

*In this integral formula*

$$a_{nm\ell} \equiv \alpha^2(\lambda_n^2 + \mu_m^2 + \nu_\ell^2),$$

$$F_{nm\ell}(t) = (L_n M_m N_\ell)^{-1} \int_0^1 \int_0^1 \int_0^1 F(x, y, z, t)X_n(x)Y_m(y)Z_\ell(z)dxdydz,$$

$$A_{nm\ell} = (L_n M_m N_\ell)^{-1} \int_0^1 \int_0^1 \int_0^1 \phi(x, y, z)X_n(x)Y_m(y)Z_\ell(z)dxdydz.$$

**Comments:** In the theorem we note:

(1) The $F_{nm\ell}(t)$'s and $A_{nm\ell}$'s computed from the integral formulas are just the generalized Fourier coefficients in the generalized Fourier series expansions of $F$ and $\phi$:

$$F(x, y, z, t) = \sum_{n=0}^{\infty} \sum_{m=0}^{\infty} \sum_{\ell=0}^{\infty} F_{nm\ell}(t)X_n(x)Y_m(y)Z_\ell(z),$$

$$\phi(x, y, z) = \sum_{n=0}^{\infty} \sum_{m=0}^{\infty} \sum_{\ell=0}^{\infty} A_{nm\ell}X_n(x)Y_m(y)Z_\ell(z).$$

The normalizing constants $L_n$, $M_m$, and $N_\ell$ are, as before, the integrals of the squares of the respective eigenfunctions:

$$L_n = \int_0^1 X_n^2(x)dx,$$

$$M_m = \int_0^1 Y_m^2(y)dy,$$

$$N_\ell = \int_0^1 Z_\ell^2(z)dz.$$

Of course, by now we have computed these for many of the different types of Sturm–Liouville problems that arise, and so we can just use these values directly.

(2) The series summations in equation (8.1) start at 0, and we use the following conventions:

$$\lambda_0 = 0, \; \mu_0 = 0, \; \nu_0 = 0.$$

Further, we take $X_0 \equiv 0$ if 0 is *not* an eigenvalue in the Sturm–Liouville problem for the $X$'s. Similarly, we agree to take $Y_0 \equiv 0$ when 0 is not an eigenvalue in SL problem for the $Y$'s, and $Z_0 \equiv 0$ when 0 is not an eigenvalue in the SL problem for the $Z$'s.

(3) The time-dependent factor $T_{nm\ell}$ is the solution of the initial value problem

$$T'_{nm\ell}(t) + a_{nm\ell} T_{nm\ell}(t) = F_{nm\ell}(t), \tag{8.3}$$

$$T_{nm\ell}(0) = A_{nm\ell}. \tag{8.4}$$

(4) Thus to write down the series solution (8.1) for a homogeneous heat problem, we must:

  (a) determine the eigenvalues and eigenfunctions for three separate Sturm–Liouville problems. The SL boundary conditions for the $X$'s and $Y$'s come from the BCs imposed on the left–right and back–front sides of the cube $\Omega$, while the SL boundary conditions on the $Z$'s come from the BCs imposed on the top–bottom of the cube;

  (b) compute the generalized Fourier coefficients for $F$ and $\phi$ and then use the integral formula (8.2) to compute the $T_{nm\ell}$'s.

**Proof of Theorem:**    To see that $u$ as given by Equation (8.1) formally satisfies the heat equation, we substitute the series in the PDE. In the calculations we interchange the differentiation and summation operations and use $X''_n = -\lambda_n^2 X_n$, $Y''_m = -\mu_m^2 Y_m$, $Z''_\ell = -\nu_\ell^2 Z_\ell$, and $T'_{nm\ell} + a_{nm\ell} T_{nm\ell} = F_{nm\ell}$ to arrive at

$$u_t - \alpha^2 \nabla^2 u$$

$$= \sum_{n=0}^{\infty} \sum_{m=0}^{\infty} \sum_{\ell=0}^{\infty} T'_{nm\ell} X_n Y_m Z_\ell - \alpha^2 T_{nm\ell} [X''_n Y_m Z_\ell + X_n Y''_m Z_\ell + X_n Y_m Z''_\ell]$$

$$= \sum_{n=0}^{\infty} \sum_{m=0}^{\infty} \sum_{\ell=0}^{\infty} [T'_{nm\ell} + a_{nm\ell} T_{nm\ell}] X_n Y_m Z_\ell$$

$$= \sum_{n=0}^{\infty} \sum_{m=0}^{\infty} \sum_{\ell=0}^{\infty} F_{nm\ell} X_n Y_m Z_\ell$$

$$= F.$$

Next, we verify that the series representation for $u$ satisfies the boundary conditions. The first heat flow BC is

$$- \kappa_1 u_x(0, y, z, t) + h_1 u(0, y, z, t)$$

$$= \sum_{n=0}^{\infty} \sum_{m=0}^{\infty} \sum_{\ell=0}^{\infty} T_{nm\ell}(t) \left[ -\kappa_1 X_n'(0) + h_1 X_n(0) \right] Y_m(y) Z_\ell(z)$$

$$= \sum_{n=0}^{\infty} \sum_{m=0}^{\infty} \sum_{\ell=0}^{\infty} T_{nm\ell}(t) \left[ 0 \right] Y_m(y) Z_\ell(z)$$

$$= 0,$$

which follows from the fact that the $X_n$'s satisfy the corresponding Sturm–Liouville BC. The verification that $u$ satisfies the other five heat flow BCs is exactly similar to the above. This finishes the proof, which is largely heuristic and formalistic since a rigorous proof would require details on convergence of the series involved and the legitimacy of interchanging the operations of partial derivatives and summations of series.

**Corollary 8.1 (The Homogeneous Problem)**    *In the special case where $F \equiv 0$, the solution (8.1) reduces to the solution of the homogeneous problem*

$$u(x, y, z, t) = \sum_{n=0}^{\infty} \sum_{m=0}^{\infty} \sum_{\ell=0}^{\infty} A_{nm\ell} \, e^{-(\lambda_n^2 + \mu_m^2 + \nu_\ell^2)\alpha^2 t} X_n(x) Y_m(y) Z_\ell(z). \qquad (8.5)$$

Thus we see that the semi-homogeneous and homogeneous heat problems in three dimensions are entirely similar to those in dimensions one and two. However, visualizing the flow of heat in the cube $\Omega$ over time is a bit more difficult, as we shall see in the examples that come later.

The theory for the steady-states is also the same as in dimensions one and two:

**Corollary 8.2 (Steady States)**    *Suppose in the semi-homogeneous problem that the internal source/sink function $F$ does not depend on $t$.*

(1) *If at least one face of the cube is not insulated, then there exists a unique steady-state, and it has series representation*

$$S(x, y, z) = \sum_{n=0}^{\infty} \sum_{m=0}^{\infty} \sum_{\ell=0}^{\infty} \frac{F_{nm\ell}}{a_{nm\ell}} X_n(x) Y_m(y) Z_\ell(z). \tag{8.6}$$

(2) *If each face of the cube is insulated, then there exists a steady-state if and only if*

$$\int_0^1 \int_0^1 \int_0^1 F(x, y, z) \, dx \, dy \, dz = 0,$$

*i.e., the Fourier coefficient $F_{000} = 0$. When this condition is met, the problem has steady-states of the form*

$$S(x, y, z) = c + \sum_{(n,m,\ell) \neq (0,0,0)} \frac{F_{nm\ell}}{a_{nm\ell}} \cos n\pi x \, \cos m\pi y \, \cos \ell\pi z, \tag{8.7}$$

*where $c$ is an arbitrary constant and the series is a triple sum over all ordered triples $(n, m, \ell)$ different from $(0, 0, 0)$. Thus there is not a unique steady-state in this case.*

**Proof**    This follows easily from the above theorem by assuming that the $T_{nm\ell}$'s do *not* depend on $t$ and using equation (8.3). The details are left as an exercise.

## 8.3.    Transforming to Homogeneous BCs

Now that we know how to solve the general semi-homogeneous problem (at least for the cube), we need to look at the technique for reducing any heat problem to one with homogeneous BCs. This technique is the same as the one we studied before in dimensions 1 and 2 and amounts to splitting the problem into two parts

$$u = S + U,$$

getting a BVP for the function $S$ (a Dirichlet, Neumann, or Robin problem), and a new, transformed heat IBVP for $U$.

**Theorem 8.2 (Transforming to Homogeneous BCs)**    *For a general region $\Omega$ in $\mathbb{R}^3$, suppose that $S$ satisfies the BVP*

$$\nabla^2 S = 0 \qquad in \ \Omega, \tag{8.8}$$

$$\kappa \nabla S \cdot n + hS = g \qquad on \ \partial \Omega, \tag{8.9}$$

*and that U satisfies the semi-homogeneous IBVP*

$$U_t = \alpha^2 \nabla^2 U + F - S_t \qquad in \ \Omega, for \ t > 0, \tag{8.10}$$

$$\kappa \nabla U \cdot n + hU = 0 \qquad on \ \partial \Omega, for \ t > 0, \tag{8.11}$$

$$U = \phi - S \qquad in \ \Omega, for \ t = 0. \tag{8.12}$$

*Then the function*

$$u = S + U$$

*satisfies the heat IBVP*

$$u_t = \alpha^2 \nabla^2 u + F \qquad in \ \Omega, for \ t > 0, \tag{8.13}$$

$$\kappa \nabla u \cdot n + hu = g \qquad on \ \partial \Omega, for \ t > 0, \tag{8.14}$$

$$u = \phi \qquad in \ \Omega, for \ t = 0. \tag{8.15}$$

**Proof**   The proof is entirely elementary and amounts merely to substituting $u = S + U$ into equations (8.13)–(8.15). For the PDE we find

$$u_t = S_t + U_t$$
$$= S_t + \alpha^2 \nabla^2 U + F - S_t$$
$$= \alpha^2 \nabla^2 U + F$$
$$= \alpha^2 (\nabla^2 S + \nabla^2 U) + F$$
$$= \alpha^2 \nabla^2 u + F.$$

Next, substituting into the BCs gives

$$\kappa \nabla u \cdot n + hu = \kappa \nabla (S + U) \cdot n + h(S + U)$$
$$= [\kappa \nabla S \cdot n + hS] + [\kappa \nabla U \cdot n + hU]$$
$$= g + 0$$

on the boundary for all times $t > 0$. Finally, we see that for $(x, y, z)$ in $\Omega$, the IC

$$u(x, y, z, 0) = S(x, y, z, 0) + U(x, y, z, 0)$$

$$= S(x, y, z, 0) + \phi(x, y, z) - S(x, y, z, 0)$$

$$= \phi(x, y, z).$$

This completes the proof.

The function $S$ used in the solution of the general non-homogeneous heat problem will, in general, be a time-dependent function $S = S(x, y, z, t)$, since the boundary conditions may be time-dependent. However, in many examples, the BCs will not involve $t$, and so $S = S(x, y, z)$ will likewise not depend on $t$.

As before, the general BVP with BCs (8.9), known as Robin boundary conditions, has two special cases that frequently occur:

**The Dirichlet Problem:**    Find a solution $S$ of Laplace's equation that has specified values on the boundary:

$$\nabla^2 S = 0 \qquad \text{in } \Omega, \tag{8.16}$$

$$S = g \qquad \text{on } \partial\Omega. \tag{8.17}$$

**The Neumann Problem:**    Find a solution $S$ of Laplace's equation that has specified normal derivative $\partial S/\partial n \equiv \nabla S \cdot n$ on the boundary:

$$\nabla^2 S = 0 \qquad \text{in } \Omega, \tag{8.18}$$

$$\nabla S \cdot n = g \qquad \text{on } \partial\Omega. \tag{8.19}$$

In equation (8.19) we require that the given function $g$ have the property that its surface integral over the boundary suface $\partial\Omega$ be zero:

**Solvability Condition**

$$\int_{\partial\Omega} g \, dA = 0.$$

This is a necessary condition in order for the Neumann problem to have a solution.

The Dirichlet problem arises from a heat problem where the temperature is held fixed at specified values on the boundary of $\Omega$, while the Neumann problem comes from a heat problem where the boundary is insulated.

## 8.4.   Examples of BVPs for the Unit Cube

Before solving some heat flow problems, we look at several BVPs that arise in transforming to homogeneous BCs. The solutions $S$ to these BVPs can be viewed as eventual temperature distributions in the cube toward which some heat flow distribution $u$ tends over time: $\lim_{t\to\infty} u(x, y, z, t) = S(x, y, z)$.

**Example 8.1   (Dirichlet BCs)**   Suppose we seek a solution $S$ of Laplace's equation in the cube that satisfies the BCs

$$S = 0 \qquad\qquad \text{on the sides and bottom,}$$

$$S(x, y, 1) = 16(x - x^2)(y - y^2) \qquad \text{on the top.}$$

At this point in your studies, you should be able to formulate immediately what the series solution will look like for a problem of this sort. However, if you need some detailed reasoning to get the correct form for the series, use a separation of variables argument as follows: For $S$ of the form $S(x, y, z) = X(x)Y(y)Z(z)$, Laplace's equation $\nabla^2 S = 0$, with the variables separated, reads

$$\frac{X''(x)}{X(x)} + \frac{Y''(y)}{Y(y)} + \frac{Z''(z)}{Z(z)} = 0.$$

This together with the boundary conditions gives two complete Sturm–Liouville problems

$$X'' = k_1 X,$$
$$X(0) = 0,$$
$$X(1) = 0,$$

and

$$Y'' = k_2 Y,$$
$$Y(0) = 0,$$
$$Y(1) = 0,$$

and one partial Sturm–Liouville problem,

$$Z'' = -(k_1 + k_2)Z,$$
$$Z(0) = 0.$$

Thus we get $X_n = \sin n\pi x$, $k_1 = -n^2\pi^2$, and $Y_m = \sin m\pi y$, $k_2 = -m^2\pi^2$ for the eigenfunctions/values of the first two problems. The partial Sturm–Liouville problem for $Z$ now reads

$$Z'' = (n^2 + m^2)\pi^2 Z, \quad Z(0) = 0.$$

Thus we take $Z = \sinh((n^2 + m^2)^{1/2}\pi z)$ as a solution to this. Using the superposition principle and the above reasoning, we are led to a general series expression

$$S(x, y, z) = \sum_{n=1}^{\infty}\sum_{m=1}^{\infty} A_{nm} \sin n\pi x \; \sin m\pi y \; \sinh r_{nm}\pi z,$$

which satisfies Laplace's equations an five of the BCs (i.e., that $S$ be 0 on the sides and the bottom of the cube). Here for convenience we are letting

$$r_{nm} = (n^2 + m^2)^{1/2}.$$

The coefficients $A_{nm}$ in the series must now be chosen such that $S$ satisfies the remaining BC, $S(x, y, 1) = 16(x - x^2)(y - y^2)$. Hence we must have

$$\sum_{n=1}^{\infty}\sum_{m=1}^{\infty} \sinh(r_{nm}\pi) A_{nm} \sin n\pi x \; \sin m\pi y = 16(x - x^2)(y - y^2).$$

Thus the factors $\sinh(r_{nm}\pi) A_{nm}$ are the Fourier sine coefficients of $16(x - x^2)(y - y^2)$, and a short computation gives

$$A_{nm} = \frac{256[1 - (-1)^n][1 - (-1)^m]}{n^3 m^3 \pi^6 \sinh(r_{nm}\pi)}.$$

This completes the analytic solution of the problem.

**Example 8.2  (Mixed Boundary Conditions)**    In this example suppose we seek a solution $S$ of Laplace's equation that satisfies the BCs

$$\nabla S \cdot n = 0 \qquad\qquad \text{on the sides,}$$

$$S = 0 \qquad\qquad \text{on the bottom,}$$

$$S(x, y, 1) = 1 + \cos \pi x \, \cos \pi y \qquad \text{on the top.}$$

When we separate variables $S(x, y, z) = X(x)Y(y)Z(z)$, in this problem, we get SL problems for $X$, $Y$ that have eigenfunctions

$$X_n = \cos n\pi x, \qquad n = 0, 1, 2, \ldots,$$

$$Y_m = \cos m\pi y, \qquad m = 0, 1, 2, \ldots,$$

and 0 is an eigenvalue for both problems. The partial SL problem for $Z$ is

$$Z'' = r_{nm}^2 \pi^2 Z, \quad Z(0) = 0,$$

where as before, we are letting $r_{mn} = (n^2 + m^2)^{1/2}$. Now in this problem, when $n = 0$ and $m = 0$, the DE reduces to $Z'' = 0$. So $Z(z) = Az + B$, and the requirement that $Z(0) = 0$ forces $B = 0$. Thus $Z(z) = Az$. For all other choices of $n$ and $m$, the problem for $Z$ has solution $Z(z) = \sinh r_{nm} \pi z$. Thus taking products of these functions $X$, $Y$, $Z$ and forming a series gives

$$S(x, y, z) = Az + \sum_{(n,m)\neq(0,0)} A_{nm} \cos n\pi x \, \cos m\pi y \, \sinh r_{nm} \pi z.$$

This is the general solution of Laplace's equation that satisfies five of the BCs (normal derivative $= 0$ on the sides of the cube and $S = 0$ on the bottom). The boundary condition on the top is

$$A + \sum_{(n,m)\neq(0,0)} \sinh(r_{nm}\pi) A_{nm} \cos n\pi x \, \cos m\pi y = 1 + \cos \pi x \, \cos \pi y,$$

and as you can see, this is one of those problems where there is no need to use the integral formulas to compute $A$, $A_{nm}$. Thus clearly, we take $A = 1$, $A_{1,1} = (\sinh(\sqrt{2}\pi))^{-1}$, and all the other $A_{nm}$'s equal to zero. The solution then of this Dirichlet problem reduces to

$$S(x, y, z) = z + (\sinh(\sqrt{2}\pi))^{-1} \cos \pi x \, \cos \pi y \, \sinh \sqrt{2}\pi z.$$

We turn now to a graphical analysis of the solution. Viewing $S$ as a steady-state temperature distribution in the cube, we wish to visualize how the temperature varies throughout the cube. There are several ways to do this. First, we can take planar cross sections (slices) through the cube and plot the temperature distribution on each section. For example, the plane $z = 1$ intersects in the top face of the cube. On this face $S(x, y, 1) = 1 + \cos \pi x \cos \pi y$, and the graph of this function is a surface showing the variation of temperature on the top face. On the other hand, the plane $z = 1/2$ gives a section through the middle of the cube with temperature variation given by $S(x, y, .5) = 0.5 + 0.107192 \cos \pi \, x \cos \pi y$.

The plot of this is a similar surface to the one we just considered but with smaller amplitudes of variation. Continuing with such sections down to the bottom of the cube gives such surfaces that become more and more flat (because the bottom is held at zero degrees, which is the lowest temperature). The plots of these surfaces are shown in Figure 8.2.

*Note*: These surfaces, for clarity, are plotted immediately over the respective sections for which they are the temperature profiles. From these plots you can discern that the highest temperature in the cube is 2 degrees, which occurs on the top at the points $(0, 0, 1)$ and $(1, 1, 1)$. Also from the plot of the temperature distribution on the top, or from its analytic expression $S(x, y, 1) = 1 + \cos \pi x \cos \pi y$, you can see that the temperature is a constant 1 degree along the two lines $(s, 0.5, 1)$, $(0.5, s, 1)$, $s \in [0, 1]$. The temperature distributions on the other slices through the cube have similar properties.

A second way of visualizing the temperature variation through the cube is to plot some of the *isotherms*. An isotherm is a surface consisting of all points that have the same temperature, say temperature $c$. Mathematically, this is the plot of the surface $S(x, y, z) = c$ (i.e., a level surface for the function $S$). Plots of some of these isotherms are shown in Figure 8.3. Yet a third way to visualize the distribution of heat in the cube is to plot several of the heat flow lines, i.e., the flow for the heat flux vector field $-\nabla S$ on $\Omega$. Recall that even though the temperature at any point in the cube does not change over time, there still must be a heat flow to maintain the steady-state. The top face of the cube is constantly heated, with temperatures distributed beten 2 and 0 degrees, while the sides are

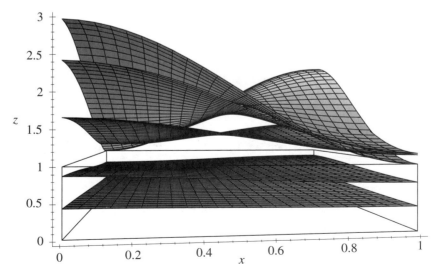

Figure 8.2.  Plots of the temperatures $S(x, y, c)$ along horizontal sections $z = c$ of the cube for $c = 0.2, 0.4, 0.7, 0.9, 1$ in Example 8.2.

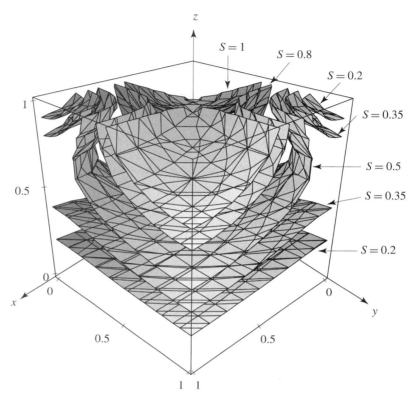

Figure 8.3. Plots of several isotherms $S(x, y, z) = c$ in the unit cube, for $c = 0.2, 0.35, 0.5, 0.8, 1$ in Example 8.2.

insulated and the bottom is maintained at 0 degrees. Given that heat flows from hot to cold, we pick several hot points on the top face and plot the heat flow lines from there to where they end up (at points either on the bottom or top). Figure 8.4 shows the resulting plots. Notice that the flow lines starting at points $C$ and $E$ end on the top of the cube, while those for the other points end on the bottom. The flow line beginning at the center of the top face is a straight line.

## 8.5. Heat Problems for the Cube

We shall consider just two examples of heat problems for the cube, each of which is non-homogeneous. Both are rather standard and simple, but the exercises that follow will deal with a wider variety of more complicated examples.

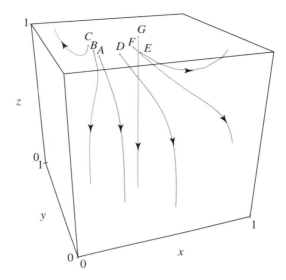

Figure 8.4.   Heat flow lines, for Example 8.2, starting at the points $A = (0.2, 0.2, 1)$, $B = (0.2, 0.3, 1)$, $C = (0.2, 0.4, 1)$, $D = (0.3, 0.2, 1)$, $E = (0.4, 0.2, 1)$, $F = (0.4, 0.3, 1)$, and $G = (0.5, 0.5, 1)$.

**Example 8.3    (Fixed Temperature on the Boundary)**    Consider the heat problem where the temperature is held at a constant value $k$ on the boundary,

$$u = k \qquad \text{on the boundary } \partial \Omega,$$

and the initial temperature is

$$\phi(x, y, z) = 4K_{1/2,3/4}(x)K_{1/2,3/4}(y)K_{1/2,3/4}(z),$$

for $(x, y, z) \in \Omega$. Here $K_{a,b}$ is the *parabolic pulse* function

$$K_{a,b}(x) = \begin{cases} 0 & \text{for } 0 \leq x < a, \\ -4(x-a)(x-b)/(b-a)^2 & \text{for } a \leq x \leq b, \\ 0 & \text{for } b < x \leq 1, \end{cases} \tag{8.20}$$

where $a < b$ are points in $[0, 1]$. Thus the initial temperature distribution is a localized heating confined to the region $\Gamma$, which is the cube of side 1/4 centered at the point (5/8, 5/8, 5/8). The Fourier sine coefficients for $K_{a,b}$ are easily computed in general, and for the case at hand they are

$$K_n = \frac{-32}{n^3 \pi^3} \left[ 8\big(\cos(3n\pi/4) - \cos(n\pi/2)\big) + n\pi \big(\sin(3n\pi/4) + \sin(n\pi/2)\big) \right].$$

Thus the Fourier sine coefficients of $\phi$ are $4K_n K_m K_\ell$.

We assume that there is no internal source/sink of heat: $F = 0$. Transforming to homogeneous BCs, it is clear that

$$u = k + U,$$

where $U$ satisfies

$$U_t = \alpha^2 \nabla^2 U \qquad \text{in } \Omega, \text{ for } t > 0,$$

$$U = 0 \qquad \text{on } \partial\Omega \text{ for } t > 0,$$

$$U = \phi - k \qquad \text{in } \Omega, \text{ for } t = 0.$$

This latter problem is easy to solve, and doing so gives the following solution to the total problem:

$$u(x, y, z, t) = k + \sum_{n=1}^{\infty} \sum_{n=1}^{\infty} \sum_{\ell=1}^{\infty} B_{nm\ell} e^{-a_{nm\ell} t} \sin n\pi x \, \sin \pi y \, \sin \pi z.$$

where $a_{nm\ell} = \alpha^2(n^2 + m^2 + \ell^2)\pi^2$ and the $B_{nm\ell}$'s are the Fourier sine coefficients of $\phi - k$. Since the sine coefficients of the constant function 1 are $g_n = 2(1 - (-1)^n)/(n\pi)$, we get that

$$B_{nm\ell} = 4K_n K_m K_\ell - k g_n g_m g_\ell.$$

From the simple physics of this problem we know that the cube, starting from the initial temperature distribution $\phi$, will heat up to a uniform temperature of $k$ degrees throughout (we assume $k > 0$):

$$\lim_{t \to \infty} u(x, y, z, t) = k.$$

However, initially, near time zero, the heat in the region $\Gamma$ will flow out of this region toward the boundary while at the same time the heat being supplied at the boundary to maintain a constant $k$ temperature there will flow inward, beginning to heat up the entire interior.

The intermediate temperatures in this heating process can be examined in various ways. We can always use the method of slices to examine, at any given time $t$, what the temperature distributions are on various horizontal slices through the cube. Figure 8.5 shows this for $t = 0.005$. Note: in the ensuing analysis we take $k = 1$ and $\alpha^2 = 1$.

Another standard technique for analyzing the heat evolution is to select several appropriate points in the cube, like the center of the initial hot spot $(0.625, 0.625, 0.625)$, and plot the evolution of temperature at these points over

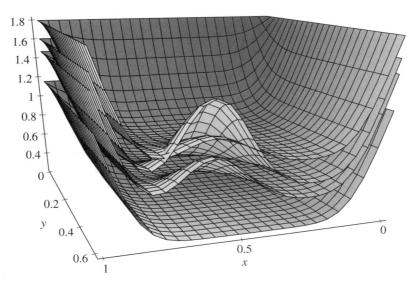

Figure 8.5.   Plots of the temperature distributions on various horizontal slices through the cube in Example 8.3. The time is $t = 0.005$, and the slices are at $z = 0.2$, $z = 0.5$, $z = 0.625$, $z = 0.8$. For clarity the plots have been shifted along the $z$-axis and are actually the graphs of $u(x, y, x, 0.005) + z$, for $z = 0.2, 0.5, 0.625, 0.8$.

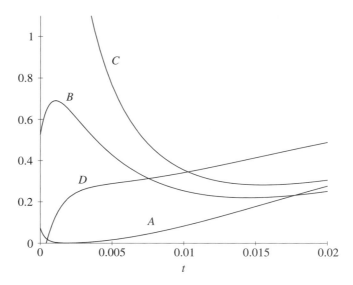

Figure 8.6.   Time evolution of the temperature at the four points $A = (0.3, 0.625, 0.625)$, $B = (0.5, 0.625, 0.625)$, $C = (0.625, 0.625, 0.625)$, $D = (0.8, 0.625, 0.625)$ in the cube for Example 8.3.

time. An example of this is shown in Figure 8.6. The graph shows that the
Point $D$, initially at temperature 0, heats up rapidly as the heat from the initial
temperature distribution reaches it, and then more slowly toward its steady-state
value 1 as the heat from the boundary reaches it. The point $C$, which is initially
the hottest spot at 4 degrees, drops rapidly in temperature as its heat diffuses away
toward the boundary. This continues until about time $t = 0.012$, when the heat
from the boundary reaches it and begins to raise its temperature. Point $B$ starts at
temperature 0.5, is heated briefly as the heat from $C$ diffuses by it, then drops in
temperature, and finally begins its rise toward 1 degree at about $t = 0.017$. Point
$A$, being farthest from the initial hot region $\Gamma$, starts a 0 degrees (the inaccuracy
in the graph is due to the approximation of $\phi$ by a truncated series) and, receiving
heat primarily from the boundary, rises steadily to 1 degree.

A third way to depict the heat content in the cube over time is to plot the
temperature profiles at various times on given line segments running through the
cube, say the segments $[x, 0.625, 0.625]$ and $[x, 0.4, 0.4]$, for $x \in [0, 1]$. These
are shown in Figure 8.7(a). The plot on the left shows in a very precise way
how the temperature drops near the initial hot spot at $(0.625, 0.625, 0.625)$,
while the temperature rises near the boundary of the cube. The rate of change
of temperature at $(.625, .625, .625)$ is very rapid relative to the rates near the
boundary. The initial heating in the region $\Gamma$ is seen to affect the longer-term
evolution toward the steady-state by the presence of skewness (lack of symmetry)
in the temperature profiles. This skewness is also present in the temperature
profile shown in Figure 8.7(b).

Plotting isotherms is not so effective for this example, so we omit this from
the analysis. However, it is sometimes instructive to combine the information
gained from the last two techniques of analysis into one plot, called a *spacetime*
plot. This is, for a given line segment in the cube, a plot of the temperature at
various points at various times. An example is shown in Figure 8.8.

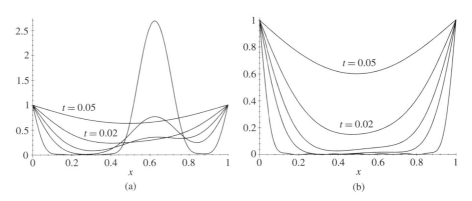

Figure 8.7.   Temperature profiles, in Example 8.3, along the line segments: (a) $[x, .625, .625]$
and (b) $[x, .4, .4]$ at times $t = .001, .005, .01, .02, .05$.

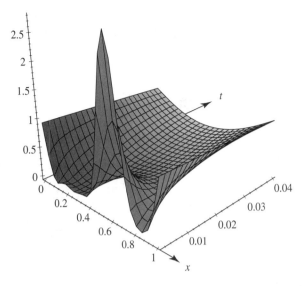

Figure 8.8.    The spacetime temperature surface for the line segment $[x, 0.625, 0.625]$ in Example 8.3. This is the graph of the function $u(x, 0.625, 0.625, t)$ for $x \in [0, 1]$ and $t \in [0, .04]$.

Notice that the curves shown in Figure 8.6 are the intersections of the surface in Figure 8.8 with planes perpendicular to the $x$-axis at $x = 0.3, 0.5, 0.625, 0.8$. Also, the curves shown in Figure 8.7(a) are the intersection of this surface with a plane perpendicular to the $t$-axis at times $t = 0.001, 0.005, 0.01, 0.02, 0.05$. Thus the spacetime surface contains all the information from these prior figures. You can also look at the orthogonal projections of this spacetime surface to obtain the curves in the prior figures.

**Example 8.4**    This problem is the analogue of the one we did for the unit square in the chapter on two-dimensional heat flow. We will use the calculations done there. Consider the heat problem where the temperature is fixed at 0 at all points along the boundary and there is an internal source of heat with density

$$F(x, y, z) = f(x)f(y)f(z),$$

where $f$ is the function defined by

$$f(x) = \begin{cases} 5 \sin 2\pi x & \text{if } x \in [0, 1/2], \\ 0 & \text{otherwise.} \end{cases} \qquad (8.21)$$

Also, assume that the initial temperature distribution is identically zero: $\phi = 0$ in $\Omega$. Thus the internal heat source $F$, which is localized around the point $(0.25, 0.25, 0.25)$ in the cube, will heat the cube up to some steady-state temperature distribution $S$. We wish to solve the problem for $u$, determine $S$, and analyze the temperature distribution in the cube represented by $S$.

With boundary conditions $u = 0$ on $\partial\Omega$, the solution has the form

$$u(x, y, z, t) = \sum_{n=1}^{\infty}\sum_{m=1}^{\infty}\sum_{\ell=1}^{\infty} T_{nm\ell}(t) \sin n\pi x \, \sin m\pi y \, \sin \ell\pi z,$$

where

$$T_{nm\ell}(t) = \frac{F_{nm\ell}}{a_{nm\ell}}[1 - e^{-a_{nm\ell}t}].$$

The quantities involved here are the analogues of those in the 2-D problem. Thus $F_{nm\ell} = f_n f_m f_\ell$, where the $f_n$'s are the Fourier sine coefficients of $f(x)$ (see the previous calculation of these). Further, $a_{nm\ell} = \alpha^2(n^2 + m^2 + \ell^2)\pi^2$. From this solution for $u$ we see that the steady-state temperature distribution $S$ is given by

$$S(x, y, z) = \lim_{t\to\infty} u(x, y, z, t)$$

$$= \sum_{n=1}^{\infty}\sum_{m=1}^{\infty}\sum_{\ell=1}^{\infty} \frac{F_{nm\ell}}{a_{nm\ell}} \sin n\pi \, x \sin m\pi y \, \sin \ell\pi z.$$

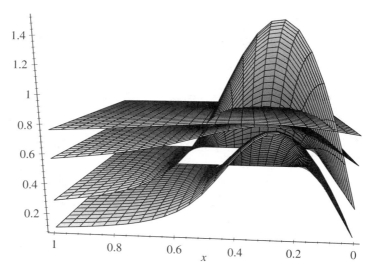

Figure 8.9.   Plots of the temperature distribution $S(x, y, c)$ on horizontal slices $c = 0.1, 0.3, 0.6,$ 0.8 in Example 8.4. For the sake of visualization, the cube has been cut by the plane $y = 0.25$ and the parts beyond this removed.

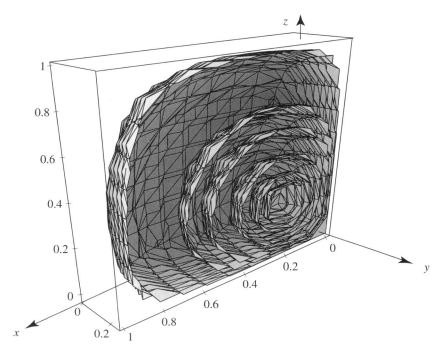

Figure 8.10.   Isotherms $S(x, y, z) = c$ of the steady-state temperature, for $c = 0.01$ (the outermost surface), $c = 0.1, 0.2, 0.5, 0.8$, and $c = 1$ (the innermost surface) in Example 8.4

To analyze the steady-state temperature distribution $S$ in the cube, we use the methods of slices and isotherms. For the slice method, we plot the temperature distribution $S(x, y, c)$ on horizontal slices $z = c$ through the cube. This is shown in Figure 8.9.

*Note*: The plots are the graphs of $c + S(x, y, c)$, which is the plot of $S(x, y, c)$ raised to the level $c$ at which the slice occurs. Also, since the heat source $F$ is centered at the point $(0.25, 0.25, 0.25)$, the cube in the figure has been cut by the plane $y = 0.25$ and the parts beyond that deleted from the picture. As an alternative to the slice method, we can plot isotherms. Figure 8.10 shows several isotherms of $S$. The evolution of the temperatures in the cube toward their steady-state values is more difficult to study. The spacetime temperature surfaces in Figure 8.11 are somewhat effective in showing what happens. The plot on the left is for the line segment running through the center of the region where the heat source is concentrated. It displays the rapid rise of the temperature at $(0.25, 0.25, 0.25)$ to its maximum value 1.2 (also the maximum temperature for the cube in its steady-state). The plot on the right is for a line segment somewhat distant from the region of the heat source and shows a lapse of about 0.04 seconds

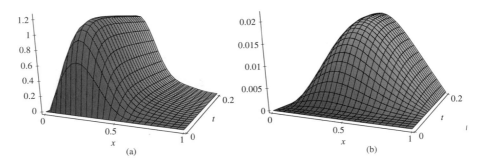

Figure 8.11.   Spacetime temperature surfaces for the line segments $[x, 0.25, 0.25]$ and $[x, 0.8, 0.8]$ in the cube from Example 8.4.

before the points along it begin to heat up and reach their steady-state values at about time $t = 0.2$.

### Exercises 8.5

1. For the solution $S$ in Example 8.2, show that the heat flux vector field $-\nabla S$ is tangent to the plane $x = y$ and also to the plane $1 - x = y$. What can you conclude from this about a heat flow line that starts at a point in one of these planes? Use a computer to plot the flow lines in these planes. Explain why the flow line starting at the point $(0.5, 0.5, 1)$ is a straight line (see Figure 8.4).

2. Solve the Dirichlet BVP that is the same as Example 8.1 except that on the top of the cube, $S$ is specified by

$$S(x, y, 1) = \sin \pi x \, \sin \pi y.$$

Note that the function here, $\sin \pi x \, \sin \pi y$, is very similar to the function $16(x - x^2)(y - y^2)$ used there, except that now the solution $S$ does not require a series for its representation. Viewing $S$ as a steady-state temperature distribution in the cube, study the nature of this distribution much as was done in Example 8.3, and indeed, compare and contrast the two. Specifically, study the temperature distribution by the method of slices, isotherms, and heat flow lines.

3. As a variation of Exercise 2, find the solution $S$ of the Dirichlet problem with BCs: $S = 0$ on the sides and bottom of the cube, and

$$S(x, y, 1) = r K_{a,b}(x) K_{a,b}(y)$$

on the top of the cube. Here $K_{a,b}$ is the parabolic pulse function from Example 8.3, and $r$ is a constant. This represents maintaining a hot spot of maximum temperature $r$ on the top face. Study how varying the location and magnitude of this hot spot affects the distribution of temperature in the cube.

4. Do a study similar to that in Exercise 3, except now use the mixed boundary conditions

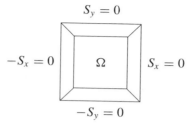

$$S_y = 0$$

$$-S_x = 0 \qquad \Omega \qquad S_x = 0$$

$$-S_y = 0$$

bottom: $S = c$ \qquad\qquad top: $S = K_{a,b}(x)K_{a,b}(y)$

5. (*Steady states*) Consider the general heat IBVP for the cube and assume that the boundary conditions are time-independent (i.e., the $g_i$, $i = 1, \ldots, 6$, do not depend on $t$) and that the heat source/sink density $F$ does not depend on $t$. Discuss the nature of the steady-states and eventual temperature distribution (if any). Do this by letting $S$ be a solution of the associated BVP, transforming to homogeneous BCs, and writing out the series solution $u = S + U$ explicity (except for $S$). Give the condition under which an eventual temperature distribution exists and determine when a steady-state exists and is unique.

6. If you did not already do so in the previous chapter, show that the solvability condition

$$\int_{\partial\Omega} g \, dA = 0$$

is a necessary condition for the existence of solutions to the Neumann BVP $\nabla^2 S = 0$ in $\Omega$ and $\nabla S \cdot n = g$ on $\partial\Omega$. Show also that for the unit cube, with $g_i$, $i = 1, \ldots, 6$, denoting the six boundary functions (see Figure 8.1), the solvability condition is

$$\sum_{i=1}^{6} \int_R g_i(s_1, s_2)ds_1ds_2 = 0.$$

Here $R = [0, 1] \times [0, 1]$ is the unit square, and the integral is the Lebesgue (double) integral over $R$.

7. Solve the Neumann problem where the flux is constant on all six sides of the cube $\Omega$:

$$S_y = d$$

$$-S_x = a \qquad \Omega \qquad S_x = c$$

$$-S_y = b$$

bottom: $-S_z = e$ \qquad\qquad top: $S_z = -(a + b + c + d + e)$

Do this by looking for a solution that is a second-degree polynomial

$$S(x, y, z) = Ax^2 + By^2 + Cz^2 + Dxy + Exz + Fyz + Gx + Hy + Iz.$$

Viewing $S$ as a steady-state temperature distribution in the cube, the isotherms of $S$ are quadric surfaces. Which types of quadric surfaces can occur?

8. Solve the following Neumann problems (using a second-degree polynomial for $S$ as in Exercise 7) and plot a sequence of isotherms for each solution. Since each isotherm is a quadric surface, you should be able to do the plots *by hand*, and in doing so you may wish to sketch the isotherms over a larger region than the unit cube. Study how the isotherms intersect the boundary surfaces (faces of the cube). Use the fact that the heat flow lines intersect each isotherm orthogonally to describe how the heat flow corresponds to the given boundary conditions. In particular, when one of the faces is insulated, discuss how the isotherms dictate heat flow that is parallel on this face (as opposed to across it).

(a)

$S_y = 2$

$-S_x = 0$    $\Omega$    $S_x = 2$

$-S_y = 0$

bottom: $-S_z = 0$        top: $S_z = -4$

(b)

$S_y = -2$

$-S_x = 0$    $\Omega$    $S_x = 4$

$-S_y = 0$

bottom: $-S_z = 0$        top: $S_z = 2$

(c)

$S_y = 2$

$-S_x = 0$    $\Omega$    $S_x = -2$

$-S_y = 0$

bottom: $-S_z = 0$        top: $S_z = 0$

(d)

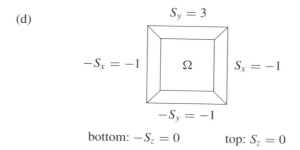

$$S_y = 3$$

$$-S_x = -1 \qquad \Omega \qquad S_x = -1$$

$$-S_y = -1$$

bottom: $-S_z = 0$        top: $S_z = 0$

9. Discuss how to solve the general Neumann problem for the cube by decomposing it into seven other Neumann problems, six of which are insulated on all but one face and one of which has constant heat flux on each face. Using this, indicate the form for the series solution of the original problem.

10. Solve the following heat IBVPs and use a computer to analyze the evolution of the temperature distribution over time, steady-states (if any), and the heat flow in the steady-states. Use some or all of the techniques (slices, isotherms, heat flow lines, temperature profiles along given line segments, spacetime surfaces) discussed in the text for this analysis or devise your own new technique. Do not feel compelled to use every possible technique, since often a technique (like the isotherms) does not help in understanding the heat distribution and flow. Use your judgment! You may take $\alpha = 1$ in these problems.

(a) Heat source $F = 0$; initial temperature distribution

$$\phi(x, y, z) = k K_{a,b}(x) K_{a',b'}(y) K_{a'',b''}(z),$$

where $K_{a,b}$ is the parabolic pulse function on $[a, b]$ and $k$ is a constant, and boundary conditions

$$u_y = 0$$

$$-u_x = 0 \qquad \Omega \qquad u_x = 0$$

$$-u_y = 0$$

top: $u = 0$        bottom: $u = c$

Here $c$ is a constant. Experiment with the location of the initial hot spot $\phi$ by using several choices for $(a, b)$, $(a', b')$, $(a'', b'')$, for example: (i) $(0.5, 0.75)$, $(0.5, 0.75)$, $(0.5, 0.75)$, and $(0.5, 0.75)$, $(0.5, 0.75)$, $(0.25, 0.5)$.

(b) Heat source

$$F(x, y, z) = p K_{c,d}(x) K_{c',d'}(y) K_{c'',d''}(z);$$

initial temperature distribution

$$\phi(x, y, z) = k K_{a,b}(x) K_{a',b'}(y) K_{a'',b''}(z),$$

where $K_{a,b}$ is the parabolic pulse function on $[a, b]$ and $k$, $p$ are constants; and boundary conditions

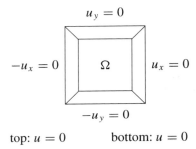

$$u_y = 0$$

$$-u_x = 0 \qquad \Omega \qquad u_x = 0$$

$$-u_y = 0$$

top: $u = 0$ \qquad bottom: $u = 0$

Experiment with the locations and magnitudes of the initial hot spot $\phi$ and the internal heat source density $F$.

# 9 Maxwell's Equations

In the previous chapters we studied in detail the heat equation and Laplace's equation together with various IBVPs and BVPs that arise by imposing appropriate initial conditions and boundary conditions on the solutions of these PDEs. These equations are of the simplest type: Each consists of a single PDE involving a single unknown function (scalar field). In this chapter we begin our study of the more complicated situation consisting of a system of several PDEs involving several unknown functions. For this we have chosen a *linear* system, Maxwell's equations, for our initial work with systems of PDEs. In later chapters we will look at important examples of *nonlinear* systems of PDEs, in particular the Navier–Stokes equations and the Euler equations from fluid mechanics.

Maxwell's equations constitute one of the nicest and most important systems of PDEs in the theory, and an enormous amount of study has been devoted to them. J.C. Maxwell formulated these equations in the late 1800s as a means of describing mathematically the various phenomena in electricity and magnetism, which previously had been investigated experimentally by Faraday, Coulomb, Ohm, and others. In fact, by means of these equations, Maxwell was able to unify the separate phenomena of electricity and magnetism into one quantity, the electromagnetic field, and predict the existence of electromagnetic waves propagating through space (with ordinary light being one such wave).

What we do here with Maxwell's equations will be quite modest, it should prove to be interesting (and teach us something about systems of PDEs, even though this system is rather special). We focus mainly on the notion of potentials for the electromagnetic field, which, by means of gauge transformations, allow Maxwell's equations to be replaced by a system of second-order, uncoupled PDEs for the potentials. The question of existence of potentials is encountered first in the electrostatic equations and leads to interesting connections with geometry, path integrals, and the topology of the underlying domain. We do not attempt to derive Maxwell's equations here (because that would be too much of a diversion), nor do we devote much space to motivating the physics of electromagnetism (hopefully, you have some crude notions about this phenomenon from everyday experience). See [Ja 75], [LC 62], [RMC 93].

The general form of Maxwell's equations that we will study is

**Maxwell's Equations:**

$$\text{div}(E) = 4\pi\rho, \qquad \text{(Coulomb's Law)} \qquad (9.1)$$

$$\text{curl}(E) + \frac{1}{c}\frac{\partial B}{\partial t} = 0, \qquad \text{(Faraday's Law)} \qquad (9.2)$$

$$\text{curl}(B) - \frac{1}{c}\frac{\partial E}{\partial t} = \frac{4\pi}{c}J, \qquad \text{(Ampére–Maxwell Law)} \qquad (9.3)$$

$$\text{div}(B) = 0. \qquad \text{(Absence of Monopoles)} \qquad (9.4)$$

In these equations, $E$, $B$ are (time-dependent) vector fields, known as the *electric* and *magnetic* fields respectively. In terms of their scalar components, these vector fields are expressed by

$$E(x, y, z, t) = \left( E^1(x, y, z, t),\ E^2(x, y, z, t),\ E^3(x, y, z, t) \right),$$

$$B(x, y, z, t) = \left( B^1(x, y, z, t),\ B^2(x, y, z, t),\ B^3(x, y, z, t) \right).$$

The function $\rho$ is a given scalar function representing the distribution of charge in space, and $J = (J^1, J^2, J^3)$ is a given vector field representing the currents (flow of charges) that are present. The constant $c$ is the speed of light. Physically, equations (9.1)–(9.4) are for an electromagnetic field in a vacuum. When an electromagnetic field propagates in a medium, Maxwell's equations as we have stated them must be modified.

Maxwell's equations comprise a system of eight scalar PDEs involving the six components of $E$ and $B$

$$E_x^1 + E_y^2 + E_z^3 = 4\pi\rho,$$

$$E_y^3 - E_z^2 + c^{-1}B_t^1 = 0,$$

$$E_z^1 - E_x^3 + c^{-1}B_t^2 = 0,$$

$$E_x^2 - E_y^1 + c^{-1}B_t^3 = 0,$$

$$B_y^3 - B_z^2 - c^{-1} E_t^1 = 4\pi c^{-1} J^1,$$

$$B_z^1 - B_x^3 - c^{-1} E_t^2 = 4\pi c^{-1} J^2,$$

$$B_x^2 - B_y^1 - c^{-1} E_t^3 = 4\pi c^{-1} J^3,$$

$$B_x^1 + B_y^2 + B_z^3 = 0.$$

This display of the scalar PDEs comprising Maxwell's equations serves to clarify the structure of these equations. However, the vector version, equations (9.1)-(9.4), of Maxwell's equations will be the most useful for the analysis.

As is indicated in equations (9.1)-(9.4), Coulomb, Ampére, and Faraday deserve credit for formulating some of the laws for electromagnetism. However, it was Maxwell who first wrote down these laws as partial differential equations, expressed in terms of the newly developed vector calculus of Gibbs and Heaviside (the $\nabla$, curl, and div operators), and used these PDEs to study and predict many aspects of electromagnetism. Maxwell's modification of Ampére's original law (from electrostatics) to give the Ampére–Maxwell law (9.3) was a crucial step in the formulation of the laws of electrodynamics.

In order to discuss solving Maxwell's equations in a region $\Omega$ of space, we would have to impose boundary conditions and initial conditions. It is not our purpose to do this here, but rather merely to examine the structure of the Maxwell equations and discuss potentials for the electromagnetic field.

One characteristic of systems of PDEs is that they *can* have no solutions at all. This can occur because the form of the system precludes it from having solutions or because the given functions (like $\rho$ and $J$) do not satisfy certain conditions derived from the structure of the system. To see how this could occur in Maxwell's equations, suppose that with $\rho$ and $J$ given the system (9.1)–(9.4) has at least one solution $E$, $B$. Then take the divergence of both sides of equation (9.3) to get

$$\frac{4\pi}{c} \operatorname{div}(J) = \operatorname{div}(\operatorname{curl}(B)) - \frac{1}{c} \operatorname{div}\left(\frac{\partial E}{\partial t}\right)$$

$$= -\frac{1}{c}\frac{\partial}{\partial t}\operatorname{div}(E)$$

$$= -\frac{4\pi}{c}\frac{\partial \rho}{\partial t}.$$

Here we have used Coulomb's Law $\operatorname{div}(E) = 4\pi\rho$ and the general identity $\operatorname{div}(\operatorname{curl}(B)) = 0$, which holds for any vector field. Thus we see that a *necessary* condition for Maxwell's equations to have a solution is that the given charge and current densities satisfy

**The Continuity Equation:**

$$\frac{\partial \rho}{\partial t} + \operatorname{div}(J) = 0.$$

From a mathematical point of view, the continuity equation is just a condition (integrability condition) that we must impose on the structure of the system of PDEs in order that there be solutions. From a physical point of view, $\rho$ and $J$ are not completely arbitrary, since the current consists of the charges in motion, and one can argue that the continuity equation is physically necessary for no charge to be created or destroyed (see the exercises).

## 9.1.  Maxwell's Equations in Empty Space

Here we wish simply to show that when the region $\Omega = \mathbb{R}^3$ and there are no charges and currents ($\rho = 0$ and $J = 0$), then any solution of Maxwell's equations also satisfies a PDE known as the wave equation.

Thus suppose $(E, B)$ is a solution of Maxwell's equations in empty space:

$$\operatorname{div}(E) = 0, \tag{9.5}$$

$$\operatorname{curl}(E) + \frac{1}{c}\frac{\partial B}{\partial t} = 0, \tag{9.6}$$

$$\operatorname{curl}(B) - \frac{1}{c}\frac{\partial E}{\partial t} = 0, \tag{9.7}$$

$$\operatorname{div}(B) = 0. \tag{9.8}$$

These are first-order PDEs, and we wish to show that $E$ and $B$ necessarily also satisfy the following second-order PDEs (which are uncoupled):

$$\frac{\partial^2 E}{\partial t^2} - c^2 \nabla^2 E = 0, \tag{9.9}$$

$$\frac{\partial^2 B}{\partial t^2} - c^2 \nabla^2 B = 0. \tag{9.10}$$

This is easy to demonstrate and relies on the following identity:

$$\operatorname{curl}(\operatorname{curl}(v)) = \nabla \operatorname{div}(v) - \nabla^2 v,$$

which holds for any vector field $v = (v^1, v^2, v^3)$. Using this, we take the curl of both sides of equation (9.6) and use (9.5) and (9.7) to get

$$0 = \text{curl}(\text{curl}(E)) + \frac{1}{c}\,\text{curl}\left(\frac{\partial B}{\partial t}\right)$$

$$= \nabla\,\text{div}(E) - \nabla^2 E + \frac{1}{c}\frac{\partial}{\partial t}\,\text{curl}(B)$$

$$= -\nabla^2 E + \frac{1}{c}\frac{\partial}{\partial t}\left(\frac{1}{c}\frac{\partial E}{\partial t}\right)$$

$$= -\nabla^2 E + \frac{1}{c^2}\frac{\partial^2 E}{\partial t^2}.$$

Rearranging this gives equation (9.9). In a similar fashion, we can easily show that $B$ satisfies equation (9.10).

The fact that solutions of Maxwell's equations in empty space also satisfy (homogeneous) wave equations was a far-reaching prediction of Maxwell's theory. The basic (homogeneous) wave equation

$$u_{tt} - c^2\nabla^2 u = 0$$

was known by mathematicians and physicists of the 1700s to model the propagation of sound in air as well as the vibrations of elastic strings and drumheads (see Chapter 11, Waves in Elastic Materials). By analogy, Maxwell was able to conclude that the electromagnetic field $(E, B)$ was indeed a wave propagating through space with speed $c$. This was soon thereafter confirmed by the experiments of Hertz.

## Exercises 9.1

1. Show that Coulomb's law, $\text{div}(E) = 4\pi\rho$, on $\Omega$ is equivalent to the statement that the integral equation

$$\int_{\partial D} E \cdot n\, dA = 4\pi \int_D \rho\, dV$$

holds for every subregion $D \subseteq \Omega$. This integral version of Coulomb's law is often interpreted as saying that the electric flux through the surface $\partial D$ is equal to the amount of charge enclosed by the surface.

2. Suppose that $E$ is the vector field defined on $\Omega = \mathbb{R}^3 - \{(0, 0, 0)\}$ by

$$E(x, y, z) = \frac{q(x, y, z)}{(x^2 + y^2 + z^2)^{3/2}},$$

where $q$ is a constant.

(a) Show that for $D$ = the ball of radius 1 centered at the origin,

$$\int_{\partial D} E \cdot n \, dA = 4\pi q$$

(assuming that the parametrization is chosen such that $n$ is the outward-directed normal at each point of the sphere $\partial D$). Interpret this result in terms of Exercise 1.

(b) Show by direct calculation that $\operatorname{div}(E) = 0$ and $\operatorname{curl}(E) = 0$ at all points of $\Omega$.

Coulomb originally phrased his law, *Like charges repel each other with a force whose magnitude is the product of the charge amounts and the reciprocal of the square of their distance apart.* Based on this, the vector field $E$ in this problem is the electric field produced by a charge of amount $q$ located at the origin. However, in terms of the way Maxwell phrased Coulomb's law, $\operatorname{div}(E) = 4\pi\rho$, there appears to be a contradiction. Namely, by part (b), $\operatorname{div}(E) = 0$. Thus the most fundamental electric field in electricity and magnetism fails to satisfy Maxwell's equations! Physicists have various means of making sense out of this apparent contradiction. However, the mathematical theory to explain precisely such situations in the theory of PDEs was invented by L. Schwartz in the 1940s. This theory is known as *distribution theory*, or the theory of *generalized functions*. See [Sch 66, 67], [Fr 63] and the next exercise set.

3. Show that Faraday's law implies that

$$\int_{\partial S} E \cdot \tau \, dL = -\frac{1}{c} \frac{\partial}{\partial t} \int_{S} B \cdot n \, dA, \tag{9.11}$$

for every surface $S$ contained in the region $\Omega$. This integral version of Faraday's law (of induction) more adequately describes the results of his original experiments. For example, Faraday found that by moving a magnet back and forth through the center of a circuit $C$ (a closed loop of wire), an electric current is induced in the circuit. Viewing $C$ as a closed curve and taking a surface $S$ whose boundary is $C$, we see how this phenomenon is contained in equation (9.11). The line integral on the left side of the equation is called the *electromotive force* around the circuit, and the surface integral on the right side is the *magnetic flux* through the surface $S$ linking the circuit. This equation says that the electromotive force induced around the circuit is proportional to the time rate of change of the magnetic flux linking the circuit.

4. Show that the continutity equation

$$\frac{\partial \rho}{\partial t} + \operatorname{div}(J) = 0$$

can be written in the integral form

$$\frac{\partial}{\partial t} \int_{D} \rho \, dV = -\int_{\partial D} J \cdot n \, dA,$$

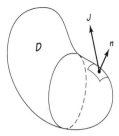

Figure 9.1.    Illustration of the integral version of the continuity equation.

for any region $D \subseteq \Omega$. Interpret this geometrically, referring to Figure 9.1, and explain how it justifies the statement that no charge is created or destroyed in $\Omega$.

5. In a fashion similar to what we did for Maxwell's equations in empty space, i.e., the homogeneous Maxwell equations, show that if $E, B$ satisfy the general, non-homogeneous Maxwell equations (9.1)–(9.4), then $E, B$ satisfy non-homogeneous wave equations of the form

$$\frac{\partial^2 E}{\partial t^2} - c^2 \nabla^2 E = F,$$

$$\frac{\partial^2 B}{\partial t^2} - c^2 \nabla^2 B = G.$$

Make sure you explicitly say what $F$ and $G$ are.

## 9.2.    Electrostatics and Magnetostatics

Another special case of Maxwell's equations, which is easier to study, arises from the assumption that the charge and curent densities $\rho$ and $J$ do *not* depend on time. Then Coulomb's law (9.1) forces $E$ to not depend on $t$, and thus from Faraday's law (9.2), $B$ does not depend on $t$ either. Consequently, Maxwell's equations reduce to

$$\text{div}(E) = 4\pi\rho, \tag{9.12}$$

$$\text{curl}(E) = 0, \tag{9.13}$$

$$\text{curl}(B) = 4\pi c^{-1} J, \tag{9.14}$$

$$\text{div}(B) = 0. \tag{9.15}$$

These are the equations of electrostatics and magnetostatics. They are a decoupled system of PDEs for the $E$ and $B$ field separately, and they therefore allow us to tackle the solution of the PDEs for $E$ separately from those for $B$. Bear in mind that there is a given region $\Omega \subseteq \mathbb{R}^3$ on which we seek solutions $E$, $B$ and that there will have to be boundary conditions imposed to specify the problem completely.

One way to find solutions of the electrostatic and magnetostatic equations is to make the following assumption:

**Assumption (Existence of Potentials):**   Assume that $E$ and $B$ have the form

$$E = -\nabla\phi, \tag{9.16}$$

$$B = \operatorname{curl}(A), \tag{9.17}$$

for some scalar field $\phi$ and vector field $A$ on $\Omega$. The fields $\phi$ and $A$ are called *potentials* for the respective $E$ and $B$ fields.

With $E$ and $B$ given by potentials, equations (9.13) and (9.15) are automatically satisfied (no matter what $\phi$ and $A$ are). This is so, since $\operatorname{curl}(E) = \operatorname{curl}(\nabla\phi) = 0$ and $\operatorname{div}(B) = \operatorname{div}(\operatorname{curl}(A)) = 0$. Furthermore, if $E$ and $B$ are to satisfy equations (9.12) and (9.14), then it is easy to see that their potentials must satisfy

$$\nabla^2\phi = -4\pi\rho, \tag{9.18}$$

$$\nabla \operatorname{div}(A) - \nabla^2 A = 4\pi c^{-1} J. \tag{9.19}$$

The first equation is Poisson's equation (i.e., the non-homogeneous version of Laplace's equation), and we can solve this given appropriate boundary conditions, like specifying the values of $\phi$ on the boundary of $\Omega$. The second equation will reduce to a vector version of Poisson's equation $\nabla^2 A = -J$, if we assume that the magnetic potential satisfies $\operatorname{div}(A) = 0$. Thus, equating components gives three separate Poisson equations to solve.

**Remark**   The minus sign in the potential equation $E = -\nabla\phi$ is there for conventional reasons. From a mathematical viewpoint, the minus sign could be incorporated in the $\phi$. From a physical viewpoint, the minus sign is put there so that the work done by moving a positive charge in an electrostatic field is positive when going from lower to higher potential (interpreted as moving against the electric force field). An analogy with heat flow is helpful here. The field $E = -\nabla\phi$ is like the heat flux vector field. It points in the direction a positive charge should move to *decrease* the work done against the field (go with the flow of the field). This is like heat flowing from hot to cold. For us, the use of the minus sign will be important only if a physical interpretation is required.

**Example 9.1**   Consider a metal tube along the $z$-axis with square cross sections $R = [0, 1] \times [0, 1]$. Assume that three of the sides (called electrodes) are grounded (held at zero potential $\phi = 0$) and the fourth side is held at constant potential $\phi = 1$. See Figure 9.2. Assume that there is no charge within the tube. It seems reasonable to seek a solution $E$ of the electrostatic equations that has the form $E = (E^1, E^2, 0)$, with $E^1$, $E^2$ being functions of $x$ and $y$ only. The potential for the field then has the form $\phi = \phi(x, y)$ and is the solution of Laplace's equation $\nabla^2 \phi = 0$ with the boundary conditions in Figure 9.2. We have already solved this problem in the chapter on boundary value problems and found that

$$\phi(x, y) = \sum_{n=1}^{\infty} \frac{2[1 - (-1)^n]}{n\pi \sinh n\pi} \sinh n\pi (1 - x) \sin n\pi y.$$

The corresponding electric field $E = -\nabla \phi$ is given by a series as well. The flow generated by the electric field $E$ (see the Introduction) consists of curves called *electric field lines*. A positive charge placed within the electric field would be forced to move along a field line toward one of the grounded electrodes. The plot of the flow for $E$, i.e., the electric field lines, is shown in Figure 9.3 and is entirely analogous to the previous heat flow problem we discussed.

There is one fine detail in the above discussion of solutions to the electrostatic/magnetostatic equations that warrants a closer look because it leads to some important modern mathematics. This detail is the assumption of existence of potentials for $E$ and $B$. Might there be solutions $E$ and $B$ of the electrostatic and magnetostatic equations that do *not* arise from potentials? The two homogeneous PDEs curl$(E) = 0$ and div$(B) = 0$ are often thought to imply the existence of potentials for $E$ and $B$, but is this really true?

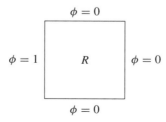

Figure 9.2.   Cross section of a tube whose sides are held at potentials $\phi = 0$ and $\phi = 1$.

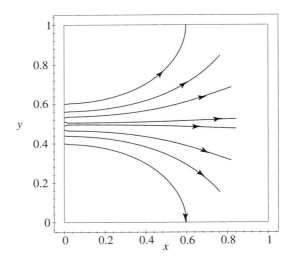

Figure 9.3.   The electric field lines for $E$ in Example 9.1. Notice that the field lines intersect the boundary orthogonally. This is so because the boundary consists of level curves for the potential function $\phi$.

**Question:**   When is it true that

$$\text{curl}(E) = 0, \text{ on } \Omega \implies E = \nabla\phi, \text{ on } \Omega \text{ for some } \phi, \qquad (9.20)$$

$$\text{div}(B) = 0, \text{ on } \Omega \implies B = \text{curl}(A), \text{ on } \Omega \text{ for some } A? \qquad (9.21)$$

We examine this question in the next section. It will be seen that the answer depends on the nature of the domain $\Omega$.

**Exercises 9.2**

1. Assuming the existence of potentials, show that if $B$ is a solution of the magnetostatic equations

$$\text{curl}(B) = 4\pi c^{-1} J,$$

$$\text{div}(B) = 0$$

in $\Omega$, then $B = \text{curl}(A)$ for some vector field $A$ that satisfies the PDEs

$$\text{div}(A) = 0,$$

$$\nabla^2 A = -4\pi c^{-1} J,$$

on $\Omega$. *Hint*: choose some arbitrary potential $A_0$ for $B$ and note that for any scalar field $\Gamma$, the vector field $A = A_0 + \nabla\Gamma$ is also a potential for $B$.

2. Solve the Dirichlet problem $\nabla^2\phi = 0$ in $R$ subject to the following boundary conditions. Plot the electric field lines for $E = -\nabla\phi$ and interpret the result. *Note:* The types of BCs for electric potentials $\phi$ are limited because of the physics. Theory and experiment show that $\phi$ must be constant on the surface of any conductor.

   (a)

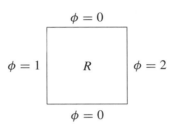

   (b)

$$\phi = 1$$

| $\phi = 1$ | $R$ | $\phi = 1$ |

$$\phi = 0$$

3. (*Generalized vector fields and scalar fields*) This exercise gives a brief introduction to the topic of generalized functions (distributions) and how they occur in the study of PDEs (see [Sch 66] for more details). This topic was alluded to in Exercise 2 in the last section, and in fact, making sense out of Coulomb's law $\operatorname{div}(E) = 4\pi\rho$ is one of the best examples to motivate the need for generalized functions (in this case generalized vector fields $E$ and generalized scalar fields $\rho$). We start with a few definitions.

   Let $\mathcal{D}(\Omega)$ denote the set of all scalar fields $\psi$ on $\Omega$ that vanish outside a compact subset $K \subset \Omega$; i.e., $\psi(x, y, z) = 0$ for all $(x, y, z) \notin K$ (we also implicitly assume that $\psi$ is infinitely differentiable). In particular, note that this requirement ensures that $\psi(x, y, z) = 0$ for all $(x, y, z)$ on the boundary $\partial\Omega$ of $\Omega$ (the other reasons for this requirement will not be needed in this heuristic discussion). A *generalized scalar field* on $\Omega$ is a function $\Gamma : \mathcal{D}(\Omega) \to \mathbb{R}$ that is linear,

$$\Gamma(a_1\psi_1 + a_2\psi_2) = a_1\Gamma(\psi_1) + a_2\Gamma(\psi_2),$$

for every $\psi_1, \psi_2 \in \mathcal{D}(\Omega)$ and every pair of scalars $a_1, a_2 \in \mathbb{R}$. There is also a requirement that $\Gamma$ be continuous in a certain sense, but without fear of spoiling the main ideas, we omit discussion of this. Note that while a scalar field $f : \Omega \to \mathbb{R}$ can be evaluated at a point $(x, y, z)$ in $\Omega$, that is, $f(x, y, z)$ makes sense, the same is not true for a generalized scalar field $\Gamma$; i.e., $\Gamma(x, y, z)$ does not makes sense. Of course, $\Gamma(\psi)$ makes sense for any $\psi \in \mathcal{D}(\Omega)$, and this is an important distinction between ordinary and generalized scalar fields.

As the name would imply, each ordinary scalar field $f$ on $\Omega$ ought also to be a generalized scalar field. This is so if, for the sake of distinction, we denote by $\tilde{f}$ the generalized scalar field defined by

$$\tilde{f}(\psi) = \int_\Omega f\psi \, dV.$$

(Certain integrability conditions need to be imposed on $f$ for this integral to exist.) You can easily check that this function is linear. Also inherent in the name is the implication that there should be some generalized scalar fields that are not ordinary scalar fields. The most famous example is the *Dirac delta function* $\delta_p$ at $p \in \Omega$. Here $p = (p_1, p_2, p_3)$ is any given point in $\Omega$, and $\delta_p$ is defined by

$$\delta_p(\psi) = \psi(p).$$

It can be shown that $\delta_p$ is not an ordinary scalar field.

The notion of generalized vector fields is similar: Let $\mathcal{D}(\Omega, \mathbb{R}^3)$ denote the set of all vector fields $w$ on $\Omega$ that vanish outside a compact subset $K \subset \Omega$; i.e., $w(x, y, z) = 0$ for all $(x, y, z) \notin K$. A *generalized vector field* on $\Omega$ is a function $\Gamma : \mathcal{D}(\Omega, \mathbb{R}^3) \to \mathbb{R}$ that is linear,

$$\Gamma(a_1 w_1 + a_2 w_2) = a_1 \Gamma(w_1) + a_2 \Gamma(w_2),$$

for every $w_1, w_2 \in \mathcal{D}(\Omega, \mathbb{R}^3)$ and every pair of scalars $a_1, a_2 \in \mathbb{R}$. Each ordinary vector field $v$ on $\Omega$ can be viewed as a generalized vector field: $\tilde{v}$ defined by

$$\tilde{v}(w) = \int_\Omega v \cdot w \, dV.$$

You can easily check that this defines a function that is linear.

There is one further definition we will need for this exercise: the definition of the divergence and the Laplacian operators acting on generalized vector and scalar fields. This definition is, Suppose $\Gamma$ is a generalized vector field. Then $\operatorname{div}(\Gamma)$ denotes the generalized scalar field defined by

$$\operatorname{div}(\Gamma)(\psi) = -\Gamma(\nabla\psi).$$

Suppose $\Gamma$ is a generalized scalar field. Then $\nabla^2\Gamma$ denotes the generalized scalar field defined by

$$(\nabla^2\Gamma)(\psi) = -\Gamma(\nabla^2\psi).$$

These definitions are motivated by the next part of this exercise:

(a) Use the identity $\operatorname{div}(\psi v) = \psi \operatorname{div}(v) + \nabla\psi \cdot v$ and Gauss's divergence theorem to show that

$$\int_\Omega \operatorname{div}(v)\psi \, dV = -\int_\Omega v \cdot \nabla\psi \, dV,$$

for $\psi \in \mathcal{D}(\Omega)$ and $v$ a vector field on $\Omega$. Discuss how this identity motivates the definition of the divergence for generalized vector fields.

(b) Suppose that $E$ is the vector field defined on $\mathbb{R}^3 - \{0\}$ by

$$E(x, y, z) = \frac{q(x, y, z)}{(x^2 + y^2 + z^2)^{3/2}},$$

where $q$ is a constant. Show that the corresponding generalized vector field $\widetilde{E}$ on $\mathbb{R}^3$ satisfies Coulomb's law

$$\operatorname{div}(\widetilde{E}) = 4\pi q \delta_0,$$

where $\delta_0$ is the Dirac delta function at the origin $0 = (0, 0, 0)$ and the equality is in the sense of generalized scalar fields on $\mathbb{R}^3$.
*Hint:* For each $h > 0$, let $D_h$ denote the region exterior to the sphere of radius $h$ centered at the origin; i.e.,

$$D_h = \{ (x, y, z) \mid x^2 + y^2 + z^2 \geq h^2 \}.$$

Then consider that for any $\psi \in \mathcal{D}(\mathbb{R}^3)$

$$\operatorname{div}(\widetilde{E})(\psi) = -\widetilde{E}(\nabla\psi)$$

$$= -\int_{\mathbb{R}^3} E \cdot \nabla\psi \, dV$$

$$= -\lim_{h \to 0} \int_{D_h} E \cdot \nabla\psi \, dV.$$

Now convert (via Gauss's divergence therem) the last integral into an integral over $D_h$ plus a surface integral over its boundary $\partial D_h$. Use the fact that $\operatorname{div}(E) = 0$ in $D_h$ to reduce to just a surface integral. Write out the surface integral using the parametrization of the sphere given in Appendix A. Finally, let $h \to \infty$ to get $4\pi q \psi(0)$.

(c) Suppose that $\phi$ is the scalar field defined on $\mathbb{R}^3 - \{(0, 0, 0)\}$ by

$$\phi(x, y, z) = \frac{q}{(x^2 + y^2 + z^2)^{1/2}}.$$

Use the techniques in the proof of part (b) to show that the corresponding generalized scalar field $\widetilde{\phi}$ on $\mathbb{R}^3$ satisfies Poisson's equation

$$\nabla^2 \widetilde{\phi} = -4\pi q \delta_0,$$

in the sense of generalized scalar fields on $\mathbb{R}^3$.

(d) Derive an alternative proof of the result in part (c) by doing the following. Devise a definition of the gradient $\nabla\Gamma$ of a generalized scalar field $\Gamma$. Use this to show that

$$\nabla\widetilde{\phi} = \widetilde{E},$$

where $E$ and $\phi$ are as in parts (b) and (c). Then use the result in (b) to get the result in (c).

4. The last exercise gave a rigorous meaning to the electrostatic field generated by a point charge of amount $q$ located at the origin $0 = (0, 0, 0)$. The charge density $\rho$ in Coulomb's law is naturally interpreted as the generalized scalar field $\rho = q\delta_0$, and the electric field is considered as a generalized vector field on $\mathbb{R}^3$. It is not hard to verify (same techniques) that for a charge of amount $q$ located at the point $p = (a, b, c)$, the charge distribution $\rho = q\delta_p$ generates an electric field

$$E(x, y, z) = \frac{q(x - a, y - b, z - c)}{((x - a)^2 + (y - b)^2 + (z - c)^2)^{3/2}},$$

with corresponding potential

$$\phi(x, y, z) = \frac{q}{((x - a)^2 + (y - b)^2 + (z - c)^2)^{1/2}},$$

*provided* that we interpret these fields in the generalized sense. We can go further with this and consider charges of amounts $q_1, \ldots, q_n$ located at distinct points $p_1 = (a_1, b_1, c_1), \ldots, p_n = (a_n, b_n, c_n)$. Then the charge density $\rho$ is the generalized scalar field

$$\rho = \sum_{j=1}^{n} q_j \delta_{p_j}.$$

This generates an electric field

$$E(x, y, z) = \sum_{j=1}^{n} \frac{q_j(x - a_j, y - b_j, z - c_j)}{((x - a_j)^2 + (y - b_j)^2 + (z - c_j)^2)^{3/2}},$$

with corresponding potential

$$\phi(x, y, z) = \sum_{j=1}^{n} \frac{q_j}{((x - a_j)^2 + (y - b_j)^2 + (z - c_j)^2)^{1/2}}.$$

This exercise deals with plotting the electric field lines for such electric fields. To keep it simple, assume that the charges are located in the $x$-$y$ plane and just consider these situations:

(a) Two charges located at at $p_1 = (0, 0, 0)$ and $p_2 = (1, 0, 0)$ with (i) $q_1 = 1, q_2 = 2$, (ii) $q_1 = 1, q_2 = -2$, and (iii) $q_1 = -1, q_2 = -2$.

(b) Three charges located at at $p_1 = (0, 0, 0)$, $p_2 = (1, 0, 0)$, and $p_3 = (0, 1, 0)$ with (i) $q_1 = 1, q_2 = 2, q_3 = 1$, (ii) $q_1 = 1, q_2 = -2, q_3 = 1$, and (iii) $q_1 = -1, q_2 = -2, q_3 = -1$.

In each situation, plot in the $x$-$y$ plane the electric field lines for $E$.

5. The electrostatic field generated by a point charge at the origin has a magnetostatic analogue. It is the magnetic field generated by a constant current flowing along an infinitely long straight wire. As in Exercise 3 above, this basic example can be easily modeled using generalized vector fields. Suppose the wire lies along the $z$-axis, which we designate by $L$, and let $\varepsilon_3 = (1, 0, 0)$ be the unit vector in the positive $z$ direction. The *Dirac delta line distribution* $\varepsilon_3\delta_L$ is the generalized vector field on $\mathbb{R}^3$ defined by

$$(\varepsilon_3\delta_L)(w) \equiv \int_{-\infty}^{\infty} w(0, 0, z) \cdot \varepsilon_3 \, dz,$$

for $w \in \mathcal{D}(\mathbb{R}^3, \mathbb{R}^3)$. Use arguments like those in Exercise 3 to do the following:

(a) The *definition* of the curl operator acting on a generalized vector field $\Gamma$ on $\Omega$ is

$$\text{curl}(\Gamma)(w) \equiv -\Gamma(\text{curl}(w)),$$

for $w \in \mathcal{D}(\Omega, \mathbb{R}^3)$. Justify that this is an extension of the curl operator acting on ordinary (differentiable) vector fields $v$ on $\Omega$ by showing that

$$\int_{\Omega} \text{curl}(v) \cdot w \, dV = -\int_{\Omega} v \cdot \text{curl}(w) \, dV,$$

for all $w \in \mathcal{D}(\Omega, \mathbb{R}^3)$.

(b) Suppose that $B$ is the vector field defined on $\mathbb{R}^3 - L$ by

$$B(x, y, z) = \left(\frac{-y}{x^2 + y^2}, \frac{x}{x^2 + y^2}, 0\right).$$

Show that the corresponding generalized vector field $\widetilde{B}$ on $\mathbb{R}^3$ satisfies Ampére's law

$$\text{curl}(\widetilde{B}) = 2\pi \varepsilon_3 \delta_L$$

as well as

$$\operatorname{div}(\widetilde{B}) = 0.$$

Thus $\widetilde{B}$ can be viewed as the solution of the magnetostatic equations with current density $J = \frac{c}{2}\varepsilon_3\delta_L$.

6. The *work* done against an electric field $E$ by moving a charge of amount $q > 0$ along a path $C$ in $\Omega$ is by definition

$$W = -\int_C E \cdot \tau \, dL.$$

Assuming that $E$ has a potential $E = -\nabla\phi$, show that $W$ is positive if $C$ starts at a point in $\Omega$ where the potential is lower and ends at a point where it is higher.

## 9.3.   Existence of Potentials

The question of existence of potentials for vector fields arises in many areas, so in this section we will not necessarily be talking about electrostatic or magnetostatic fields. Thus we will consider a given vector field on a region $\Omega$ and ask for necessary and sufficient conditions for it to be (1) the gradient of a scalar field, or (2) the curl of a vector field on $\Omega$.

We consider (1) here, leaving question (2) for the exercises. We phrase the question in terms of being able to solve the system of PDEs

$$\nabla\phi = E, \quad \text{on } \Omega,$$

where $E$ is a given vector field on $\Omega$ and $\phi$ is the unknown function. Written out in terms of components, this system is

$$\phi_x = E^1,$$

$$\phi_y = E^2,$$

$$\phi_z = E^3.$$

As often happens with systems of PDEs, this particular system need not have any solutions at all unless the given $E$ satisfies certain conditions (called the integrability conditions). To see this, note that *if* $\phi$ is a solution, then necessarily $\operatorname{curl}(E) = \operatorname{curl}(\nabla\phi) = 0$. Thus a necessary condition for solutions to exist is the following:

**Integrability Conditions (necessary conditions):**

$$\mathrm{curl}(E) = 0,$$

or in component form,

$$E^3_y - E^2_z = 0, \tag{9.22}$$

$$E^1_z - E^3_x = 0, \tag{9.23}$$

$$E^2_x - E^1_y = 0. \tag{9.24}$$

The following example illustrates why these are called the integrability conditions and also indicates a technique for solving $\nabla\phi = E$, which works for special cases of $\Omega$.

**Example 9.2**   Suppose $\Omega = \mathbb{R}^3$. A naive way to solve the potential equations is to start with the first PDE, $\phi_x = E^1$, and integrate partially with respect to $x$:

$$\phi = \int E^1 dx.$$

Here the integration of $E^1$ is with respect to $x$, with $y$ and $z$ held fixed. More precisely, we let

$$\phi(x, y, z) = \int_0^x E^1(s, y, z)ds + g(y, z), \tag{9.25}$$

where $g$ is an arbitrary function of $y$ and $z$ (which is like adding an arbitrary constant of integration to an indefinite integral). You can check that $\phi_x(x, y, z) = E^1(x, y, z)$ (by the fundamental theorem of calculus) for any choice of $g$. We now must determine $g$ such that $\phi$ satisfies the other two PDEs in the potential equations. Thus calculating $\phi_y$ and using the integrability condition $E^1_y = E^2_x$ (and the fundamental theorem of calculus), we get

$$\phi_y(x, y, z) = \int_0^x E^1_y(s, y, z)ds + g_y(y, z)$$

$$= \int_0^x E^2_x(s, y, z)ds + g_y(y, z)$$

$$= E^2(x, y, z) - E^2(0, y, z) + g_y(z, y).$$

In order to have $\phi_y = E^2$, the above calculation shows that we must choose $g$ such that

$$g_y(y, z) = E^2(0, y, z).$$

To find $g$, we integrate partially with respect to $y$ and add on an arbitrary function $h$ of integration:

$$g(y, z) = \int_0^y E^2(0, s, z)ds + h(z). \tag{9.26}$$

The function $h$ is arbitrary so far, but we now want to choose it such that $\phi$ satisfies the third PDE, $\phi_z = E^3$, of the potential equations. If we calculate $\phi_z$ from its definition in equation (9.25), use the expression (9.26) for $g$, and use the integrability conditions $E_z^1 = E_x^3$, $E_z^2 = E_y^3$, we get

$$\phi_z(x, y, z) = \int_0^x E_z^1(s, y, z)ds + g_z(y, z)$$

$$= \int_0^x E_z^1(s, y, z)ds + \int_0^y E_z^2(0, s, z)ds + h'(z)$$

$$= \int_0^x E_x^3(s, y, z)ds + \int_0^y E_y^3(0, s, z)ds + h'(z)$$

$$= E^3(x, y, z) - E^3(0, y, z) + E^3(0, y, z) - E^3(0, 0, z) + h'(z).$$

If we want to have this reduce to just $E^3(x, y, z)$, we need to choose $h$ such that $h'(z) = E^3(0, 0, z)$. Integrating this gives the following expression for $h$:

$$h(z) = \int_0^z E^3(0, 0, s)ds + c, \tag{9.27}$$

where $c$ is a constant of integration. There are no further requirements by which we can determine $c$, and indeed, the reasoning shows that for any value of $c$, the $\phi$ we have constructed will satisfy $\nabla\phi = E$. For convenience, we take $c = 0$. If we put formulas (9.26)–(9.27) back into formula (9.25), we get the following explicit integral formula for a potential for the vector field $E$:

**A Potential Formula:**

$$\phi(x, y, z) = \int_0^x E^1(s, y, z)ds + \int_0^y E^2(0, s, z)ds + \int_0^z E^3(0, 0, s)ds. \tag{9.28}$$

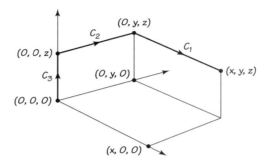

Figure 9.4.    A polygonal curve $C = C_1 \cup C_2 \cup C_3$, joining the origin $(0, 0, 0)$ to the point $(x, y, z)$.

It is important to note that this formula is *exactly* the expression for the line integral of $E$ along the curve $C = C_1 \cup C_2 \cup C_3$ shown in Figure 9.4 (exercise).

While the integrability conditions are necessary for there to exist solutions of the potential equations, and the above example illustrates how to use them to construct a potential in the case that $\Omega = \mathbb{R}^3$, these conditions are by no means sufficient, as the following standard example shows:

**Example 9.3**    Let $\Omega = \mathbb{R}^3 - \{$the $z$-axis$\}$, and consider the vector field $E$ on $\Omega$ given by

$$E(x, y, z) = (\frac{-y}{x^2 + y^2}, \frac{x}{x^2 + y^2}, 0).$$

It is easy to verify that $\text{curl}(E) = 0$ in $\Omega$, and so $E$ satisfies the integrability conditions. As you can see, $E$ does not depend on $z$ and is not defined at points on the $z$-axis. Thus we cannot use the potential formula (9.28) to construct a potential for $E$, since that formula requires an integration along the $z$-axis. Attempts to modify this formula or construct a potential by other means will fail because a potential for this vector field does not exist! To be more specific, we claim that there does not exist a scalar field $\phi$ on $\Omega$ such that: $\nabla \phi = E$, on $\Omega$. To verify this assertion, assume to the contrary that such a $\phi$ does exist. Then let $C$ be the curve $\gamma : [0, 2\pi] \to \mathbb{R}^3$ given by

$$\gamma(t) = (\cos t, \sin t, 0).$$

This is a closed curve, $\gamma(0) = (1, 0, 0) = \gamma(2\pi)$, which lies entirely in $\Omega$. Then note that

$$0 = \phi(\gamma(2\pi)) - \phi(\gamma(0))$$

$$= \int_0^{2\pi} \frac{d}{dt}\left[\phi(\gamma(t))\right] dt$$

$$= \int_0^{2\pi} \nabla\phi(\gamma(t)) \cdot \gamma'(t)\, dt$$

$$= \int_0^{2\pi} E(\gamma(t)) \cdot \gamma'(t)\, dt$$

$$= \int_0^{2\pi} 1\, dt = 2\pi.$$

This is a contradiction! Thus no such potential exists. While this argument shows that $E$ has no potential on *all* of $\Omega$, it is possible that $E = \nabla\phi$ for some $\phi$ on a subregion $D \subset \Omega$. For example, it is easy to check that $\phi(x, y, z) = \tan^{-1}(y/x)$ is a potential for $E$ restricted to $D = \{(x, y, z) \in \Omega \,|\, x > 0\}$. Thus the question about the existence of potentials is intimately related to the region on which $E$ is defined.

The idea of using line integrals in the above example turns out to be central to establishing necessary and sufficient conditions for the existence of potentials for $E$. *Note*: See Appendix A for the definition of the line integral $\int_C E \cdot \tau\, dL$ and other related concepts. Here $\tau$ represents a unit tangential vector field to $C$.

**Lemma 9.1**    *Suppose $E$ is a vector field on $\Omega$. Then the following assertions are equivalent:*

(1)  *For every closed curve $C$ in $\Omega$,*

$$\int_C E \cdot \tau\, dL = 0.$$

(2)  *For any two curves $C_1, C_2$ in $\Omega$ that have the same boundary,*

$$\int_{C_1} E \cdot \tau\, dL = \int_{C_2} E \cdot \tau\, dL.$$

**Proof**    To prove that (1) $\Longrightarrow$ (2), suppose $C_1 = \{\alpha_1, \ldots, \alpha_m\}$ and $C_2 = \{\alpha_{m+1}, \ldots, \alpha_k\}$ are two curves in $\Omega$ with the same boundary. Supposing the domain of $\alpha_j$ is $[a_j, b_j]$, define the inverse curve

$$\tilde{\alpha}_j(s) = \alpha_j\left(\frac{(s - a_j)a_j + (b_j - s)b_j}{b_j - a_j}\right),$$

for $s \in [a_j, b_j]$. This is the same curve, except that it is traversed in the opposite direction (with the same speed). For convenience, let $D_j = \{\alpha_j\}$ and $\tilde{D}_j = \{\tilde{\alpha}_j\}$. It is easy to see (exercise) that the inverse to a curve has the properties

$$\int_{\tilde{D}_j} E \cdot \tau \, dL = -\int_{D_j} E \cdot \tau \, dL,$$

$$\int_{\partial \tilde{D}_j} f = -\int_{\partial D_j} f,$$

for any vector field $E$ and scalar field $f$. Now let $C$ be the curve

$$C = \{\alpha_1, \ldots, \alpha_m, \tilde{\alpha}_{m+1}, \ldots, \tilde{\alpha}_k\}.$$

Then we get, for any scalar field $f$,

$$\int_{\partial C} f = \sum_{j=1}^{m} \int_{\partial D_j} f + \sum_{j=m+1}^{k} \int_{\partial \tilde{D}_j} f$$

$$= \sum_{j=1}^{m} \int_{\partial D_j} f - \sum_{j=m+1}^{k} \int_{\partial D_j} f$$

$$= \int_{\partial C_1} f - \int_{\partial C_2} f$$

$$= 0.$$

The last equation follows because $C_1$ and $C_2$ have the same boundary. Thus $C$ is a closed curve. Hence, by assumption, the integral of $E$ around $C$ is zero:

$$0 = \int_C E \cdot \tau \, dL$$

$$= \int_{C_1} E \cdot \tau \, dL - \int_{C_2} E \cdot \tau \, dL.$$

Thus (2) holds.

(2) $\Longrightarrow$ (1): Suppose $C = \{\alpha_1, \ldots, \alpha_m\}$ is a closed curve in $\Omega$. Let $[a_1, b_1]$ be the domain of $\alpha_1$ and define $\beta : \mathbb{R} \to \Omega$ by

$$\beta(s) = \alpha_1(a_1),$$

for every $s$. Then $\beta$ is a constant curve (and so is also a closed curve). Letting $C_1 = C$ and $C_2 = \{\beta\}$, we have, from the definition of closed curves,

$$\int_{\partial C_1} f = 0 = \int_{\partial C_2} f,$$

for every scalar field $f$. Thus $C_1 = C$ and $C_2$ have the same boundary, since the integral of any scalar field around their boundaries is the same. So by hypothesis,

$$\int_C E \cdot \tau \, dL = \int_{C_2} E \cdot \tau \, dL = 0.$$

The latter equation comes from the fact that $C_2$ is a constant curve. This proves the lemma.

The following theorem characterizes those vector fields on $\Omega$ that have potentials. Alternatively, the theorem can be interpreted as giving necessary and sufficient conditions for existence of solutions to the PDE $\nabla \phi = E$.

**Theorem 9.1**   *Suppose $E$ is a vector field on a region $\Omega$. Then the following are equivalent:*

(1) *There is a scalar field $\phi$ such that $\nabla \phi = E$ at each point of $\Omega$.*

(2) *For every closed curve $C$ in $\Omega$,*

$$\int_C E \cdot \tau \, dL = 0.$$

**Proof**   (1) $\Longrightarrow$ (2): Suppose $E = \nabla \phi$ on $\Omega$ for some scalar field $\phi$. Then for any closed curve $C$, we have by the gradient theorem (Appendix A)

$$\int_C E \cdot \tau \, dL = \int_C \nabla \phi \cdot \tau \, dL = \int_{\partial C} \phi = 0.$$

(2) $\Longrightarrow$ (1): Suppose the line integral of $E$ around every closed curve in $\Omega$ is zero, or equivalently, by the last lemma, the line integrals of $E$ along two curves with the same boundary are the same. Using this, we construct a potential $\phi$

for $E$ as follows. Fix some point $(x_0, y_0, z_0)$ in $\Omega$. Since $\Omega$ is a region (an open connected set) in $\mathbb{R}^3$, it is path connected; i.e., for each point $(x, y, z)$ in $\Omega$ there is a continuous curve $C = \{\alpha_1, \ldots, \alpha_m\}$ in $\Omega$, with $\alpha_j : [a_j, b_j] \to \Omega$, such that $\alpha_1(a_1) = (x_0, y_0, z_0)$ and $\alpha_m(b_m) = (x, y, z)$. Continuity of the curve means that $\alpha_j(b_j) = \alpha_{j+1}(a_{j+1})$, for $j = 1, \ldots, m - 1$. As $(x, y, z)$ varies over $\Omega$, we choose one such curve $C = C_{(x,y,z)}$. See Figure 9.5. Using this, we define $\phi$ on $\Omega$ by

$$\phi(x, y, z) = \int_{C_{(x,y,z)}} E \cdot \tau \, dL.$$

Otherwise stated, the value of $\phi$ at $(x, y, z)$ is defined as the value of the line integral $E$ along the chosen curve $C_{(x,y,z)}$ joining $(x, y, z)$ to the fixed point $(x_0, y_0, z_0)$.

We now need to show that $\nabla\phi = E$. For this, we first show that $\phi_x = E^1$. So suppose $(x, y, z)$ is some point in $\Omega$ and choose $r > 0$ such that the open ball of radius $r$ and center $(x, y, z)$ is completely contained in $\Omega$. For any $0 < h < r$, let

$$C_{(x,y,z),h} = C_{(x,y,z)} \cup \{\gamma_h\},$$

where $\gamma_h : [0, h] \to \Omega$ is the curve (line segment) $\gamma_h(s) \equiv (x + s, y, z)$. The curve $C_{(x,y,z),h}$ runs along $C_{(x,y,z)}$ from $(x_0, y_0, z_0)$ to $(x, y, z)$ and then along a straight line segment from $(x, y, z)$ to $(x + h, y, z)$. See Figure 9.5. This curve and the curve $C_{(x+h,y,z)}$ clearly have the same boundary,

$$\int_{\partial C_{(x,y,z),h}} f = f(x + h, y, z) - f(x_0, y_0, z_0) = \int_{\partial C_{(x+h,y,z)}} f,$$

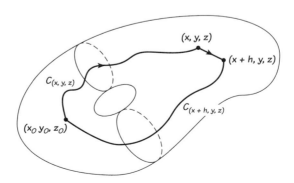

Figure 9.5.    Curves $C_{(x,y,z)}$ and $C_{(x+h,y,z)}$ joining the origin to the points $(x, y, z)$ and $(x+h, y, z)$.

for every scalar field $f$. By the previous lemma, the line integral of $E$ along each curve is the same. Thus

$$\int_{C_{(x+h,y,z)}} E \cdot \tau \, dL = \int_{C_{(x,y,z)}} E \cdot \tau \, dL + \int_0^h E^1(x+s, y, z) \, ds.$$

Consequently, we get

$$\frac{1}{h}[\phi(x+h, y, z) - \phi(x, y, z)] = \frac{1}{h}\int_0^h E^1(x+s, y, z) \, ds.$$

Since this holds for any $h < r$, letting $h \to 0$ and using the fundamental theorem of calculus gives

$$\phi_x(x, y, z) = E^1(x, y, z).$$

The proofs that $\phi_y = E^2$ and $\phi_z = E^3$ at each point of $\Omega$ are similar and so will be omitted.

The theorem not only completely characterizes when a vector field $E$ has a scalar potential $E = \nabla\phi$, but it also exhibits a way of constructing $\phi$ from line integrals. The corresponding results for the existence of vector potentials $B = \operatorname{curl}(A)$ for a given vector field $B$ are discussed in the exercises.

The following diagram may help clarify the relationships and similarities involved between these two types of potentials:

$$\left\{ \begin{array}{c} \text{scalar} \\ \text{fields} \end{array} \right\} \xrightarrow{\nabla} \left\{ \begin{array}{c} \text{vector} \\ \text{fields} \end{array} \right\} \xrightarrow{\text{curl}} \left\{ \begin{array}{c} \text{vector} \\ \text{fields} \end{array} \right\} \xrightarrow{\text{div}} \left\{ \begin{array}{c} \text{scalar} \\ \text{fields} \end{array} \right\}$$

Note that in the diagram the composition of any two of the differential operators in the sequence (in order) gives the identically zero operator:

$$\operatorname{curl} \circ \nabla = 0, \quad \operatorname{div} \circ \operatorname{curl} = 0.$$

In linear algebra terminology, this says that (1) the image of the gradient operator is contained in the kernel (null space) of the curl operator, and (2) the image of the curl operator is contained in the kernel of the divergence operator. The question of existence of potentials amounts to whether a vector field that is in the kernel is also in the image. Thus the previous theorem can be interpreted as a characterization of the image of the gradient operator $\nabla$. The interested student can learn how these basic ideas are generalized and clarified by consulting texts on homology and cohomology theory.

## 9.3.1.  Simply Connected Regions

From a practical point of view, it is rather difficult to check if a given vector field $E$ has a potential by applying the criterion from Theorem 9.1. Namely, integrate $E$ along each closed curve in $\Omega$ and see if you get 0. Checking the integrability condition $\text{curl}(E) = 0$ is easy, but this condition, as we've seen, is necessary but not sufficient to guarantee the existence of a potential. However, there is a broad category of regions $\Omega$ for which the integrability condition is also sufficient for existence of potentials. These are called *simply connected regions*.

**Definition 9.1**    Suppose $\Omega$ is a region in $\mathbb{R}^3$, i.e., an open, connected subset of $\mathbb{R}^3$. Then $\Omega$ is called *simply connected* if it has the property that for any two smooth curves $C_1 = \{\alpha\}, C_2 = \{\beta\}$ in $\Omega$ with common domain $\alpha, \beta : J = [a, b] \to \Omega$ and the same boundary $\alpha(a) = \beta(a), \alpha(b) = \beta(b)$, there exists a smooth map $h : J \times [0, 1] \to \Omega$ such that

$$h(s, 0) = \alpha(s) \qquad \text{for all } s \in J, \tag{9.29}$$

$$h(s, 1) = \beta(s) \qquad \text{for all } s \in J, \tag{9.30}$$

$$h(a, t) = p_0 \qquad \text{for all } t \in [0, 1], \tag{9.31}$$

$$h(b, t) = p_1 \qquad \text{for all } t \in [0, 1]. \tag{9.32}$$

Here $p_0 = \alpha(a) = \beta(a)$ and $p_1 = \alpha(b) = \beta(b)$ are the common endpoints of the two curves. A region $\Omega$ that is not simply connected is called *multiply connected*.

The map $h$ in the above definition is called a *homotopy* between the two curves $\alpha, \beta$. It can be thought of kinematically as a smooth deformation of one curve onto the other as $t$ varies over $[0, 1]$. See Figure 9.6.

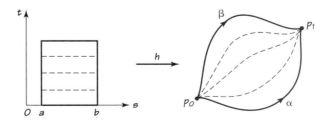

Figure 9.6.    A homotopy $h$ of the curve $C_1 = \{\alpha\}$ onto the curve $C_2 = \{\beta\}$.

An alternative way to think of a homotopy $h$ is as a surface $S = \{h\}$ in $\mathbb{R}^3$ that lies completely inside $\Omega$ and that has boundary

$$\partial S = \{\alpha, p_1, \widetilde{\beta}, p_0\}. \tag{9.33}$$

Here, the points $p_0$, $p_1$ denote constant curves, and $\widetilde{\beta}$ denotes the inverse of $\beta$ (see the exercises). Thus each homotopy of the curve $C_1$ onto the curve $C_2$ is a smooth surface $S$ in $\Omega$ with $C_1, C_2$ comprising its boundary, and (by Stokes's theorem) for any vector field $E$ on $\Omega$,

$$\int_S \mathrm{curl}(E) \cdot n \, dA = \int_{C_1} E \cdot \tau \, dL - \int_{C_2} E \cdot \tau \, dL. \tag{9.34}$$

From this, Lemma 9.1, and Theorem 9.1, we easily get the following corollary:

**Corollary 9.1**    *Suppose $\Omega$ is a simply connected region and $E$ is a vector field on $\Omega$. Then $E$ has a potential if and only if it satisfies the integrability conditions.*

**Proof**    Exercise.

Simply connected regions $\Omega$ are *intuitively* easy to identify using the remarks preceding the above corollary: Any two smooth curves in $\Omega$ with the same boundary must comprise the boundary of some surface in $\Omega$. Equivalently, each closed curve in $\Omega$ must be the boundary of some surface in $\Omega$. Thus the region $\Omega = \mathbb{R}^3 - \{z\text{-axis}\}$ in the example prior to Lemma 9.1 is *not* simply connected. This is so since the unit circle $C$ in the $x$-$y$ plane is a closed curve in $\Omega$, and any surface with $C$ as its boundary must intersect the $z$-axis (and thus not lie entirely in $\Omega$).

**Exercises 9.3**

1. Show that the expression on the right side of the potential formula (9.28) is precisely the line integral along the curve $C$ shown in Figure 9.4.

2. (*Inverse curves*) Suppose $C = \{\alpha\}$ is a smooth curve, with $\alpha : [a, b] \to \mathbb{R}^3$. Define the *inverse curve* $\widetilde{C} = \{\widetilde{\alpha}\}$ by

$$\widetilde{\alpha}(s) = \alpha\left(\frac{(s-a)a + (b-s)b}{b-a}\right),$$

for $s \in [a, b]$. Show that for any vector field $E$ and and scalar field $f$,

$$\int_{\tilde{C}} E \cdot \tau \, dL = -\int_{C} E \cdot \tau \, dL,$$

$$\int_{\partial \tilde{C}} f = -\int_{\partial C} f.$$

3. Prove that equations (9.33)–(9.34) hold. Then prove Corollary 6. For the latter, you may use the fact that any two points in a region $\Omega$ can be connected by a *smooth* curve.

4. (*Vector potentials*) The problem of finding a *vector* potential $A$ for a given vector field $B$ on a region $\Omega \subseteq \mathbb{R}^3$, amounts to solving the system of PDEs

$$\text{curl}(A) = B,$$

or written out in terms of components,

$$A_y^3 - A_z^2 = B^1, \tag{9.35}$$

$$A_z^1 - A_x^3 = B^2, \tag{9.36}$$

$$A_x^2 - A_y^1 = B^3. \tag{9.37}$$

The integrability condition (necessary condition) for this system to have a solution is that the given vector field $B$ have divergence zero: $\text{div}(B) = 0$; i.e.,

**Integrability Condition:**    $B_x^1 + B_y^2 + B_z^3 = 0$.

For $\Omega = \mathbb{R}^3$, construct a formula giving a potential $A$ in terms of integrals of the components of $B$. This will be similar to what was done in Example 9.2. There are several ways to proceed on this, so to be specific, do the following.

(a) Note that if $A_0$ is any vector potential for $B$, then so is $A = A_0 - \nabla \Gamma$ for any scalar field $\Gamma$. Use this to show that if $B$ has a potential, then it has one with $A^3 = 0$.

(b) Based on part (a), it seems reasonable to look for a potential $A$ for which $A^3 = 0$. Then equations (9.35)–(9.37) reduce to

$$-A_z^2 = B^1,$$

$$A_z^1 = B^2,$$

$$A_x^2 - A_y^1 = B^3.$$

Solve the first two equations by partial integration, adding on arbitrary functions of integration. Then use the third equation to determine these functions of integration.

5. (*Alternative formulas for potentials*) Suppose $\Omega = \mathbb{R}^3$, and for convenience of notation, we change to the subscripting notation for the coordinates of points in $\mathbb{R}^3$:

$$\vec{x} = (x_1, x_2, x_3).$$

This exercise presents some alternative formulas for constructing potentials $\phi$, $A$ for vector fields $E$, $B$. These formulas are

$$\phi(\vec{x}) = \int_0^1 E(s\vec{x}) \cdot \vec{x} \, ds, \qquad (9.38)$$

$$A(\vec{x}) = \int_0^1 s[B(s\vec{x}) \times \vec{x}] ds. \qquad (9.39)$$

Assuming that $\text{curl}(E) = 0$ and $\text{div}(A) = 0$, show that the $\phi$ and $A$ given by these formulas are potentials: $\nabla\phi = E$ and $\text{curl}(A) = B$. Interpret formula (9.38) as a line integral and relate it to formula (9.28) for scalar potentials derived in the text. Formulas (9.38)–(9.39) are sometimes easier to use than the other formulas (formula (9.28) from the text and the formula you found in the last exercise).

6. (*Poincaré (chain) homotopies*) Consider the following reformulation of the results in the last exercise. The fields $\phi$, $A$ given by formulas (9.38)–(9.39) depend on $E$, $B$, and to exhibit this explicitly in the notation, we rewrite the formulas as

$$h_1(E)(\vec{x}) = \int_0^1 E(s\vec{x}) \cdot \vec{x} \, ds, \qquad (9.40)$$

$$h_2(B)(\vec{x}) = \int_0^1 s[B(s\vec{x}) \times \vec{x}] ds. \qquad (9.41)$$

Also, for each scalar field $f$ on $\mathbb{R}^3$ define a new vector field $h_3(f)$ by

$$h_3(f)(\vec{x}) = \int_0^1 s^2 f(s\vec{x})\vec{x} \, ds. \qquad (9.42)$$

We view $h_1, h_2, h_3$ as linear maps with domains as shown in the following diagram:

$$\left\{ \begin{array}{c} \text{scalar} \\ \text{fields} \end{array} \right\} \mathrel{\mathop{\rightleftarrows}^{h_1}_{\nabla}} \left\{ \begin{array}{c} \text{vector} \\ \text{fields} \end{array} \right\} \mathrel{\mathop{\rightleftarrows}^{h_2}_{\text{curl}}} \left\{ \begin{array}{c} \text{vector} \\ \text{fields} \end{array} \right\} \mathrel{\mathop{\rightleftarrows}^{h_3}_{\text{div}}} \left\{ \begin{array}{c} \text{scalar} \\ \text{fields} \end{array} \right\}$$

The maps $h_1, h_2, h_3$ are sometimes called *Poincaré* (or chain) *homotopies*. With this notation set, prove the following identites:

$$\nabla(h_1(E)) + h_2(\text{curl}(E)) = E, \qquad (9.43)$$

$$\text{curl}(h_2(B)) + h_3(\text{div}(B)) = B, \qquad (9.44)$$

for any two vector fields $E$, $B$ on $\mathbb{R}^3$. Discuss how this generalizes the result in the previous exercise.

7. For each of the following vector fields $v$ on $\mathbb{R}^3$ determine whether $v$ has a scalar or a vector potential by checking whether $\mathrm{curl}(v) = 0$ or $\mathrm{div}(v) = 0$. When possible, determine an actual potential for $v$ by using one of the integral formulas that have been presented in the text and exercises.

(a) $v = (2xy^3z + z, 3x^2y^2z, x^2y^3 + x)$.

(b) $v = (yz\cos(xz), \sin(xz), xy\cos(xz))$.

(c) $v = (3x^2ze^{-y}, 2y - zx^3e^{-y}, x^3e^{-y})$.

(d) $v = (xy^2 - xz^2, yz^2 - yx^2, zx^2 - zy^2)$.

(e) $v = (-xy\cos z, xy\cos z, (y - x)\sin z)$.

(f) $v = (xy(3z^2 + z^3)e^{-z}, -3y^2z^2e^{-z}, yz^3e^{-z})$.

(g) $v = (x(z^2 - y^2)\cos(yz), x^2y\cos(yz) - z\sin(yz), y\sin(yz) - x^2z\cos(yz))$.

## 9.4.   Potentials and Gauge Transformations

The use of potentials in the electrostatic and magnetostatic equations can be extended in a similar fashion to the time-varying case for solutions of the general Maxwell equations. In essence, the idea is to transform the problem of solving the first-order PDEs comprising Maxwell's equations into the problem of solving second-order wave equations for the potentials. Part of this leads naturally to the idea of gauge transformations.

In order to guarantee the existence of potentials, we assume that the spatial domain $\Omega$ for Maxwell's equations is simply connected.

We begin with a motivational discussion before stating the precise results. Thus suppose $(E, B)$ is a solution of Maxwell's equations

$$\mathrm{div}(E) = 4\pi\rho, \tag{9.45}$$

$$\mathrm{curl}(E) + \frac{1}{c}\frac{\partial B}{\partial t} = 0, \tag{9.46}$$

$$\mathrm{curl}(B) - \frac{1}{c}\frac{\partial E}{\partial t} = \frac{4\pi}{c}J, \tag{9.47}$$

$$\mathrm{div}(B) = 0. \tag{9.48}$$

The last equation, together with our assumption on the domain, imply that $B$ is the curl of some vector field, say

$$B = \text{curl}(A).\tag{9.49}$$

Substituting this into Faraday's law (9.46) gives

$$0 = \text{curl}(E) + \frac{1}{c}\frac{\partial}{\partial t}\text{curl}(A)$$

$$= \text{curl}\left(E + \frac{1}{c}\frac{\partial A}{\partial t}\right).$$

This implies that the quantity in the parentheses must be the gradient of some scalar field, say

$$E + \frac{1}{c}\frac{\partial A}{\partial t} = -\nabla\phi.$$

The minus sign here is traditional for physical reasons, although mathematically is not necessary. Rewriting this gives the following expression for $E$:

$$E = -\nabla\phi - \frac{1}{c}\frac{\partial A}{\partial t}.\tag{9.50}$$

Thus we see that an electromagnetic field $(E, B)$ arises from a scalar and vector potential $(\phi, A)$ in the manner specified by equations (9.49)–(9.50). If we then substitute the expression (9.50) for $E$ into Coulomb's law (9.45), we get

$$-\nabla^2\phi - \frac{1}{c}\frac{\partial}{\partial t}\text{div}(A) = 4\pi\rho.\tag{9.51}$$

This is one equation that the potentials $\phi, A$ must satisfy. Another equation that they must satisfy comes from substituting $B = \text{curl}(A)$ and $E = -\nabla\phi - c^{-1}\partial A/\partial t$ into the Ampére–Maxwell law (9.47). This gives (after rearranging slightly)

$$\frac{1}{c^2}\frac{\partial^2 A}{\partial t^2} - \nabla^2 A + \nabla\left(\frac{1}{c}\frac{\partial\phi}{\partial t} + \text{div}(A)\right) = \frac{4\pi}{c}J\tag{9.52}.$$

The reasoning so far is that any solution $(E, B)$ of Maxwell's equations comes from a pair of potentials $(\phi, A)$, and that $(\phi, A)$ must satisfy the PDEs (9.51)–(9.52). This is all fairly direct and obvious. Now we do something that is not quite so obvious. We assume that $(\phi, A)$ also satisfies

**The Lorentz Condition:**

$$\frac{1}{c}\frac{\partial \phi}{\partial t} + \mathrm{div}(A) = 0. \tag{9.53}$$

As we shall see, by using what are known as gauge transformations, we can always select potentials for the electromagnetic field that satisfy this condition. The nice part about having the potentials satisfy the Lorentz condition is that the PDEs (9.51)–(9.52) decouple into a pair of wave equations:

$$\frac{\partial^2 \phi}{\partial t^2} - c^2 \nabla^2 \phi = 4\pi c^2 \rho,$$

$$\frac{\partial^2 A}{\partial t^2} - c^2 \nabla^2 A = 4\pi c J.$$

**Theorem 9.2** *(Lorentz Potential Equations) On a simply connected spatial region, the vector fields E, B are a solution of Maxwell's equations if and only if*

$$E = -\nabla\phi - \frac{1}{c}\frac{\partial A}{\partial t}, \tag{9.54}$$

$$B = \mathrm{curl}(A), \tag{9.55}$$

*for some scalar field $\phi$ and vector field A that satisfy the Lorentz potential equations*

$$\frac{1}{c}\frac{\partial \phi}{\partial t} + \mathrm{div}(A) = 0, \tag{9.56}$$

$$\frac{\partial^2 \phi}{\partial t^2} - c^2 \nabla^2 \phi = 4\pi c^2 \rho \tag{9.57},$$

$$\frac{\partial^2 A}{\partial t^2} - c^2 \nabla^2 A = 4\pi c J. \tag{9.58}$$

**Proof**   Suppose first that $E, B$ is a solution of Maxwell's equations. We repeat some of the above arguments because we have to change the notation slightly. You will see why shortly. Thus, since $\mathrm{div}(B) = 0$, there exists a vector field $A_0$ such that $\mathrm{curl}(A_0) = B$. Substituting this expression for $B$ into Faraday's law gives $\mathrm{curl}(\partial A_0/\partial t + E) = 0$. Thus there exists a scalar field $\phi_0$ such that $-\nabla\phi_0 = \partial A_0/\partial t + E$. Rearranging this gives $E = -\nabla\phi_0 - \partial A_0/\partial t$. Thus $E$ and $B$ are given by potentials $\phi_0$ and $A_0$ in the form of equations (9.54)–(9.55).

This was our argument prior to the theorem. The potentials $\phi_0$, $A_0$ we selected (perhaps constructed as indicated in the last section) do *not* necessarily satisfy the Lorentz condition, but that's OK, since we can always transform them by a gauge transformation into new potentials that *do*. Here is how this works. Suppose $\Gamma$ is any scalar function on $\Omega$, and let $\phi$ and $A$ be defined by

$$\phi = \phi_0 - \frac{1}{c}\frac{\partial \Gamma}{\partial t}, \tag{9.59}$$

$$A = A_0 + \nabla\Gamma. \tag{9.60}$$

This is called a *gauge transformation* of $(\phi_0, A_0)$ by the *gauge* $\Gamma$. The key feature of such gauge transformations is that they transform potentials for an electromagnetic field into potentials for the *same* field. Specifically, since $\phi_0$, $A_0$ are potentials for $(E, B)$, we get

$$-\nabla\phi - \frac{1}{c}\frac{\partial A}{\partial t} = -\nabla\left(\phi_0 - \frac{1}{c}\frac{\partial \Gamma}{\partial t}\right) - \frac{1}{c}\frac{\partial}{\partial t}\left(A_0 + \nabla\Gamma\right)$$

$$= -\nabla\phi_0 - \frac{1}{c}\frac{\partial A_0}{\partial t}$$

$$= E$$

and

$$\text{curl}(A) = \text{curl}(A_0 + \nabla\Gamma) = \text{curl}(A_0) = B.$$

This freedom to transform potentials by gauges $\Gamma$ allows us to choose a particular $\Gamma$ such that the $\phi$, $A$ given by equations (9.59)–(9.60) satisfy the Lorentz condition. To see this, let

$$f_0 = \frac{1}{c}\frac{\partial \phi_0}{\partial t} + \text{div}(A_0).$$

If $f_0 = 0$, then $\phi_0$, $A_0$ already satisfy the Lorentz condition, and so we take $\Gamma = 0$. If $f_0 \neq 0$, then we find

$$\frac{1}{c}\frac{\partial \phi}{\partial t} + \text{div}(A) = \frac{1}{c}\frac{\partial}{\partial t}\left(\phi_0 - \frac{1}{c}\frac{\partial \Gamma}{\partial t}\right) + \text{div}\left(A_0 + \nabla\Gamma\right)$$

$$= \nabla^2\Gamma - \frac{1}{c^2}\frac{\partial^2 \Gamma}{\partial t^2} + f_0.$$

Thus if we want $\phi$, $A$ to satisfy the Lorentz condition, we need to choose the gauge $\Gamma$ such that it satisfies the PDE

$$\frac{1}{c^2}\frac{\partial^2 \Gamma}{\partial t^2} - \nabla^2 \Gamma = f_0$$

on $\Omega$. This is always possible. So choose such a $\Gamma$. Then, as we've seen, $E = -\nabla\phi - c^{-1}\partial A/\partial t$ and $B = \text{curl}(A)$, and so by the arguments prior to the theorem, we get that $\phi$ and $A$ satisfy the wave equations (9.57)–(9.58). This completes the first half of the proof.

For the other half of the proof, suppose $\phi$, $A$ are fields that satisfy the Lorentz potential equations (9.56)–(9.58). Let $E = \nabla\phi - c^{-1}\partial A/\partial t$ and $B = \text{curl}(A)$. Then it is straightforward to verify that $E$ and $B$ satisfy Maxwell's equations (exercise).

# 10 Fluid Mechanics

In this chapter we apply the continuum mechanics from Appendix B to the study of fluids, which are just particular types of continua. The system of PDEs that govern the motion of any continuum are

**The Continuum Equations (Eulerian Version):**

$$\frac{\partial \rho}{\partial t} + \nabla_v \rho + \rho \operatorname{div}(v) = 0, \tag{10.1}$$

$$\rho \left[ \frac{\partial v}{\partial t} + \nabla_v v \right] = \rho F + \operatorname{div}(T), \tag{10.2}.$$

$$T^{ij} = T^{ji} \qquad \text{for every } i, j. \tag{10.3}$$

This is a system of PDEs on a region $\Omega \subseteq \mathbb{R}^3$ and an interval of times $(0, b)$. It is customary to refer to $\Omega$ as the continuum. The first of these equations is the *conservation of mass equation* (*continuity equation*), and $\rho$ is the mass density function, which is one of the unknowns in this system of PDEs. The second equation is the *linear momentum equation* and is the continuum version of Newton's second law, $m\dot{v} = f$. The force on the right side of equation (10.2) consists of an external force $F$ (a given, time-dependent vector field on $\Omega$) and an internal force $\operatorname{div}(T)$ that models the force arising from internal stress in the continuum. Here $T$ is a time-dependent, second rank tensor field on $\Omega$ that, as we will see, depends on $v$ in a specific way dictated by the nature of the continuum. The third equation (10.3) is a consequence of the law requiring the balance of of angular momentum with external and internal torques.

We will need to specialize these equations to model the motion of fluids. This amounts to specifying what the stress tensor $T$ for a fluid should be. Actually, we will study several different types of fluids: incompressible fluids, perfect fluids, viscous fluids, etc. In each case the stress tensor $T$ is chosen to reflect the different properties of the fluids.

We begin with some definitions that apply to any continuum $\Omega$ undergoing a motion governed by the flow $\phi_t$ generated by the velocity vector field $v$.

**Definition 10.1**    The flow $\phi_t$, or motion, of $\Omega$ is called

(1) *Incompressible* if $\mathrm{div}(v) = 0$.

(2) *Irrotational* if $\mathrm{curl}(v) = 0$.

(3) *Steady* if $\partial v / \partial t = 0$.

Notice that we have phrased the above concepts in terms of $v$ rather than in terms of the flow. This is for convenience and reflects the preference we are giving the Eulerian description of the motion as opposed to the Lagrangian description. However, the meaning and geometric understanding of these concepts is most clearly understood from the Lagrangian viewpoint. Thus an incompressible flow $\phi_t$ is one that preserves volumes of each piece $B$ of $\Omega$; i.e., the volume of $\phi_t(B)$ is the same for all times $t$ (see the exercises for the details). Steady flows are particularly important in fluid mechanics: Their velocity vector fields $v = v(x, y, z)$ are independent of time, which means that the velocity of the fluid flowing through the point $(x, y, z)$ does not fluctuate with time.

## 10.1.  The Stress Tensor for a Fluid

Throughout this chapter, $T$ refers to the stress tensor in the Eulerian description.

To determine the form of the stress tensor $T$ for a fluid, we could proceed deductively: laying down postulates on how a fluid responds to deformations and how stresses are built up internally, i.e., how one part of the fluid exerts force across its boundary on an adjacent part. From these postulates we could then deduce the form of $T$. This deductive approach is quite detailed, forming a large part of the science of continuum mechanics, and discussing it here would take us too far afield. Instead, we just take the end result of this approach as a definition and then explore how such stress tensors appear to model various types of fluids appropriately.

**Definition 10.2**    A continuum $\Omega$ is called a *fluid* if in response to the flow generated by a velocity vector field $v$, the stress tensor assumes the following form:

$$T = [-p + \lambda \, \mathrm{div}(v)] I + 2\mu D. \qquad (10.4)$$

In this definition the various quantities on the right-hand side are as follows: $v$ is the velocity vector field, as usual, and if we let

$$v' = \begin{bmatrix} v_x^1 & v_y^1 & v_z^1 \\ v_x^2 & v_y^2 & v_z^2 \\ v_x^3 & v_y^3 & v_z^3 \end{bmatrix} \tag{10.5}$$

be the Jacobian matrix, with $v'^*$ the transpose of this Jacobian matrix, then $D$ is the symmetric matrix

$$D \equiv \frac{1}{2}(v' + v'^*) \tag{10.6}$$

$$= \begin{bmatrix} v_x^1 & \frac{1}{2}(v_y^1 + v_x^2) & \frac{1}{2}(v_z^1 + v_x^3) \\ \frac{1}{2}(v_y^1 + v_x^2) & v_y^2 & \frac{1}{2}(v_z^2 + v_y^3) \\ \frac{1}{2}(v_z^1 + v_x^3) & \frac{1}{2}(v_z^2 + v_y^3) & v_z^3 \end{bmatrix} \tag{10.7}$$

The matrix $D$ is a tensor field, known as the *stretching tensor field*. The $I$ in equation (10.4) is the $3 \times 3$ identity matrix, and $\lambda$, $\mu$ are nonnegative constants. The $p$ is a scalar field, known as the *pressure function*:

$$p(x, y, z, t) = \left\{ \begin{array}{c} \text{The pressure at the point } (x, y, z) \\ \text{in the fluid at time } t \end{array} \right\}. \tag{10.8}$$

The pressure is measured in units of force per unit area. The constant $\mu$ is called the *(dynamical) shear viscosity* constant.

Special types of fluids are defined as follows. The fluid is called

(1) *Viscous* if $\mu \neq 0$.

(2) *Navier–Stokes* if it is viscous and incompressible. These fluids have stress tensors of the form

$$T = -pI + 2\mu D$$

(3) *Inviscous* if $\mu = 0$.

(4) *Perfect* if it is inviscous and $\lambda = 0$.

(5) *Ideal* if it is inviscous and incompressible. These fluids have stress tensors of the form

$$T = -pI.$$

The meaning and interpretation of the various constituent parts of the expression (10.4) for $T$ will be explored below (and in the exercises), but for now, we concentrate on how this expression influences the form of the equations of motion for the fluid.

*Note:* Each of these $T$'s satisfies the symmetry condition $T^{ij} = T^{ji}$.

## 10.2.   The Fluid Equations

With the above model for a fluid, expressed in terms of the stress tensor, we now can specialize the continuum equations (10.1)–(10.2) to obtain a system of PDEs that governs the motion of any fluid.

A straightforward calculation of the divergence of the stress tensor in equation (10.4) gives

$$\text{div}(T) = -\nabla p + (\lambda + \mu)\nabla \, \text{div}(v) + \mu \nabla^2 v. \tag{10.9}$$

If we now insert this into the continuum equations (10.1)–(10.2), we get the following equations governing the motion of a fluid:

**The Fluid Equations:**

$$\frac{\partial \rho}{\partial t} + \nabla_v \rho + \rho \, \text{div}(v) = 0, . \tag{10.10}$$

$$\rho \left[ \frac{\partial v}{\partial t} + \nabla_v v \right] = \rho F - \nabla p + (\lambda + \mu)\nabla \, \text{div}(v) + \mu \nabla^2 v. \tag{10.11}$$

These fluid equations constitute a system of four PDEs involving five unknown functions: the mass denity $\rho$, the components of velocity $v^1$, $v^2$, $v^3$, and the pressure $p$. Generally, there will be other equations involving these unknown functions. We will study only two prominent special cases of the fluid equations: (1) ideal fluids, i.e., fluids that are inviscous and incompressible, and (2) Navier-Stokes fluids, i.e., viscous, incompressible fluids. The equations governing the flow of such fluids are

**The Euler Equations (for an Ideal Fluid):**

$$\frac{\partial \rho}{\partial t} + \nabla_v \rho = 0, \tag{10.12}$$

$$\rho \left[ \frac{\partial v}{\partial t} + \nabla_v v \right] = \rho F - \nabla p, \tag{10.13}$$

$$\text{div}(v) = 0; \tag{10.14}$$

**The Navier–Stokes Equations:**

$$\frac{\partial \rho}{\partial t} + \nabla_v \rho = 0, \tag{10.15}$$

$$\rho \left[ \frac{\partial v}{\partial t} + \nabla_v v \right] = \rho F - \nabla p + \mu \nabla^2 v, \tag{10.16}$$

$$\mathrm{div}(v) = 0. \tag{10.17}$$

The incompressibility condition $\mathrm{div}(v) = 0$ not only simplifies the fluid equations considerably, but also adds a fifth equation to the system, so that both the Euler equations and the Navier–Stokes equations are systems of five equations for five unknown functions $\rho$, $v^1$, $v^2$, $v^3$, $p$. Because of this we can hope, with suitable initial conditions and boundary conditions, that there will exist solutions of the system of PDEs, and that these solutions will be unique. The initial conditions simply state that at some initial time, say $t = 0$, we know the density, velocity, and pressure throughout the vessel

**Initial Conditions:**

$$\rho(x, y, z, 0) = \rho_0(x, y, z),$$

$$v(x, y, z, 0) = v_0(x, y, z), \tag{10.18}$$

$$p(x, y, z, 0) = p_0(x, y, z),$$

for all $x \in \Omega$. Here $\rho_0$, $v_0$, and $p_0$ are given functions.

The boundary conditions just specify what happens with the fluid at the boundary $\partial \Omega$ of $\Omega$. Several examples of appropriate boundary conditions (BCs) are

**Rigid Walls BC:**

$$v \cdot n = 0 \qquad \text{on } \partial \Omega, \forall t, \tag{10.19}$$

where $n$ is the outward-directed unit normal to the boundary surface $\partial \Omega$. One says in this case that $v$ (and hence the fluid flow) is parallel, or tangent, to the boundary at all points along the boundary. See Figure 10.1

The rigid walls BC says only that the velocity $v$ is tangential at the boundary and says nothing about the magnitude of $v$ there. One could require $v$ to match a prescribed (tangential) velocity at each boundary point.

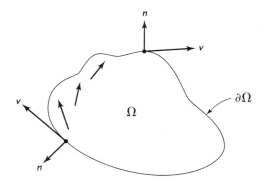

Figure 10.1.   The rigid walls BC.

**Prescribed Tangential Velocity BC:**

$$v \cdot n = 0 \qquad \text{on } \partial\Omega, \forall t,$$

$$v = h \qquad \text{on } \partial\Omega, \forall t. \tag{10.20}$$

A particular case of this is

**No Slip BC:**

$$v = 0 \qquad \text{on } \partial\Omega, \forall t. \tag{10.21}$$

This requires the motion of the fluid to vanish at the boundary walls, and one conceives of the boundary walls as stationary (not in motion). Of course, the more general BC (10.20) can be thought of as a no slip BC, if one interprets the boundary walls as moving with velocity $h$. One important example of this is Taylor–Couette flow. This is for fluid contained between two concentric cylinders, each rotating with a uniform (constant) speed. Requiring the fluid to match the velocity of the boundary walls is interpreted as a no slip BC. The fluid near the walls gets dragged along by the motion of the walls.

For the rest of the chapter we will concentrate our study on the Euler equations for an ideal fluid and the Navier–Stokes equations for a viscous incompressible fluid. Note that these equations differ merely by the term $\mu\nabla^2 v$ in the Navier–Stokes equation (10.16). This term accounts for the viscosity of the fluid, and in the absence of viscosity ($\mu = 0$), the Navier–Stokes equations reduce to the Euler equations. In the ensuing discussion, we will explain more fully the meaning of viscosity and other physical properties/behaviors of fluids.

We should mention here, at the beginning, that many mathematical (and physical) aspects of the theory for these fluid equations are still incomplete at this point in time, even though the equations have been around for hundreds of years. There is much current interest in and research on completing much of the theory.

**Exercises 10.2**

1. If you did not already do so in Chapter 1, prove the following identities. Here $f$, $g$ are scalar fields, and $u$, $v$, $w$ are vector fields.

   (a) $\nabla(fg) = g\nabla f + f\nabla g$.

   (b) $\text{curl}(fv) = \nabla f \times v + f\,\text{curl}(v)$.

   (c) $\text{div}(fv) = \nabla f \cdot v + f\,\text{div}(v)$.

   (d) $\nabla_v(fw) = (\nabla_v f)w + f\nabla_v w$.

   (e) $\nabla_{fv}w = f\nabla_v w$.

   (f) $\frac{\partial}{\partial x_i}(u \times v) = \frac{\partial u}{\partial x_i} \times w + u \times \frac{\partial w}{\partial x_i}$.

   (g) $\nabla_v(u \times w) = (\nabla_v u) \times w + u \times (\nabla_v w)$.

   (h) $\text{curl}(v \times w) = \nabla_w v - \nabla_v w + \text{div}(w)v - \text{div}(v)w$.

   (i) $\text{div}(v \times w) = \text{curl}(v) \cdot w - v \cdot \text{curl}(w)$.

   (j) $\text{div}(\nabla f \times \nabla g) = 0$.

   (k) $\nabla_v v = \nabla(\frac{|v|^2}{2}) + \text{curl}(v) \times v$.

2. Use the identity in Exercises 1(c) to show that the continuity equation (10.1) can be written as

$$\frac{\partial \rho}{\partial t} + \text{div}(\rho v) = 0.$$

Use Gauss's divergence theorem from vector analysis to show that the above version of the continuity equation follows from the hypothesis that

$$\frac{d}{dt}\int_B \rho\,dV = -\int_{\partial B} \rho v \cdot n\,dA,$$

for every subregion $B \subseteq \Omega$. This formulation of the continuity equation in terms of integrals is the clearest geometrically. The term on the left side of the equation is the time rate of change of the mass inside $B$, while the term on the right side is minus the flux of mass outward across the boundary surface to $B$. See Figure 10.2. Comment on how to interpret this in terms of an infinitesimal portion of the mass flowing across the boundary.

3. Suppose $v : \Omega \to \mathbb{R}^3$ is a differentiable vector field on $\Omega$, and let

$$D = \frac{1}{2}(v' + v'^*),$$

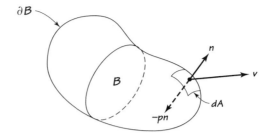

Figure 10.2.   A subregion $B$ of the fluid. For a point on the boundary, $n$ is the outward-directed normal and $dA$ is an infinitesimal element of boundary surface area.

where * denotes matrix transpose. Show that

$$\mathrm{tr}(D) = \mathrm{div}(v),$$

$$\mathrm{div}(\mathrm{div}(v)I) = \nabla \,\mathrm{div}(v),$$

$$\mathrm{div}(2D) = \nabla^2 v + \nabla \,\mathrm{div}(v).$$

From this show that if

$$T = [-p + \lambda \,\mathrm{div}(v)]I + 2\mu D,$$

then

$$\mathrm{div}(T) = -\nabla p + (\lambda + \mu)\nabla \,\mathrm{div}(v) + \mu \nabla^2 v.$$

4. For vector fields $v$, $w$, interpret the covariant derivative $\nabla_v w$ as the rate of change of $w$ along the flow $\phi$ generated by $v$. Specifically, show that

$$\left. \frac{d}{dt} w(\phi_t(x, y, z)) \right|_{t=0} = (\nabla_v w)(x, y, z),$$

for $x \in \Omega$. See Figure 10.3.

5. Let $\phi_t$ be the flow generated by a velocity vector field $v$ on $\Omega$. A subregion $B \subseteq \Omega$ is continuously deformed over time into the subregion $\phi_t(B)$ at time $t$, and this has volume given by the volume integral $\int_{\phi_t(B)} 1 \, dV$. Use the transport theorem from Appendix B to show

$$\frac{d}{dt} \int_{\phi_t(B)} 1 \, dV = \int_{\phi_t(B)} \mathrm{div}(v) \, dV.$$

Use this to show that for an incompressible fluid, the volume of $B$ does not change when transported by the flow.

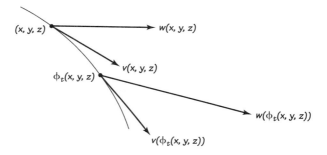

Figure 10.3.   Geometric significance of the covariant derivative $\nabla_v w$ of a vector field $w$ along a vector field $v$.

6. Suppose the stress tensor for the continuum has the form

$$T = -pI,$$

where $p = p(x, y, z, t)$ is a scalar field and $I$ is the $3 \times 3$ identity matrix. The function $p$ is the *pressure function* (with units of force per unit area). Assuming also incompressibility, the continuum is then an *ideal fluid*. This exercise explores the nature of the stress tensor and statics for such a fluid.

The force acting on an infinitesimal planar surface of area $dA$ is $Tn\, dA = -pn\, dA = -(p\, dA)n$. This force is always perpendicular to the surface, has magnitude $p\, dA$, and is directed oppositely to $n$. For a subregion $B$, the total force on $B$ due to pressure on its boundary is the sum of all these infinitesimal forces:

$$\mathcal{F}_{\text{int}}(B) = -\int_{\partial B} pn\, dA.$$

See Figure 10.2.

It is clear that the stress tensor is symmetric (and so the balance of the angular momenta with the applied torques holds). Also, the linear momentum equation is

$$\rho\left[\frac{\partial v}{\partial t} + \nabla_v v\right] = \rho F - \nabla p.$$

Consider the case where the continuum is at rest, $v = 0$; the material is of constant density, $\rho = \text{constant}$; and the external force density is due solely to gravity, $F = (0, 0, -g)$, with $g$ the acceleration of gravity (near the surface of the earth).

(a) Show that the linear momentum equation is

$$0 = -p_x,$$

$$0 = -p_y,$$

$$0 = -\rho g - p_z.$$

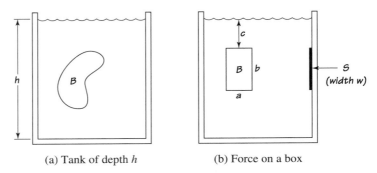

(a) Tank of depth $h$              (b) Force on a box

Figure 10.4.    Fluid at rest in a tank and hydrostatic bouancy force on a submerged object.

(b) Assume that the fluid is in a rectangular tank of depth $h$, as shown in Figure 10.4(a). Determine the pressure function $p$, assuming that the pressure is zero at the top surface.

(c) Suppose $B$ is a subregion of the tank. Show that the total force on $B$ due to fluid pressure acting across its boundary is

$$\mathcal{F}_{\text{int}}(B) = (0, 0, \rho g \text{ vol } (B)).$$

This is *Archimedes's principle*: The bouancy force on a submerged object is equal to the weight of the water it displaces. Also verify this principle by direct calculation when $B$ is a rectangular box oriented as shown in Figure 10.4(b).

(d) Calculate the force exerted by the fluid on the rectangular piece $S$ of the side of the tank shown in Figure 10.4. By definition, this force is $\mathcal{F} = -\int_S T n \, dA$.

(e) Assume the same situation as described above, except now allow the fluid to be *density stratified*; i.e., suppose $\rho$ varies in a linear fashion with the depth:

$$\rho = \rho(z) = k(h - z).$$

Thus the fluid is more dense at the bottom of the tank. Do parts (b), (c), and (d) now, with the appropriate changes.

7. The model for a Navier–Stokes fluid,

$$T = -pI + 2\mu D,$$

reduces to that for an ideal fluid when $\mu = 0$. But for $\mu > 0$ the extra term in the stress tensor accounts for forces due to viscosity as one part of the fluid attempts to slide past an adjacent part. For perfect fluids these forces are absent, and only the forces due to hydrostastic pressure are present. This exercise deals with the effect of viscosity on the motion.

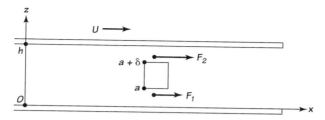

Figure 10.5.   A tank of depth $h$ with the top surface in motion with constant speed $U$.

Consider a Navier–Stokes fluid in a tank $\Omega = \mathbb{R}^2 \times [0, h]$ of depth $h$ and infinite extent in the $x$, $y$ directions (see Figure 10.5). Assume that at the top surface of the tank ($z = h$), the fluid is maintained at a constant speed $U$ in the $x$ direction, so that on this part of the boundary, $v = (U, 0, 0)$. Suppose also that $v = 0$ on the bottom surface of the tank. In this exercise we seek a solution of the Navier–Stokes equations that has the form

$$v = (v^1(z), 0, 0).$$

This will be a steady flow.

(a) Show that $\mathrm{div}(v) = 0$ and $\nabla_v v = 0$.

(b) Assume that $\rho = $ constant and $F = (0, 0, -g)$. Show that the Navier–Stokes equations reduce to

$$0 = -p_x + \mu v^1_{zz},$$
$$0 = -p_y,$$
$$0 = -\rho g - p_z.$$

Solve these PDEs for $v$ and $p$ assuming that the pressure vanishes on the top surface of the tank.

(c) Plot the velocity profile on a planar cross section $x = c$ through the tank.

(d) Compute the stress tensor $T$ explicitly in terms of the $v$ and $p$ you found in part (b). For the small cubical subregion $B$ in the tank shown in Figure 10.5, compute the stress force density $Tn$ on the bottom and top surfaces ($z = a$ and $z = a + \delta$). Call these $F_1$ and $F_2$ respectively. Plot these on their respective boundary surfaces and explain why they account for the shearing motion of the viscous fluid.

## 10.3.   Special Solutions of the Euler Equations

The appropriate (and simplest) place to start in discussing solutions $(\rho, v, p)$ of the Euler equations is with steady flows. That is, we look for solutions where the velocity $v$ does not depend on the time $(\partial v/\partial t = 0)$. We will also assume that the external forces arise from a potential and that the fluid is homogeneous. Thus, throughout this section we will assume the following

**Assumptions for This Section:**

(1)  The flow is steady: $\partial v/\partial t = 0$.

(2)  The external forces arise from a potential $F = -\nabla\Psi$ (with $\Psi$ not depending on $t$).

(3)  The fluid is *homogeneous*; i.e., $\rho$ is a constant function.

Note that assumption (3) means that the conservation of mass equation $\partial\rho/\partial t + \nabla_v\rho = 0$ is automatically satisfied, and the other assumptions reduce the Euler equations to

$$\rho\nabla_v v = -\nabla(p + \rho\Psi), \qquad (10.22)$$

$$\mathrm{div}(v) = 0. \qquad (10.23)$$

Equation (10.22) implies that the pressure $p$ does not depend on the time either, and suggests the strategy of manufacturing solutions by first choosing an appropriate divergence-free vector field $v$ and then trying to determine $p$ from (10.22). This strategy relies on the following basic identity:

**Lemma 10.1**   *For any vector field $v$, the following identity holds:*

$$\nabla_v v = \nabla\left(\frac{|v|^2}{2}\right) + \mathrm{curl}(v) \times v. \qquad (10.24)$$

Using this identity, we see that if $v$ is curl free, $\mathrm{curl}(v) = 0$, then $\nabla_v v = \nabla(\frac{|v|^2}{2})$, and thus we get the following special types of solutions to the Euler equations:

**Proposition 10.1**   *Suppose $v$ is any vector field on $\Omega$ that is divergence-free and curl-free: $\mathrm{div}(v) = 0$ and $\mathrm{curl}(v) = 0$. Let $\rho > 0$ and $k$ be any constants, and define a function $p$ by*

$$p \equiv -\rho\frac{|v|^2}{2} - \rho\Psi + k. \qquad (10.25)$$

*Then $(\rho, v, p)$ is a solution of the Euler equations whose flow is irrotational.*

**Proof**   The proof is rather elementary, in that it just follows from all the above observations. Since $\rho$ is a constant, the continuity equation is trivially satisfied. By assumption, the incompressibility condition is satisfied. Finally, we have defined the pressure function $p$ such that the linear momentum equation (10.22) holds:

$$-\nabla(p + \rho\Psi) = \nabla\left(\rho\frac{|v|^2}{2} - k\right) = \rho\nabla_v v,$$

where we have used the curl-free assumption and identity (10.24).

*Note*: The proposition gives us a way of constructing solutions of the Euler equations that are steady ($\partial v/\partial t = 0$), irrotational (curl$(v) = 0$), and homogeneous ($\rho = $ constant). The pressure function is defined (up to an arbitrary constant) in terms of the veclocity as in equation (10.25), and this gives an important relationship between pressure and velocity. Conversely, we have

**Proposition 10.2**   (*Bernoulli's Law*) *If $(\rho, v, p)$ is a solution of the Euler equations that is steady, irrotational, and homogeneous (and the external force density $F = -\nabla\Psi$), then there exists a constant $k$ such that*

$$\rho\frac{|v|^2}{2} + p + \rho\Psi = k. \tag{10.26}$$

In essence, Bernoulli's law says that at places in the fluid where the velocity is high, the pressure is low (and conversely, where the velocity is low, the pressure is high).

**Example 10.1**   Let $v$ be the vector field defined on $\mathbb{R}^3 - \{z \text{ axis}\}$ by

$$v(x, y) = \left(\frac{-y}{x^2 + y^2}, \frac{x}{x^2 + y^2}, 0\right). \tag{10.27}$$

By direct calculation, we see that div$(v) = 0$ and curl$(v) = 0$. Thus, we get a solution of the Euler equations if we take

$$p(x, y, z) = \frac{-\rho}{x^2 + y^2} - \rho g z + k, \tag{10.28}$$

where $\rho$ and $k$ are constants. Here we have taken $\Psi = gz$ as the potential for the gravitational field density. If we restrict attention to a fluid in the region $\Omega$

between two concentric cylinders (see Figure 10.6), then the above velocity and pressure represent an ideal, homogeneous fluid, circulating in $\Omega$ with a flow that is steady and irrotational. The velocity profile is shown in Figure 10.6. Note that the velocity $v$ is tangential to the boundary cylinders, and that $|v| = R_i^{-1}$, $i = 1, 2$, on the these cylinders (where $R_1, R_2$ are the radii of the bases of these cylinders). Thus, viewing the cylinders as revolving with these constant speeds, we get that the solution $(\rho, x, p)$ satisfies the prescribed tangential velocity BC.

As a means of producing more examples, let us look back at Proposition 10.1 and rephrase it slightly. It says that any solution $v$ of the system of PDEs

$$\mathrm{curl}(v) = 0,$$

$$\mathrm{div}(v) = 0 \tag{10.29}$$

gives rise to a solution $(\rho, v, p)$ of the Euler equations via the prescription outlined there. The next two subsections give general methods for finding solutions of the system (10.29).

## 10.3.1.  Potential Flows and Harmonic Functions

As we have noted in other chapters, the identity

$$\mathrm{curl}(\nabla \Lambda) = 0$$

holds for any scalar field $\Lambda$ and gives a class of vector fields that are curl free, namely the class of vector fields $v$ that are gradients of scalar fields:

$$v = \nabla \Lambda. \tag{10.30}$$

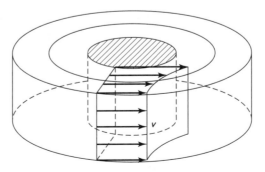

Figure 10.6.    Flow between concentric cylinders.

In the present setting, such vector fields generate flows that are called *potential flows*. Thus any potential flow is irrotational. We have also noted before that if curl($v$) = 0 holds on $\Omega$, and $\Omega$ is a fairly reasonable domain (technically, a simply connected domain), then $v$ is the gradient of some scalar field $v = \nabla\Lambda$ on $\Omega$. This is not always the case, as the last example shows. Try to convince yourself that there is *no* scalar field $\Lambda$ on $\Omega = \mathbb{R}^3 - \{z\text{-axis}\}$ such that $\nabla\Lambda = v$ for the $v$ given there (note that this $\Omega$ is not simply connected). Thus an irrotational flow need not be a potential flow.

In this subsection we will consider potential flows and assume that $v = \nabla\Lambda$ for some $\Lambda$, so that the flow for $v$ is automatically irrotational. To satisfy the incompressibility condition, recall the identity div($\nabla\Lambda$) = $\nabla^2\Lambda$, and hence we have

**Proposition 10.3** (*Potential Flows*) *Suppose* $\Lambda$ *is a scalar function that is harmonic, i.e., satisfies Laplace's equation*

$$\nabla^2\Lambda = 0$$

*on* $\Omega$. *Define a velocity and pressure by*

$$v = \nabla\Lambda, \tag{10.31}$$

$$p = -\rho\frac{|\nabla\Lambda|^2}{2} - \rho\Psi + k, \tag{10.32}$$

*where* $\rho > 0$ *and* $k$ *are constants, and* $\Psi$ *is a potential for the external force density. Then* $(\rho, v, p)$ *is a solution of the Euler equations, and the fluid flow is steady and potential.*

This proposition gives us a large class of fluid motions that we can analytically determine by the methods developed in the previous chapters. Recall that the solutions of Laplace's equation are called *harmonic functions*, and the proposition says that each harmonic function $\Lambda$ generates a fluid flow with velocity $v = \nabla\Lambda$ and pressure determined by Bernoulli's law. Note that since $\nabla\Lambda$ is perpendicular to each level surface of $\Lambda$, it follows that the streamlines of the flow are perpendicular to each level surface of $\Lambda$. See Figure 10.7.

The geometric fact that the streamlines for a potential flow are perpendicular to the level surfaces of $\Lambda$ is interesting and often useful in constructing the flow from the known level surfaces. Some of the exercises deal with this technique. On the other hand, by using computer graphics packages, we can directly draw the flow lines generated by the vector field $v$. This will also be covered in the exercises. The next section gives yet another method for drawing the streamlines by using elementary facts from complex analysis.

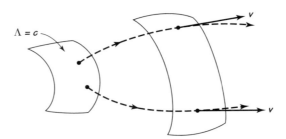

Figure 10.7.    Streamlines of a potential flow.

As we shall see, a great number of potential flows are *planar flows*. The exact meaning of this latter term is as follows:

**Definition 10.3 (Planar Flows)**    The flow generated by a vector field $v$ is called *planar* if the third component of $v$ is identically zero, $v^3 = 0$, and the other two components do not depend on $z$. Thus

$$v(x, y) = \left( v^1(x, y),\ v^2(x, y),\ 0 \right). \tag{10.33}$$

This means that there is no component of flow in the $z$-direction, and the velocity profile is the same on each plane $z = c$, for any constant $c$.

One can easily generalize this notion to planar flows (exercise), but the above definition will suffice for the most part here.

In a number of situations, we can combine several standard harmonic functions to easily obtain potential flows that satisfy the rigid walls BC. This is illustrated in the following example (as well as in the exercises).

**Example 10.2**    The most basic potential flow is a *uniform flow*, where all the streamlines are straight lines parallel to a given direction. If the direction is along the $x$-axis, then the potential and the velocity are

$$\Lambda_1(x, y, z) = Ux,$$

$$v_1(x, y, z) \equiv \nabla\Lambda_1(x, y, z) = (U, 0, 0), \tag{10.34}$$

where $U$ is a constant (the speed of the flow). The flow is planar, and so to picture it, we need only draw the streamlines in the $x$-$y$ plane. See Figure 10.8(a). A more interesting harmonic function and its corresponding velocity vector field are

$$\Lambda_2(x, y, z) = \frac{bx}{x^2 + y^2},$$

$$v_2(x, y, z) = \left( \frac{b(y^2 - x^2)}{(x^2 + y^2)^2}, \frac{-2bxy}{(x^2 + y^2)^2}, 0 \right). \qquad (10.35)$$

It is easy to check that $\Lambda_2$ is a harmonic function. Again, the velocity vector field $v_2$ is planar, so we need only draw the picture of the flow in the $x$-$y$ plane. The corresponding fluid flow is known as a *source/sink* because of its flow pattern, which is shown in Figure 10.8(b).

Note that $v_2$ has a singularity at the origin (it is not defined there). The flow pattern illustrated in Figure 10.8(b) appears as if the fluid is flowing out from the left side of this singularity, curving around, and disappearing back into it on the right side. To draw the flow pattern by hand, we use the fact that the streamlines are perpendicular to the level surfaces $\Lambda_2 = c$. Explicitly, this equation is

$$\frac{bx}{x^2 + y^2} = c,$$

or (with the appropriate algebra, and completing the square)

$$\left( x - \frac{b}{2c} \right)^2 + y^2 = \frac{b^2}{4c^2}.$$

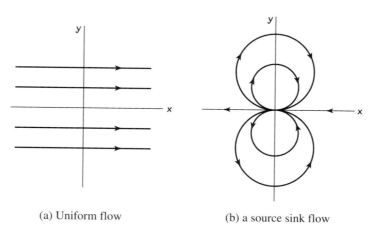

(a) Uniform flow                    (b) a source sink flow

Figure 10.8.   Pictures of two basic potential flows.

The plot of this in the $x$-$y$ plane is a circle with center $(b/2c, 0)$ and radius $b/2|c|$. (The actual level surface $\Lambda_2 = c$ is obtained by translating this circle parallel to the $z$-axis.) Drawing a number of these circles will then enable you to construct the flow lines, which will be orthogonal to each such circle. This should give the flow picture shown in Figure 10.9(b).

Now both of these flows $v_1$, $v_2$ give rise to a much more interesting combined flow: $v \equiv v_1 + v_2$; i.e.,

$$v(x, y, z) = \left( U + \frac{b(y^2 - x^2)}{(x^2 + y^2)^2}, \frac{-2bxy}{(x^2 + y^2)^2}, 0 \right). \tag{10.36}$$

This is a potential flow, whose potential is $\Lambda = \Lambda_1 + \Lambda_2$; i.e.,

$$\Lambda(x, y, z) = Ux + \frac{bx}{x^2 + y^2}. \tag{10.37}$$

This flow models a steady, irrotational, planar Euler flow past a cylinder. Namely, if we draw the streamlines of (10.36) for fluid particles on the exterior of the cylinder $x^2 + y^2 = b/U$, represented as a circle of radius $r = \sqrt{b/U}$ in the $x$-$y$ plane, then we get the picture shown in Figure 10.9. The choice of this radius $r$ is special, since it is the only value for which the vector field $v$ in (10.36) is tangential to the circle $x^2 + y^2 = r^2$, at each of its points. Thus for this value of $r$ the flow on the region $\Omega = \{(x, y, z) | x^2 + y^2 \geq r^2\}$ will satisfy the rigid walls

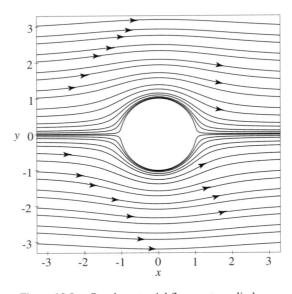

Figure 10.9.    Steady potential flow past a cylinder.

boundary condition. To see this, note that the outward unit normal to the circle $x^2 + y^2 = r^2$ is $n = (x/r, y/r, 0)$, and so

$$v \cdot n = \left( U - \frac{b}{x^2 + y^2} \right) \frac{x}{r},$$

and this is zero only when $x^2 + y^2 = r^2 = b/U$ (and on the $y$-axis).

A further topic that this example illustrates is the notion of *stagnation points*. These are points in the fluid where the velocity vanishes: $v = 0$. A particle placed at such a point will not be disturbed by the fluid flow but will remain at this point forever. In the present example there are two stagnation points, $(\pm\sqrt{b/U}, 0, 0)$, which are found by solving $v = 0$, with $v$ given by (10.36). Explicitly, this system of equations is

$$U + \frac{b(y^2 - x^2)}{(x^2 + y^2)^2} = 0,$$

$$\frac{-2bxy}{(x^2 + y^2)^2} = 0.$$

These stagnation points are also clearly visible in Figure 10.9.

It should be mentioned that Figure 10.9 was produced by *Maple*, which does not have the facility of plotting the individual streamlines in real time, so that the velocities of movement along the streamlines and a feeling for the overall motion of the fluid past the cylinder are readily apparent. The static drawing shown here can not capture the dynamics of the motion. We have augmented *Maple*'s code with an elementary procedure called DDEplot which will draw the flow lines dynamically (see CD-ROM).

In some cases the construction of potential flows can be accomplished by solving a Neumann problem

$$\nabla^2 \Lambda = 0 \qquad \text{in } \Omega,$$

$$\nabla \Lambda \cdot n = h \qquad \text{on } \partial\Omega,$$

where $h$ is an appropriate function that also must satisfy the condition

$$\int_{\partial\Omega} h \, dA = 0$$

for the solvability of the problem. In terms of the velocity vector field $v = \nabla\Lambda$, the boundary condition is

$$v \cdot n = h \qquad \text{on } \partial\Omega.$$

Thus if we require the rigid walls BC on the entire boundary, $v \cdot n = 0$ on $\Omega$, then as you know, the solution of the Neumann problem is $\Lambda = c$, where $c$ is a constant. Hence $v = \nabla\Lambda = 0$, and the fluid is at rest. To get more interesting solutions than this, we will need to allow the given function $h$ in the Neumann problem to be nonzero on parts of the boundary. This amounts to prescribing the normal component of the velocity there.

**Example 10.3**    Consider the Neumann problem on the rectangle $R = [0, 2] \times [0, 1]$, with BCs as shown in Figure 10.10. The solution of this problem is

$$\Lambda(x, y) = 0.1 \cosh \pi x \, \cos \pi y.$$

If we let $v = \nabla\Lambda$, then

$$v(x, y) = (0.1\pi \sinh \pi x \, \cos \pi y, -0.1\pi \cosh \pi x \, \sin \pi y, 0).$$

We can consider this as a fluid flow in a rectangular tank, say $\Omega = R \times [0, 1]$, where $v$ is parallel to the boundary on the top, bottom, and three sides and has specified normal component on the fourth side. The flow is planar again, so just looking at the flow pattern in the $x$-$y$ plane suffices to understand the motion. Figure 10.11 is a static picture of the flow (you should have a computer draw a dynamic picture to fully appreciate the speeds of the fluid flow along the various streamlines). *Note*: Any problem like this can also be interpreted as a steady-state heat flow problem. Then $\Lambda$ represents the temperature in the plate $R$, and $-v$ is the heat flux vector, pointing in the direction of greatest temperature increase.

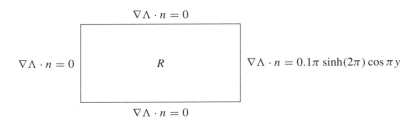

Figure 10.10.    Boundary conditions for a Neumann problem in Example 10.3.

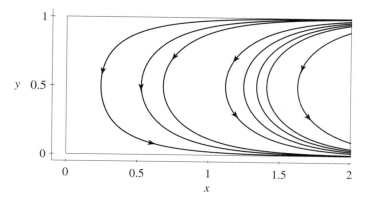

Figure 10.11.   The flow lines for Example 10.3.

Reversing the arrows in Figure 10.11 gives a picture of the *heat* flow lines (going from hot to cold).

**Exercises 10.3.1**

1. Find a potential flow that is planar,

$$v = (v^1(x, y), v^2(x, y), 0),$$

and that satisfies the boundary conditions shown in Figure 10.12. Here $R = [0, 2] \times [0, 1]$. The tank $\Omega = R \times [0, h]$ is a rectangular box, and the view shown in Figure 10.12 is looking down on it along the $z$-axis. Once you find $v$, use a computer to draw its flow lines. Also plot $v$ (as a direction field) in a separate picture.

3. Let $\Lambda$ be the following polynomial:

$$\Lambda = x^4 - 6x^2y^2 + y^4.$$

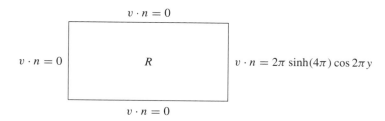

Figure 10.12.   A potential flow with $v$ tangential to three sides of the tank and prescribed normal component on the fourth.

(a) Show that $\Lambda$ is a harmonic function and calculate the potential flow $v$ corresponding to $\Lambda$.

(b) Show that $v$ is tangential to the lines $y = 0$ and $y = x$. Thus restricting $v$ to the region $R$ bounded by these lines in the first quadrant gives a planar potential flow that satisfies the rigid walls BCs. Use a computer to graph the flow lines in $U$. The flow pattern models the flow of an ideal fluid past an inside corner of 45 degrees.

(c) Find all other lines to which $v$ is tangential.

4. Formulate a generalization of the definition of a *planar flow* that also covers the case when, say,

$$v = (0, v^2(y, z), v^3(y, z)).$$

5. For finding planar, potential flows that can satisfy BCs that are related to circular geometry, it is useful to transform $\Lambda = \Lambda(x, y)$ to a function $K = K(r, \theta)$ in polar coordinates:

$$K(r, \theta) = \Lambda(r \cos \theta, r \sin \theta).$$

If $\Lambda$ is harmonic, then $K$ satisfies Laplace's equation in polar coordinates:

$$K_{rr} + \frac{1}{r^2} K_{\theta\theta} + \frac{1}{r} K_r = 0.$$

In Section 12.8 on BVPs in polar coordinates, it is shown that the general solution of this PDE is a series, some of whose terms have the forms

$$K = (Ar^m + Cr^{-m}) \cos m\theta$$

and

$$K = (Br^m + Dr^{-m}) \sin m\theta.$$

This exercise examines special flows that arise from such potentials.

(a) Show that the above functions $K$ satisfy Laplace's equation in polar coordinates.

(b) Show that the vector field (10.36) for the planar flow past a circular cylinder (with $U = 1 = b$) comes from the polar coordinate potential

$$K = (r + r^{-1}) \cos \theta.$$

(c) Determine the vector field $v$ with polar coordinate potential

$$K = (r^2 + r^{-2}) \cos 2\theta,$$

and plot $v$ (or at least its direction field) on the rectangle $R = [-2, 2] \times [-2, 2]$. Show that $v$ is tangential to the unit circle $x^2 + y^2 = 1$ and plot the flow on the part of $R$ exterior to this circle. Find all the stagnation points.

(d) Do what was asked in part (c) for the polar coordinate potential

$$K = (r^2 + r^{-2}) \sin 2\theta$$

What is the difference here from part (c)?

## 10.3.2. Planar Flows and Analytic Functions

This subsection presents another broad class of special solutions to the Euler equations and exhibits a fundamental connection between planar fluid flow and complex analysis. Thus we will use some elementary facts about analytic functions to construct planar flows

$$v(x, y) = \left( v^1(x, y), \ v^2(x, y), \ 0 \right) \tag{10.38}$$

for ideal fluids.

A planar flow is thought of as a 2-dimensional flow, i.e., as existing in the $x$-$y$ plane only, and practically is often realized with fluids in a shallow tank, circlulating at small speeds. The last section gave examples of several instances of planar flows.

If we consider a planar flow with velocity vector field (10.38) and write out the PDEs $\mathrm{div}(v) = 0$, $\mathrm{curl}(v) = 0$ in component form, we get

$$v^1_x = -v^2_y, \tag{10.39}$$

$$v^1_y = v^2_x. \tag{10.40}$$

According to Proposition 10.1, each solution of ths system gives us a stationary, irrotational (and in this case planar) solution of the Euler equations. However, the system (10.39)–(10.40) is, in essence, the well-known *Cauchy–Riemann equations* from complex analysis, which serve as necessary and sufficient conditions for complex functions to be analytic. We discuss a little background on this and then use it to construct planar fluid flows by the prescription we have indicated.

Let $\mathbb{C} = \mathbb{R}^2$ denote the set of complex numbers, with a complex number denoted by $z = x + yi = (x, y)$, where $x, y \in \mathbb{R}$ are real numbers and $i = \sqrt{-1} = (0, 1)$ denotes the complex imaginary unit. The complex conjugate of $z$ is denoted by $\bar{z} = x - yi$. In this setting, one studies complex-valued functions of a complex variable, i.e., functions $f : \mathbb{C} \to \mathbb{C}$. More generally, we require only that the domain of $f$ be an open subset of $\mathbb{C}$. Each such function has component expression

$$f(z) = f^1(z) + f^2(z)i = \left( f^1(z), f^2(z) \right),$$

where the component functions $f^1(z) = f^1(x, y)$ and $f^2(z) = f^2(x, y)$, called the *real* and *imaginary parts* of $f$ respectively, are real-valued functions of two real variables (or of a single complex variable). Primary examples of such functions are polynomial functions,

$$p(z) = c_0 + c_1 z + \cdots + c_n z^n,$$

and rational functions,

$$r(z) = \frac{p(z)}{q(z)},$$

where $p$ and $q$ are polynomial functions. In addition, most of the functions from real analysis have extensions (continuations) to their analogues as complex functions. A complex function $f$ is said to be *analytic* (or complex differentiable) at $z$ if the limit

$$\lim_{h \to 0} \frac{f(z+h) - f(z)}{h} \equiv f'(z)$$

exists. Also, $f$ is called analytic on an open subset $R \subseteq \mathbb{C}$ if it is defined and analytic at each point of $R$. It is an elementary exercise to show that if $f$ is analytic on $R$, then its real and imaginary parts $f^1$, $f^2$ satisfy the *Cauchy–Riemann* equations

$$f_x^1 = f_y^2, \tag{10.41}$$

$$f_y^1 = -f_x^2 \tag{10.42}$$

on $R$. The converse is also true. Furthermore, the derivative of $f$ is given by

$$f' = f_x^1 + f_x^2 i = \frac{1}{i}(f_y^1 + f_y^2 i). \tag{10.43}$$

(See the exercises for a proof.)

An important geometric fact that follows from the Cauchy–Riemann equations is that if $f = f^1 + f^2 i$ is analytic, then the gradients $\nabla f^1 = (f_x^1, f_y^1)$, $\nabla f^2 = (f_x^2, f_y^2)$ of its real and imaginary parts are orthogonal at each point:

$$\nabla f^1 \cdot \nabla f^2 = 0. \tag{10.44}$$

Otherwise stated, any two level curves for $f^1$ and $f^2$,

$$f^1(x, y) = c_1, \quad f^2(x, y) = c_2,$$

are orthogonal at their point of intersection. As a good example of this, consider

$$f(z) = e^{-z} = e^{-x-yi} = e^{-x} e^{-yi}$$
$$= e^{-x} \cos y - (e^{-x} \sin y) i,$$

where we have used Euler's formula on $e^{-yi}$. The real and imaginary parts of $f$ are $f^1(x, y) = e^{-x} \cos y$ and $f^2(x, y) = -e^{-x} \sin y$, and thus

$$\nabla f^1(x, y) = (-e^{-x} \cos y, -e^{-x} \sin y),$$
$$\nabla f^2(x, y) = (e^{-x} \sin y, -e^{-x} \cos y).$$

These clearly satisfy the orthogonality relation (10.44). The level curves for these functions are the curves with equations

$$e^{-x} \cos y = c_1, \quad -e^{-x} \sin y = c_2,$$

and they may be expressed as graphs of functions $y$ by solving for $x$:

$$x = \ln |c_1 \cos y|, \quad x = \ln |c_2 \sin y|.$$

This reveals the periodic structure of the level curves in the $y$ direction and allows us to obtain rough sketches by hand. Figure 10.13 shows computer-drawn sketches of these families of level curves. *Note*: Because the drawing uses different scales on the $x$- and $y$-axes, the orthogonality of the curves at each point of intersection is distorted.

An equally important fact, which follows from the Cauchy–Riemann equations (10.41)–(10.42), is that if $f = f^1 + f^2 i$ is an analytic function on an open

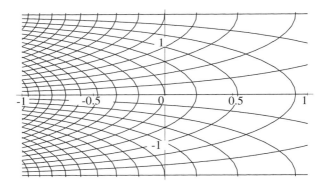

Figure 10.13.   The level curves for the real and imaginary parts $f^1$, $f^2$ of the analytic function $f(z) = e^{-z}$. Note: the use of different scales on the axes distorts the fact that these curves intersect orthogonally.

set $R$, then its real and imaginary parts are harmonic functions,

$$f^1_{xx} + f^1_{yy} = 0,$$

$$f^2_{xx} + f^2_{yy} = 0,$$

on $R$ (exercise).

This completes the brief discussion of what we need from complex analysis. We now return to fluid mechanics and note immediately that equations (10.39)–(10.40) are precisely the Cauchy–Riemann equations for the complex function

$$\bar{v} = (v^1, -v^2) = v^1 - v^2 i.$$

Here, the bar $^-$ denotes complex conjugation. Since $\bar{\bar{v}} = v$, we see that $v$ satisfies equations (10.39)–(10.40); i.e., $\text{div}(v) = 0$ and $\text{curl}(v) = 0$ if and only if $v$ is the conjugate of an analytic function. This then gives us a method of generating fluid flows from analytic functions:

**Proposition 10.4**   *Suppose $f = f^1 + f^2 i$ is an analytic function on the open set $R \subseteq \mathbb{C}$. Define a velocity and pressure function on $R$ by*

$$v = \bar{f} = (f^1, -f^2), \tag{10.45}$$

$$p = -\rho \frac{|f|^2}{2} - \rho \Psi + k, \tag{10.46}$$

*where $\rho > 0$ and $k$ are constants, and $\Psi$ is a potential for the external force density. Then $(\rho, v, p)$ is a solution of the Euler equations whose flow is steady, planar, and irrotational.*

*Furthermore, suppose $g = g^1 + g^2 i$ is an analytic function on $R$, and take $f = g'$ in the above prescription. Then, since*

$$\bar{g}' = \nabla g^1, \tag{10.47}$$

*the prescription (10.45)–(10.46) for $v$ and $p$ becomes*

$$v = \nabla g^1, \tag{10.48}$$

$$p = -\rho \frac{|\nabla g^1|^2}{2} - \rho \Psi + k. \tag{10.49}$$

*Then this $(\rho, v, p)$ is a solution of the Euler equations whose flow is steady, planar, and potential. Moreover, the streamlines of the flow are the level curves*

$$g^2(x, y) = c_2$$

*of the imaginary part of g.*

**Proof**   The first assertion about constructing solutions of the Euler equations from $f$ should be clear from the previous discussion. The construction involving $g$ will be more useful, as we shall see. The assertions about $g$ need to be checked. First, to see that the identity (10.47) holds, use the part $-g_x^2 = g_y^1$ of the Cauchy–Riemann equations for $g$ to get

$$\bar{g}' = \overline{g_x^1 + g_x^2 i} = g_x^1 - g_x^2 i$$
$$= g_x^1 + g_y^1 i = \nabla g^1.$$

You should also note that since $g$ is analytic, so is $g'$, and so we can use the first construction with $f = g'$. Finally, since $v \equiv \nabla g^1$ is tangent to each streamline, and $\nabla g^2$ is perpendicular to the level curves of $g^2$ as well as to $v$, it follows that the level curves of $g^2$ coincide with the streamlines of the flow.

**Remark**   The second prescription in the proposition is more special because a given analytic function $f$ on $R$ may not have an antiderivative on $R$. For simply connected regions $R$, this is the case; i.e., there exists an analytic function $g$ on $R$ such that $f = g'$. In this case both prescriptions in the proposition are the same. The prescription in terms of $g$ is *always* preferred because (1) the flow is potential and (2) the streamlines of the flow are the level curves of the imaginary part $g^2$.

**Definition 10.4**   Suppose $v$ is a planar vector field on $\Omega$ and $g = g^1 + g^2 i$ is an analytic function on $\Omega$ such that

$$v = \overline{g'} = \nabla g^1$$

on $\Omega$. Then $g$ is called a *complex potential* for $v$, while $g^1$ is called a *real potential*. The function $g^2$ is called the *stream function*.

Note that since $g^1$ is the real part of the analytic function $g$, it is harmonic

$$g^1_{xx} + g^1_{yy} = 0,$$

on $R$, and so flows arising from the $g$-prescription are special cases of the flows discussed in the last section. Note also that by the same reasoning, the imaginary part $g^2$ is also harmonic, and we can interchange the roles of $g^1$ and $g^2$ in generating potential flows. Specifically, the the velocity vector field

$$v = \nabla g^2$$

generates a steady, planar, potential fluid flow whose streamlines are the level curves

$$g^1(x, y) = c_1$$

of the real part of $g$.

**Example 10.4**   We return to the last example but change the notation to fit the above discussion. So we consider $g(z) = e^{-z}$, whose real and imaginary parts are $g^1(x, y) = e^{-x} \cos y$ and $g^2(x, y) = -e^{-x} \sin y$, which have level curves as shown in Figure 10.13. By the above discussion, we can use either of the vector fields

$$v(x, y) = \nabla g^1(x, y) = (-e^{-x} \cos y, -e^{-x} \sin y),$$

$$v(x, y) = \nabla g^2(x, y) = (e^{-x} \sin y, -e^{-x} \cos y)$$

as a model for a planar flow. By restricting the region under consideration and using a pair of streamlines to form the boundary walls, we get two models for the flow, as shown in Figure 10.14.

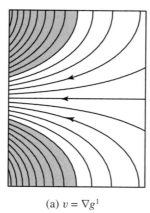

(a) $v = \nabla g^1$

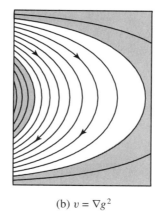

(b) $v = \nabla g^2$

Figure 10.14.    Flows arising from $g(z) = e^{-z}$.

## Exercises 10.3.2

1. Suppose $f = f^1 + f^2 i$ is an analytic function on $R \subseteq \mathbb{C}$. Show that

   (a) The (complex) derivative $f'$ of $f$ is given by

   $$f' = f_x^1 + f_x^2 i = \frac{1}{i}(f_y^1 + f_y^2 i).$$

   (*Hint:* In the limit, $\lim_{h \to 0}(f(z + h) - f(z))/h$, let $h = \varepsilon$ and $h = \varepsilon i$ with $\varepsilon$ real and tending to zero.

   (b) Use part (a) to show that if $f$ is analytic, then its real and imaginary parts satisfy the Cauchy–Riemann equations and furthermore, that each of them is a harmonic function.

2. As in Example 10.4, construct models for planar flows from the real and imaginary parts of the following complex functions. Use a computer to draw the orthogonal trajectories and use this to mark the boundary walls and flow lines for the two planar flows arising from each complex function.

   (a) $g(z) = \sinh z$. This is the complex version of the hyperbolic sine (defined as you would suspect). Comment on how $g^1$ and $g^2$ are related to Dirichlet BVPs that we have previous considered.

   (b) $g(z) = e^{-z^2}$.

3. (*A mathematical fiction*) Many books on fluid mechanics consider the function $g(z) = \ln z$ as *the* antiderivative of the function $f(z) = 1/z$. We will also, for convenience,

use this fictitious assertion here, but the student should understand the extent to which it is true or false. The validity of $g' = f$ depends on the domains of definitions for these functions. Thus $f(z) = 1/z$ is defined on $\mathbb{C} - \{0\}$ (and any subset thereof), while $g(z) = \ln z$ is a multiple-valued function whose principal branch is defined by

$$\ln z = \ln |z| + \arg(z)i,$$

for $z \in R = \mathbb{C} - \{\text{the positive } x\text{-axis}\}$. Now, it *is* true that $g' = f$ on $R$, but extending beyond this is not possible, as part (a) below shows.

(a) Show that there does *not* exist an analytic function $g$, defined on $D = \mathbb{C} - \{0\}$, such that $g' = f$ on $D$. *Hint:* To do this, show that if this were possible, then the complex path integral

$$\int_\gamma f(x)\,dz \equiv \int_a^b f(\gamma(t))\gamma'(t)dt$$

would be zero for every curve $\gamma : [a, b] \to \mathbb{R}^2$ that lies in $D$. To get a contradiction, choose a particular $\gamma$ for which the path integral is not zero.

(b) *Source/sink:* For a nonzero real number $m$, the analytic function

$$f(z) = \frac{m}{z}$$

gives a flow

$$v = \bar{f}$$

on $\mathbb{C} - \{0\}$ that is called a *source/sink* of strength $|m|$ at the origin. Calculate $v$ explicitly and draw the flow lines for $v$. This flow is planar and irrotational but, by part (a), is *not* a potential flow. Many books list this flow as arising from the complex potential $g(z) = m \ln z$, and while this is *not* true, it will be convenient in the sequel to maintain this mathematical fiction. Since only the derivative of $g$ enters into the calculation of the corresponding velocity vector field $v = \bar{g}'$, this will not lead to any problems.

**Basic Planar Flows:**

The following problems deal with flows

$$v = \bar{g}'$$

generated by the $g$-prescription, even though, as discussed in the last problem, the flows may not be potential. In each problem, make sure you use a computer software package that will allow you to view the plot of flow dynamically, in real time, with the various flow lines being plotted at their respective speeds of flow (see the CD-ROM for how to do this in *Maple*). See [Kir 85] and [Rou 65] for the static pictures.

4. *Uniform flow:* A uniform flow with speed $R$ and making an angle $\alpha$ with the $x$-axis is generated by

$$g(z) = U e^{-\alpha i} z.$$

This is a potential flow. Calculate $v$ explicitly and verify that it makes an angle $\alpha$ with the $x$-axis. The particular case $g(z) = Uz$, which is a uniform flow parallel to the $x$-axis, is the one we will use most often.

5. *Flow in a corner:* The function

$$g(z) = z^n$$

generates a flow $v$ that when restricted to an appropriate sector of the plane (bounded by rays to which $v$ is tangential), models potential flow in a corner (the sector).

(a) For $n = 5, 6$ calculate $v$ explicity and determine a sector in the first quadrant for which $v$ is tangential to the boundary. Use a computer to draw the flow lines in this sector. Compare the speeds along the respective flow lines.

(b) Generalize the first part of part (a) to an arbitrary $n$.

6. *Sources/sinks:* The function

$$g(z) = m \ln(z - w),$$

generates a flow $v$ called a source/sink at the point $w = a + bi$ with strength $|m|$. See Exercise 3 above. Such a basic flow is rather simple, but if we combine several sources/sinks and add in a uniform flow, the resulting flow can be quite useful.

(a) Investigate the direction field and flow pattern for a combination of two sources/sinks

$$g(z) = \ln(z + 1) + m \ln(z - 1)$$

located at $(-1, 0)$, $(1, 0)$ and with $m = 1, 0.1, -0.1$ respectively.

(b) *Flow past a half-body:* For the flow $v$ generated by

$$g(z) = Uz + \ln z,$$

with $U = 1$ and $U = 10$, draw the direction field and flow patterns. Notice how the uniform flow sweeps past the source flow at $(0, 0)$ to produce the combined flow. Find the stagnation point.

(c) *Rankine oval:* Let $v$ be the vector field generated by

$$g(z) = Uz + m \ln \frac{z+b}{z-b},$$

with $U, m, b$ positive real numbers. The flow for $v$ is a combination of a uniform flow past a source, and a sink at $(-b, 0)$ and $(b, 0)$ respectively. For this flow, there is a closed curve, called the *Rankine Oval*, that passes through the two stagnation points of $v$. For $U = 1, m = 1, b = 1$, draw the direction field for $v$ and flow lines exterior to the Rankine oval. Find the stagnation points. Also investigate how increasing the speed $U$ changes the size and shape of the oval as well as the location of the stagnation points relative to the source and sink.

(d) As a generalization of parts (b) and (c), investigate the flow for the vector field $v$ generated by

$$g(z) = Uz + m \ln(z+b) - n \ln(z-w) - n \ln(z-\bar{w}),$$

where $w = b + \varepsilon i$.

6. *Vortices:* The vector field generated by

$$g(z) = ik \ln(z-w)$$

models a flow called a *vortex* centered at $w = a + bi$ and strength $k$ (which is a real number). This is an irrotational flow, but it is *not* potential.

(a) Draw the direction field and flow for a vortex centered at $w = 0$ and strength $k = 1$. What is the difference if $k = -1$?

(b) Consider the flow generated by a combination of two vortices

$$g(z) = ik_1 \ln(z-w_1) + ik_2 \ln(z-w_2),$$

where, say, $w_1 = (0, 0)$ and $w_2 = (1, 0)$. Plot the direction fields and flow patterns for the cases (i) $k_1 = 1 = k_2$; (ii) $k_1 = 1, k_2 = 0.1$; and (iii) $k_1 = 1, k_2 = -0.1$.

(c) *Vortex near a wall:* The model for the flow of a vortex near a wall is generated by

$$g(z) = ik \ln \frac{z+b}{z-b},$$

where $k$ and $b$ are real numbers. Explain why this is so, and draw the direction field and flow for the model.

(d) Plot the direction field and flow for the three-vortex model

$$g(z) = i \ln z + 0.2i \ln(z-1) - 0.4i \ln(z - 0.5 - i).$$

7. *Doublet:* The potential flow $v$ with complex potential

$$g(z) = \frac{m}{z},$$

where $m$ is real, is called a *doublet* because of its flow pattern. The following parts deal with this.

(a) *Flow past a cylinder:* Combining the potential $Uz$ for a uniform flow with that of the doublet and relabeling constants gives a complex potential

$$g(z) = U\left(z + \frac{c}{z}\right)$$

for a potential flow that models flow past a cylinder (the cylinder is perpendicular to the $x$-$y$ plane and intersects this plane in a circle. Show that the vector field $v$ with this complex potential is the same as that in Example 10.2 (with $b = Uc$).

(b) *Curve ball:* Consider the vector field $v$ generated by

$$g(z) = U\left(z + \frac{c}{z}\right) + ik\ln z,$$

which adds a vortex flow to that in part (a). For this $v$, (i) determine the stagnation points and a boundary circle to which $v$ is everywhere tangential; (ii) draw the direction fields and flows for the cases $U = 1, c = 1, k = 1$ and $U = 10, c = 1, k = 1$.

## 10.4. Special Solutions of the Navier–Stokes Equations

The Euler equations that we have been studying in the previous sections illustrate many features of fluid flow and also serve as a good example of a system of PDEs. However, the Euler equations can only model flows whose internal stresses are due simply to fluid pressure, and many effects of real fluids arise from stresses due to viscosity. Historically, Euler's equations for a perfect fluid were studied extensively before the modification of these equations to account for viscosity took place.

To study the effect of viscosity, we must use the Navier–Stokes equations

$$\frac{\partial \rho}{\partial t} + \nabla_v \rho = 0, \tag{10.50}$$

$$\rho\left[\frac{\partial v}{\partial t} + \nabla_v v\right] = \rho F - \nabla p + \mu \nabla^2 v, \tag{10.51}$$

$$\operatorname{div}(v) = 0. \tag{10.52}$$

Mathematically, these equations contain the Euler equations as a special case ($\mu = 0$), but for $\mu > 0$, the presence of the second-order derivatives in the term $\nabla^2 v$ in the linear momentum equation models the effect of viscosity in the fluid. From a physical point of view, the appropriate boundary conditions on $v$ are the

**No Slip BC:**   $v = 0$ on $\partial\Omega$

(viscosity precludes the fluid from sliding along the boundary), and from a mathematical point of view, the solvability of the equations requires this too. More generally, when the boundary is in motion, the appropriate BC is that $v$ be tangential to the boundary and coincide with the prescribed velocity of the boundary.

In this section we will look at special types of solutions to the Navier–Stokes equations. To simplify the discussion, the first two assumptions we need are:

(1) $\rho = $ constant,

(2) $F = -\nabla\Psi$.

Because of (1), the continuity equation is automatically satisfied, and with (2), we can combine the pressure function $p$ and the external force potential $\Psi$ to simplify the linear momentum equation. Thus we introduce a function known as

**The Piezometric Head:**   $H = \Psi + p/\rho$.

Using this, we can rewrite the equations of motion as follows:

$$\frac{\partial v}{\partial t} + \nabla_v v = -\nabla H + \alpha\nabla^2 v,$$

$$\operatorname{div}(v) = 0.$$

Here we have introduced the constant $\alpha = \mu/\rho$, which is called the *kinematical viscosity constant* and is usually denoted by $\nu$.

Even with these assumptions, the Navier–Stokes equations do not have simple analytic solutions. To simplify further, we specialize the way the velocity vector $v$ depends on the spatial variables:

**Assumption:**   Throughout this section we assume that $v$ has the form

$$v = (v^1(y, z, t), v^2(x, z, t), v^3(x, y, t)). \tag{10.53}$$

This assumption guarantees that $v$ automatically satisfies the incompressibility condition $\mathrm{div}(v) = 0$. Thus we need only consider the linear momentum equation, which reduces to the following three scalar equations:

$$v_t^1 + v^2 v_y^1 + v^3 v_z^1 = -H_x + \alpha(v_{yy}^1 + v_{zz}^1), \tag{10.54}$$

$$v_t^2 + v^1 v_x^2 + v^3 v_z^2 = -H_y + \alpha(v_{xx}^2 + v_{zz}^2), \tag{10.55}$$

$$v_t^3 + v^1 v_x^3 + v^2 v_y^3 = -H_z + \alpha(v_{xx}^3 + v_{yy}^3). \tag{10.56}$$

In addition to the time derivatives $v_t^i$ on the left sides of these equations, there are the terms that come from $\nabla_v v$. These are called the *convective terms*, since they arise from the transport of the fluid by the flow (see the transport theorem in Appendix B). The convective terms are what make the Navier–Stokes equations *nonlinear*. Some of the initial special solutions we now look at will be such that these terms vanish identically: $\nabla_v v = 0$.

**Example 10.5   (Viscous Flow in a Conduit)**   In this example we consider flows that are of the above form and parallel to the $x$-axis: $v^2 = 0$, $v^3 = 0$. Thus $v$ has the form

$$v = (v^1(y, z, t), 0, 0),$$

and the Navier–Stokes equations reduce to

$$v_t^1 = -H_x + \alpha(v_{yy}^1 + v_{zz}^1),$$

$$0 = -H_y,$$

$$0 = -H_z.$$

First, observe that the latter two equations tell us that $H$ is a function of $x$ and $t$ only. With this understood, examination of the first equation reveals that all the terms except $-H_x$ depend only on $y$, $z$, and $t$. Thus, $H_x$ must be a function of $t$ only, say $-H_x = a(t)$. Consequently, we find that

$$H = -a(t)x + b(t),$$

for two unknown functions $a, b$. The pressure $p$ is then given (up to these unknown functions) by

$$p(x, y, z, t) = -\rho \Psi(x, y, z, t) - \rho a(t)x + \rho b(t).$$

With $a(t)$ given, we see that the PDE for $v^1$

$$v_t^1 = a(t) + \alpha(v_{yy}^1 + v_{zz}^1), \qquad (10.57)$$

is just a two-dimensional non-homogeneous heat equation, and this we know how to solve, given appropriate BCs and IC. To be more specific, assume that the fluid flows in a region of the form $\Omega = \mathbb{R} \times D$, where $D \subseteq \mathbb{R}^2$ is a region in the $y$-$z$ plane (see Figure 10.15). The region $\Omega$ is viewed as a conduit, or pipe, with $D$ representing a typical cross section through the pipe. At time zero, an initial velocity in the $x$ direction is prescribed, say $v(y, z, 0) = (g(y, z), 0, 0)$, for $(y, z) \in D$. For all later times $t > 0$, the velocity is required either to be zero on the boundary of the conduit or, if part of the boundary is in motion, to have a prescribed velocity matching that of the boundary, say $v(y, z, t) = (f(y, z, t), 0, 0)$, for $(y, z) \in \partial D$ and $t > 0$. Thus a solution $v^1$ of the heat equation (10.57) must be chosen to satisfy

$$v^1 = f \qquad \text{on } \partial D, \text{ for } t > 0, \qquad (10.58)$$

$$v^1 = g \qquad \text{in } D, \text{ for } t = 0. \qquad (10.59)$$

Then for any constant $\rho$ and arbitrary functions $a(t), b(t)$, we get a solution $(\rho, v, p)$ of the Navier–Stokes equations in $\Omega$ that satisfies the BCs and IC. We can use our understanding of heat flow to interpret the corresponding fluid flow. Thus, for example, requiring $v^1$ to be zero on $\partial D$ means holding the temperature at zero there, and if $a(t) = $ a constant in the heat equation (10.57) for $v^1$, then we know that regardless of the initial temperature distribution,

$$\lim_{t \to \infty} v^1(y, z, t) = S(y, z),$$

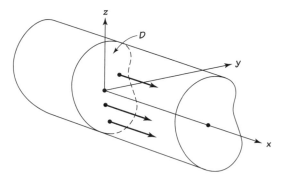

Figure 10.15.    Viscous flow in the $x$-direction in a conduit $\Omega = \mathbb{R} \times D$.

where $S$ is a steady-state temperature distribution. Hence, in terms of fluid flow, if the boundary is stationary and the gradient of the piezometric head is constant, then the eventual velocity in the conduit will be $(S(y, z), 0, 0)$. The most famous example of this is

**Poiseuille Flow:**    Suppose the typical cross section $D$ of the pipe is a disk of radius $r_0$ with center at the origin. We take $a = $ constant, require $v^1 = 0$ on the boundary $\partial D$ of the disk, and assume that the initial velocity is zero throughout the pipe. The heat equation for $v^1$ is more naturally solved by transforming to polar coordinates. Thus define $w$ by

$$w(r, \theta) = v^1(r \cos \theta, r \sin \theta).$$

Then $w$ must satisfy the heat equation in polar coordinates (see Chapter 12)

$$w_t = a + \alpha \left( w_{rr} + \frac{1}{r} w_r + \frac{1}{r^2} w_{\theta\theta} \right).$$

The solution of this for the disk is actually rather complicated in general, but we can simplify it considerably if we look for a solution $w$ that does not depend on $\theta$. This seems reasonable, since the BC and IC do not depend on $\theta$, and in general this will always work in such cases. The PDE for $w$ is non-homogeneous, and so we separate out the steady-state solution $K$ explicitly:

$$w(r, t) = K(r) + u(r, t),$$

requiring that $K$ satisfy

$$0 = a + \alpha \left( K'' + \frac{1}{r} K' \right).$$

This differential equation can be written as

$$(r K')' = \frac{-a}{\alpha} r,$$

which upon integrating once yields

$$r K' = \frac{-a}{2\alpha} r^2 + c,$$

and one further integration gives

$$K(r) = \frac{-a}{4\alpha}r^2 + c \ln r + d.$$

Now, $K$ must be defined for $r = 0$, so we must take $c = 0$. Then $d$ is determined by the boundary condition $K(r_0) = 0$. Thus we find that the steady-state solution is

$$K(r) = \frac{a}{4\alpha}(r_0^2 - r^2).$$

The other part $u$ of the total solution must satisfy the *homogeneous* heat equation

$$u_t = \alpha \left( u_{rr} + \frac{1}{r}u_r \right)$$

and homogeneous BCs. The solution is given by a series involving the Bessel function $J_0$ (see Chapter 12):

$$u(r, t) = \sum_{n=1}^{\infty} A_{0n} e^{-\alpha \lambda_{0n}^2 t} J_0 \left( \frac{\lambda_{0n} r}{r_0} \right).$$

The $\lambda$'s are the zeros of $J_0$, and the $A$'s are the Fourier–Bessel coefficients of $-K$ (because we need $0 = w(r, 0) = K(r) + u(r, 0)$).

This gives the solution of the heat problem, and the fluid flow is determined by this: $v = (v^1, 0, 0)$ and $v^1(r \cos\theta, r \sin\theta, t) = K(r) + u(r, t)$. The fluid flow approaches the steady-state flow determined by $K$. The steady-state flow is known as *Poiseuille Flow*. Figure 10.16 gives two views of the velocity distribution on a typical cross section of the pipe. The speed of the flow has

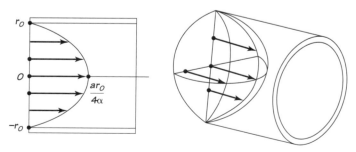

Figure 10.16.   Two views of the distribution of velocity for a Poiseuille flow at each cross section of a pipe.

a maximum of $ar_0/4\alpha$ at the center of the pipe. Figure 10.17 shows the time evolution of the flow from being initially zero throughout the pipe until the approximate time at which it reaches the Poiseuille flow distribution.

**Example 10.6   (Flow Between Two Parallel Plates)**   In this example we consider flows of the form

$$v = (v^1(y, z, t), v^2(x, z, t), 0),$$

with neither of $v^1$, $v^2$ being zero. For such a velocity vector field, equations (10.54)–(10.56) reduce to

$$v^1_t + v^2 v^1_y = -H_x + \alpha(v^1_{yy} + v^1_{zz}), \qquad (10.60)$$

$$v^2_t + v^1 v^2_x = -H_y + \alpha(v^2_{xx} + v^2_{zz}), \qquad (10.61)$$

$$0 = -H_z. \qquad (10.62)$$

Such flows are parallel to the $x$-$y$ plane, and because of the way $v^1$, $v^2$ depend on $x$, $y$, $z$, the only way $v$ can match the velocity of the boundary is for there to be *no* boundaries in the $x$ or $y$ directions; i.e., the region for the flow must be of the form

$$\Omega = \mathbb{R}^2 \times [0, h].$$

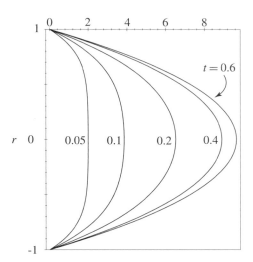

Figure 10.17.   Time evolution of the flow in a pipe from initally zero velocity at each section until it reaches the Poiseuille velocity distribution.

This is the region between two parallel plates of infinite extent and separated by a distance $h$. The form of equations (10.60)–(10.62) dictates that the piezometric head necessarily is of the form

$$H = -B(t)xy - a(t)x - b(t)y + d(t)$$

(see the exercises). The first two equations, (10.60)–(10.61), are nonlinear equations and are coupled to each other through the convective terms. Each equation can be interpreted as a 2-D heat equation with convection. (See problem 3 in Exercises B.1). Rather than discuss this general case, we specialize again.

**Further Assumption:**

$$v = (v^1(y, z, t), v^2(z, t), 0).$$

The equations for $v^1$ and $v^2$ then become

$$v_t^1 + v^2 v_y^1 = a(t) + \alpha(v_{yy}^1 + v_{zz}^1), \tag{10.63}$$

$$v_t^2 = b(t) + \alpha v_{zz}^2. \tag{10.64}$$

The second equation here is a nonhomogenous heat equation for $v^2$, which can be determined analytically once the boundary conditions from the fluid flow problem are specified. With $v^2$ known, equation (10.63) becomes a linear PDE: the heat equation with convection for $v^1$.

**Exercises 10.4**

1. Show that the piezometric head that occurs in equations (10.54)–(10.56) must necessarily have the form

$$H = -Axyz - Bxy - Cxz - Dyz - ax - by - cz + d,$$

where $A, B, C, D, a, b, c, d$ are functions of $t$ only.

2. Consider viscous fluid flow through a conduit, or pipe, with rectangular cross section $\Omega = \mathbb{R} \times D$, where $D = [0, c] \times [0, k]$. Assume (1) the piezometric head has constant gradient $\nabla H = (-a, 0, 0)$, where $a$ is a positive constant; (2) $v = 0$ on the boundary of the pipe; and (3) initially, at time zero, $v = (g(y, z), 0, 0)$.

   (a) Find a solution of the Navier–Stokes equations that has the form

$$v = (v^1(y, z, t), 0, 0)$$

and satisfies the stated conditions. Do this by the standard series method for solving the corresponding non-homogeneous heat equation (10.57). Write down the series solution and then separate into two series: one for the steady-state and the other for the transient.

(b) Let $c = 2, k = 1, \alpha = 1, a = 1$, and for the two choices of initial velocity distributions (1) $g(y, z) = 0$ and (2) $g(y, z) = \sin \pi y \sin \pi z$, do the following:

(i) Graph the steady-state velocity distribution.

(ii) Graph the evolution of the velocity distribution from $t = 0$ until the approximate time at which it reaches the steady-state. What is this approximate time?

3. Rework Exercise 2, but now assume that the gradient of the piezometric head is $\nabla H = (-a(t), 0, 0)$, with $a(t) = 1 - \cos(\omega t)$. Do only one initial condition, $g(y, z) = 0$, and take $\omega = 1$ in the graphical part. Interpret your answer.

4. Consider the fluid flow in the region between two parallel planes (infinite in extent), $\Omega = \mathbb{R}^2 \times [0, h]$. Assume (1) the piezometric head has constant gradient $= (-a, 0, 0)$; (2) $v = 0$ on the boundary of $\Omega$; and (3) $v = 0$ initially throughout $\Omega$.

(a) Find solution of the Navier–Stokes equations that has the form

$$v = (v^1(z, t), 0, 0)$$

and satisfies the stated conditions. *Note:* This is a special case of that discussed in Example 10.5, except with a different geometry for $\Omega$. Do this problem by first separating $v^1$ into steady and transient parts,

$$v^1(z, t) = S(z) + u(z, t),$$

requiring that $S$ satisfy the inhomgeneous PDE+BCs and $u$ satisfy the homogeneous PDE + BCs. You can express $S$ in closed form (without a series). The fluid flow corresponding to $S$ is also called *Poiseuille flow*.

(b) For $h = 1, \alpha = 1, a = 1$, do the following:

(i) Graph the steady-state velocity distribution.

(ii) Graph the evolution of the velocity distribution from $t = 0$ until the approximate time at which it reaches the steady-state. What is this approximate time?

5. Rework Execise 4, with everything the same, except now let the top boundary plane be in motion with constant speed $U$. Specifically, the BCs are $v = (U, 0, 0)$ for $z = h$, and $v = 0$ for $z = 0$. In the graphical parts you can take $U = 1$.

6. Consider a fluid flow in a conduit consisting of the space between two concentric cylinders of radii $r_1 < r_2$. Thus $\Omega = \mathbb{R} \times D$, where the typical cross section $D$ is an

annulus. Use the polar coordinate version of the heat equation to find the steady-state fluid flow in such a conduit that satisfies $v = 0$ on $\partial\Omega$. Graph the velocity profile for this flow.

7. As a special case of the flow discussed in Example 10.6, consider a fluid between two parallel plates: $\Omega = \mathbb{R}^2 \times [0, h]$, and assume that $v$ has the special form

$$v = (v^1(z, t), v^2(z, t), 0).$$

In this case, equations (10.63)–(10.64) simplify even further to the following pair of uncoupled, non-homogeneous heat equations for $v^1$ and $v^2$:

$$v_t^1 = a(t) + \alpha v_{zz}^1, \tag{10.65}$$

$$v_t^2 = b(t) + \alpha v_{zz}^2. \tag{10.66}$$

Suppose the top plate is fixed, while the bottom plate moves horizontally with velocity $(f^1(t), f^2(t), 0)$. Thus the boundary conditions are

$$v^1(h, t) = 0, \quad v^2(h, t) = 0,$$

$$v^1(0, t) = f^1(t), \quad v^2(0, t) = f^2(t).$$

Assume that the fluid is initially at rest: $v = 0$ throughout $\Omega$. In this exercise we explore how the motion of the bottom plate is imparted to the fluid. Since the IBVPs for $v^1$, $v^2$ have the same form, we concentrate on the one for $v^1$.

(a) The IBVP for $v^1$ has non-homogeneous BCs. Transform to one with homogeneous BCs by letting

$$v^1 = \frac{f^1(t)}{h}(h - z) + u^1,$$

where $u^1 = u^1(z, t)$ is chosen to satisfy the homogeneous BCs $u^1(0, t) = 0, u^1(h, t) = 0$.

(b) Suppose $f^1(t) = \sin \omega t$ and $a(t) =$ a constant. Calculate the series solution for $v^1$. Separate the series into two series, one for the transient part and one for the nontransient part.

(c) Suppose $f^2(t) = 0$, so that the bottom plate oscillates back and forth periodically in the $x$-direction only. Take $b = 0$, so that $v^2 = 0$, and the motion takes place in the $x$-direction only: $v = (v^1(z, t), 0, 0)$, with $v^1$ the function you found in part (b). For various fixed values of $t$, plot $v^1(z, t), 0 \le z \le h$. This will give pictures of the velocity distribution between the plates at the chosen times. Also, you might want to animate these plots as a function of time to get a feeling for the way the oscillating boundary plate establishes a simliar oscillation in the fluid. For the graphs, use values $h = 1, b = 1, \omega = 1$.

8. Rework Exercise 7, but now assume that *both* the top and bottom plates oscillate periodically in the $x$-direction with velocities $\sin 2t$ and $\sin t$ respectively.

## 10.5.   Solution of the Navier–Stokes Equations

The special solutions to the Navier–Stokes equations presented in the last section give some indication of the nature of these equations, but those were indeed very limited cases with special geometry and mostly without the convective term $\nabla_v v$ that makes the equations nonlinear.

To exhibit analytic representations of solutions of the Navier–Stokes equations is, with the exception of special solutions (like those in the last section), not possible. This situation, which is typical of most PDEs of interest, necessitates an examination of the theory about existence of solutions to the given IBVP, their uniqueness, and numerical methods for approximating these solutions. In this section, we give a very elementary (and rudimentary) description of this theory for the Navier–Stokes equations, written in the form

$$\frac{\partial v}{\partial t} + \nabla_v v - \alpha \nabla^2 v - F = \nabla P \qquad \text{in } \Omega,$$

$$\text{div}(v) = 0 \qquad \text{in } \Omega,$$

$$v = 0 \qquad \text{on } \partial\Omega.$$

We are assuming that the mass density $\rho$ is constant, and for convenience, we are letting $P = -p/\rho$. The region $\Omega$ is assumed to be bounded.

A precise discussion is found in Temam [Te 77] and requires a good background in functional analysis and some understanding of differential geometry. We can not assume this here, so the following discussion is intended to give just the essence of the theory.

### 10.5.1.   Preliminary Theorems

To discuss the theory, we need a couple of results that are useful in their own right.

**Theorem 10.1 (Helmholtz Representation)**   *Suppose $E$ is a vector field on $\Omega$. Then $E$ can be represented as*

$$E = u + \nabla P,$$

*where u is a divergence-free vector field* $(\mathrm{div}(u) = 0$ *in* $\Omega)$ *that is tangential to the boundary* $(u \cdot n = 0$ *on* $\partial\Omega)$, *and P is a scalar field on* $\Omega$. *Furthermore, this representation is unique in the sense that if also* $E = \bar{u} + \nabla\bar{P}$, *with* $\bar{u}$ *divergence-free and tangential to the boundary, then* $\bar{u} = u$ *and* $\bar{P} = P + c$ *for some constant c.*

**Proof**   The proof relies on the existence and uniqueness aspects of the Neumann problem (see Chapter 13). Thus, we choose the scalar field $P$ to be a solution of the BVP

$$\nabla^2 P = \mathrm{div}(E) \qquad \text{in } \Omega, \tag{10.67}$$

$$\nabla P \cdot n = E \cdot n \qquad \text{on } \partial\Omega. \tag{10.68}$$

There is not a unique solution to this problem, but any two solutions differ by a constant. Having chosen one such solution $P$, we then let $u$ be the vector field *defined* by

$$u = E - \nabla P.$$

Then clearly, $\mathrm{div}(u) = 0$ in $\Omega$ and $u \cdot n = 0$ on $\partial\Omega$, and so we obtain the desired Helmholtz representation of $E$. Next suppose that we can also write $E$ as $E = \bar{u} + \nabla\bar{P}$, where $\bar{u}$ is divergence-free and tangential to the boundary. Then

$$\nabla^2 \bar{P} = \mathrm{div}(\nabla\bar{P})$$

$$= \mathrm{div}(E - \bar{u})$$

$$= \mathrm{div}(E).$$

Furthermore,

$$\nabla\bar{P} \cdot n = (E - \bar{u}) \cdot n = E \cdot n$$

on $\partial\Omega$. Hence $\bar{P}$ also satisfies the Neumann BVP (10.67)–(10.68). Consequently, $P$ and $\bar{P}$ differ by a constant, say $\bar{P} = P + c$. Then since the gradients of $P$, $\bar{P}$ are the same, it follows that $\bar{u} = E - \nabla\bar{P} = E - \nabla P = u$.

Helmholtz's representation is often stated as saying that a vector field $E$ on $\Omega$ can be written in the form

$$E = \mathrm{curl}(w) + \nabla P,$$

for some vector field $w$ with $\operatorname{curl}(w) \cdot n = 0$ on $\partial\Omega$ and some scalar field $P$. This is *not* necessarily true. It depends on the region $\Omega$. As in the theorem, we can always write $E = u + \nabla P$, with $\operatorname{div}(u) = 0$, and if $\Omega$ is simply connected, then there is a vector field $w$ such that $\operatorname{curl}(w) = u$ on $\Omega$. Thus we obtain the above representation. But this is not possible in general.

**Lemma 10.2**  *The identity*

$$\int_{\Omega} \nabla f \cdot w \, dV = \int_{\partial\Omega} f \, w \cdot n \, dA$$

*holds for every scalar field $f$ and every vector field $w$ with $\operatorname{div}(w) = 0$.*

The proof of this is left as an exercise.

**Theorem 10.2 (de Rham)**  *Suppose $E$ is a vector field on $\Omega$ such that*

$$\int_{\Omega} E \cdot w \, dV = 0, \tag{10.69}$$

*for every vector field $w$ with (1) $\operatorname{div}(w) = 0$ in $\Omega$ and (2) $w = 0$ on $\partial\Omega$. Then there exists a scalar field $P$ such that*

$$E = \nabla P$$

*on $\Omega$.*

   *Conversely, if $E = \nabla P$ for some $P$, then equation (10.69) holds for every $w$ that is divergence-free and zero on the boundary of $\Omega$.*

**Proof**  The first assertion is the difficult one to prove (see [Rh 60], [Te 77]) and requires more depth than is expected in this book. However, for pedagogical purposes, we prove a weaker version of that assertion. Namely, assume that $E$ is such that equation (10.69) holds for every divergence-free vector field $w$ (not just the ones that are also 0 on the boundary). Then the existence of $P$ is proved as follows. Note that by the Helmholtz representation theorem, we can write

$$E = u + \nabla P,$$

where $u$ is divergence-free and tangential to the boundary. We wish to show that $u = 0$. To show this, observe that

$$\int_\Omega |u|^2 dV = \int_\Omega u \cdot u \, dV$$

$$= \int_\Omega u \cdot (E - \nabla P) dV$$

$$= -\int_\Omega u \cdot \nabla P \, dV$$

$$= -\int_{\partial\Omega} P \, u \cdot n \, dA$$

$$= 0.$$

The next-to-last equation comes from the lemma above. The last equation comes from the fact that $u$ is tangential to $\partial\Omega$.

The other assertion of the theorem is left as an exercise.

## 10.5.2.   Weak Solutions

One method for developing the theory of existence of solutions to PDEs involves rewriting the PDE of interest in terms of integrals (often called the variational form of the PDE) and then seeking solutions of the corresponding integral equation. These latter solutions will often not be differentiable functions, and so they are referred to as *weak solutions* of the original PDE. This is the approach we will outline here for the Navier–Stokes equations.

De Rham's theorem is instrumental in simplifying the variational formulation of the Navier–Stokes equations. It essentially allows us to eliminate the unknown function $P$ from the equations and concentrate on finding $v$ by itself. This is more or less what we found in the previous sections in dealing with special solutions of the Euler or the Navier–Stokes equations: The pressure $p$ was determined after finding the velocity $v$. To see how this applies here, we first introduce the following definition:

**Definition 10.5**   For convenience of notation, we introduce the following set of vector fields on $\Omega$:

$$\mathcal{V} = \{w \mid \operatorname{div}(w) = 0 \text{ in } \Omega, \ w = 0 \text{ on } \partial\Omega\}.$$

Now suppose $(v, P)$ is a solution of

$$\frac{\partial v}{\partial t} + \nabla_v v - \alpha \nabla^2 v - F = \nabla P \qquad \text{in } \Omega,$$

$$\text{div}(v) = 0 \qquad \text{in } \Omega,$$

$$v = 0 \qquad \text{on } \partial\Omega.$$

Bear in mind that $v$ is a time-dependent vector field on $\Omega$, and it will be convenient to let $v(t)$, for each time $t$, denote the vector field

$$v(t)(x, y, z) = v(x, y, z, t).$$

Then the last two equations above just say that

$$v(t) \in \mathcal{V},$$

for each $t$. Furthermore, the requirement in the first equation (the linear momentum equation) can be rephrased as follows: By de Rham's theorem, $\int_\Omega \nabla P \cdot w \, dV = 0$ for every $w \in \mathcal{V}$, and so taking the dot product of both sides of the linear momentum equation with such a $w$ and integrating over $\Omega$ gives

$$\int_\Omega \left[ \frac{\partial v}{\partial t} + \nabla_v v - \alpha \nabla^2 v - F \right] \cdot w \, dV = 0. \tag{10.70}$$

Thus $v(t)$ is an element of $\mathcal{V}$ that satisfies this integral equation for any $w \in \mathcal{V}$. Conversely, if we can find a $v$ with $v(t) \in \mathcal{V}$, for each $t$, that also satisfies the integral equation for every $w \in \mathcal{V}$, then by de Rham's theorem, there exists a $P$ such that

$$\frac{\partial v}{\partial t} + \nabla_v v - \alpha \nabla^2 v - F = \nabla P.$$

Then $(v, P)$ will be a solution of the Navier–Stokes equations.

Based on these observations, the problem of solving the Navier–Stokes equations is reduced to solving the integral equation (10.70) for a function $v$ taking its values in the space $\mathcal{V}$. The boundary condition $v = 0$ on $\partial\Omega$ is built into the space $\mathcal{V}$, as well as the incompressibility condition $\text{div}(v) = 0$. However, the initial values of $v$ at time $t = 0$ will need to specified.

To simplify the form of the integral equation (10.70), we introduce the following notation for some of the integral expressions:

**Definition 10.6**   Suppose $u$, $v$, $w$ are vector fields on $\Omega$. Define a function $\langle \cdot, \cdot \rangle$ of two variables $u$, $w$ by:

$$\langle u, w \rangle \equiv \int_\Omega u \cdot w \, dV.$$

In a similar fashion define a function

$$\langle\langle u, w \rangle\rangle \equiv \sum_{i=1}^{3} \langle \nabla u^i, \nabla w^i \rangle$$

$$= \sum_{i=1}^{3} \int_\Omega \nabla u^i \cdot \nabla w^i \, dV.$$

Finally, we define a function of three vector fields by

$$B(u, v, w) \equiv \int_\Omega (\nabla_u v) \cdot w \, dV.$$

**Comment:**   It is important to note that $\langle \cdot, \cdot \rangle$ is a *bilinear function*; i.e., it has the following properties:

For any vector fields $u$, $\bar{u}$, $w$, $\bar{w}$ and constants $a$, $b$:

(1) $\langle au + b\bar{u}, w \rangle = a \langle u, w \rangle + b \langle \bar{u}, w \rangle$.

(2) $\langle u, aw + b\bar{w} \rangle = a \langle u, w \rangle + b \langle u, \bar{w} \rangle$.

This is easy to verify (exercise). Also $\langle\langle \cdot, \cdot \rangle\rangle$ is bilinear. The function $B$ is what is known as a *trilinear function*.

We need one further result to aid in simplifying the integral equation (10.70). It is proof is a simple exercise:

**Lemma 10.3**   *For vector fields $v$, $w$ on $\Omega$, with $w$ vanishing on the boundary, $w = 0$ on $\partial\Omega$, the following identity holds:*

$$\int_\Omega \nabla^2 v \cdot w \, dV = -\sum_{i=1}^{3} \int_\Omega \nabla v^i \cdot \nabla w^i \, dV = -\langle\langle v, w \rangle\rangle.$$

Using this lemma and all of the above notation, we can write the integral equation (10.70) as

**Integral (Weak) Version of the Navier–Stokes Equation:** The velocity vector field $v$ is a time-dependent vector field with $v(t) \in V$ and satisfies

$$\frac{d}{dt}\langle v, w \rangle + B(v, v, w) + \alpha\langle\langle v, w \rangle\rangle = \langle F, w \rangle, \tag{10.71}$$

for all $w \in V$.

We now indicate how one would go about finding a solution of the integral version of the Navier–Stokes equation. The method for this is called the Galerkin method, and it actually suggests a numerical method that can be used in practice to approximate the theoretical solution.

## 10.5.3.   The Galerkin Method

To prove the existence of a solution $v$ of equation (10.71), which will be a weak solution of the Navier–Stokes equations, we first show the existence of approximate solutions of this equation.

Thus suppose $\{w_n\}_{n=1}^N$ is a linearly independent subset of vector fields in $V$, and let $V_N \subseteq V$ denote the subspace spanned by these vector fields. We consider the problem of finding a solution $v = v^N$ of (10.71) that takes values in $V_N$; i.e., $v^N(t) \in V_N$ for $t$ in some interval $I$. This amounts to finding scalar functions $b_n = b_n(t), n = 1, \ldots, N$, so that the vector field defined by

$$v^N(t) \equiv \sum_{n=1}^N b_n(t) w_n \tag{10.72}$$

satisfies equation (10.71) for every $w \in V_N$. Because equation (10.71) is linear in $w$, it will be satisfied for every $w$ in $V_N$ if it is satisfied for each $w_k, k = 1, \ldots, N$. Substituting the expression (10.72) for $v$ and substituting $w_k$ for $w$ in equation (10.71), and isolating the time dervatives, we get the following equations:

$$\sum_{n=1}^N b'_n(t)\langle w_n, w_k \rangle \tag{10.73}$$

$$= \langle F, w_k \rangle - \sum_{n=1}^N \sum_{m=1}^N b_n(t) b_m(t) B(w_n, w_m, w_k) - \alpha \sum_{n=1}^N b_n(t)\langle\langle w_n, w_k \rangle\rangle,$$

for $k = 1, \ldots, N$. These constitute a system of ordinary differential equations involving the unknown functions $b_1(t), \ldots, b_N(t)$. (All the other quantities are known constants, computed from the given functions $w_n$ and $F$ according to the above definitions.) We get one of these DE's for each $k = 1, \ldots, N$, and the whole collection constitutes a system of $N$ differential equations for $N$ unknown functions. Using linear algebra, we can convert this system of DE's into the normal form for a first-order system (exercise), and then we can apply the general existence and uniqueness theorem (see the Introduction), which says that for given initial values $b_n(0) = \beta_n$, $n = 1, \ldots, N$, there is one and only one set $b_1(t), \ldots, b_N(t)$ of solutions that satisfy the system of DE's and the given initial values.

In the present circumstance, since we are seeking to approximate $v(t)$ with $v^N(t)$, and the initial velocity $v(0)$ is assumed given, the initial values $\beta_n$, $n = 1, \ldots, N$, will be chosen so that $v^N(0)$ is a *projection* of $v(0)$ onto the subspace $\mathcal{V}_N$. (See the exercises for how to do this.)

This shows how to find an *approximate* (weak) solution to the Navier–Stokes equations on $\mathcal{V}_N = \text{span}\{w_1, \ldots, w_N\}$, given linearly independent vector fields $w_n \in \mathcal{V}, n = 1, \ldots, N$. To use this to find a weak solution $v(t)$ (at least theoretically), one would construct, for each $N = 1, 2, \ldots,$, an approximate solution $v^N(t)$ as above and then obtain $v(t)$ as a limit

$$v(t) = \lim_{N \to \infty} v^N(t).$$

To discuss the details of this in a rigorous and precise way requires an understanding of certain topics in functional analysis, which is not assumed as prior knowledge for this book. However, the above discussion (and following exercises) should give the reader an intuitive idea of some aspects of the theory for existence of solutions to the Navier–Stokes equations.

## 10.5.4.  Steady Solutions of the Navier–Stokes Equations

The existence of steady solutions to the Navier–Stokes equations, i.e., where $v$ does not depend on the time, is a corollary to the above discussion. Since this case is considerably simpler, it may be instructive to see what the discussion above reduces to for steady solutions.

Assuming that $F$ does not depend on time, the integral version (10.71) of the Navier–Stokes equations when restricted to a vector field $v$ that does not depend on $t$ becomes

$$B(v, v, w) + \alpha \langle\langle v, w \rangle\rangle = \langle F, w \rangle,$$

for all $w \in \mathcal{V}$. Then the Galerkin method for solving this by approximations,

$$v^N \equiv \sum_{n=1}^{N} b_n w_n,$$

leads to the *algebraic* equations

$$\sum_{n=1}^{N}\sum_{m=1}^{N} b_n b_m B(w_n, w_m, w_k) + \alpha \sum_{n=1}^{N} b_n \langle\langle w_n, w_k \rangle\rangle = \langle F, w_k \rangle,$$

for $k = 1, \ldots, N$. Each of these is a second-degree polynomial equation for the unknown constants $b_1, \ldots, b_N$. There are well-known methods for determining the existence of solutions to such algebraic equations and also for numerically approximating these solutions. The uniqueness of solutions $b_1, \ldots, b_N$ is not guaranteed unless something more is known about $\alpha$ and $\langle F, w_k \rangle$. This can lead to nonuniqueness of solutions of the Navier–Stokes equations.

## Exercises 10.5

1. Use Gauss's divergence theorem to prove Lemma 10.2.

2. For a vector field $u$ and a scalar field $f$, prove the identity

$$\int_{\Omega} u \cdot \nabla f \, dV = \int_{\partial\Omega} f \, u \cdot n \, dA - \int_{\Omega} f \, \text{div}(u) dV.$$

   Use this to prove that in the Helmholtz representation, the decomposition of $E$ is an *orthogonal sum*; i.e., $\langle u, \nabla P \rangle = 0$ with respect to the inner product $\langle \cdot, \cdot \rangle$ (see Exercise 5 below for the definition of an inner product). Also prove Lemma 10.3.

3. Show that $\langle \cdot, \cdot \rangle$ and $\langle\langle \cdot, \cdot \rangle\rangle$ are bilinear and that $B(\cdot, \cdot, \cdot)$ is trilinear.

4. Derive the system of DE's (10.73) from the integral version (10.71) of the Navier–Stokes equations, using the assumptions in the text. Write this system in the normal form

$$b'_j(t) = H_j(b(t), t),$$

   $j = 1, \ldots, N$, for a first-order system. You may use the fact that the matrix $G$ with entries $G_{nk} = \langle w_n, w_k \rangle$ is invertible (see the next exercise).

5. Suppose $W$ is a vector space (over the field of real numbers). An *inner product* on W is a symmetric, bilinear function $\langle \cdot, \cdot \rangle$ of the vectors in $W$ that has the additional property

$$\langle v, v \rangle \geq 0$$

   for every $v \in W$, and $\langle v, v \rangle = 0$ if and only if $v$ is the zero vector. An inner product gives rise to a corresponding *norm* on $W$ defined by

$$\|v\| = \langle v, v \rangle^{1/2}.$$

The norm $\|v\|$ of a vector $v$ is thought of as the length of the vector, and $(W, \langle \cdot, \cdot \rangle)$, which is called an *inner product space*, is thought of as an abstraction of the vector space $\mathbb{R}^3$ with the usual inner product (or dot product) of vectors.

(a) Show that the bilinear functions $\langle \cdot, \cdot \rangle$ and $\langle \langle \cdot, \cdot \rangle \rangle$ defined in the text are inner products on the space $\mathcal{V}$.

(b) Suppose $(W, \langle \cdot, \cdot \rangle)$ is an abstract inner product space and $\{w_n\}_{n=1}^N$ is a linearly independent subset of $W$. Let $G$ be the $N \times N$ matrix

$$G = \{\langle w_n, w_m \rangle\}_{n,m=1}^N.$$

Show that $G$ is invertible (i.e., if $Gb = 0$ for some $N$-tuple $b = (b_1, \ldots, b_N)$, then $b = 0$). With $G_{nm}$ denoting the $n$-$m$ entry of $G$, it is customary to denote the $n$-$m$ entry of $G^{-1}$ by $G^{nm}$. Using this notation, define for each $k = 1, \ldots, N$ a vector $w_k^*$ by

$$w_k^* = \sum_{n=1}^N G^{kn} w_n.$$

Show that:

(i) The $w_k^*$'s satisfy

$$\langle w_k^*, w_m \rangle = \delta_{km}$$

for each $k, m$. Here $\delta_{km}$ is the *Kronecker delta* symbol, which is equal to 1 if $k = m$ and is equal to zero if $k \neq m$.

(ii) A vector $w \in W$ is in $\text{span}\{w_1, \ldots, w_N\}$ if and only if

$$w = \sum_{n=1}^N \langle w_n^*, w \rangle w_n.$$

The vectors $w_1^*, \ldots, w_N^*$ are often called the *reciprocal vectors* relative to the given linearly independent set of vectors. *Note:* we are not assuming that the span of the vectors $w_1, \ldots, w_N$ is all of $W$. Relative to these vectors, we can define a a projection $P$ from $W$ onto $\text{span}\{w_1, \ldots, w_N\}$ by

$$Pv = \sum_{n=1}^N \langle w_n^*, v \rangle w_n.$$

Show that:

(iii) $P$ is a linear map and $P^2 = P$.

(iv) In the integral version (10.71) of the Navier–Stokes equations, suppose an initial value $v(0) \in \mathcal{V}$ for $v(t)$ is specified. In applying the Galerkin method, it is standard to take, for each $N$, the initial values $\beta_n = \langle w_n^*, v(0) \rangle$, $n = 0, \ldots, N$, for the system of DE's (10.73). With this choice of initial values, show that $v^N(0) = Pv(0)$. Thus (at least heuristically) $\lim_{N \to \infty} v^N(0) = v(0)$.

6. (*The Stokes equations*) In the case where the magnitude of the velocity $v$ is small in relation to $\alpha = \nu/\rho$, the convective term $\nabla_v v$ in the Navier–Stokes equations can, for practical purposes, be neglected. If we omit the term $\nabla_v v$, the resulting equations are

$$\frac{\partial v}{\partial t} - \alpha \nabla^2 v - F = \nabla P \qquad \text{in } \Omega,$$

$$\mathrm{div}(v) = 0 \qquad \text{in } \Omega,$$

$$v = 0 \qquad \text{on } \partial\Omega.$$

Thus is a system of linear PDEs (and BC) known as the *Stokes equations*, and it approximates the flow of a very slow, highly viscous fluid.

(a) Show that the integral version of the Stokes equations is

$$\frac{d}{dt}\langle v, w \rangle + \alpha \langle\langle v, w \rangle\rangle = \langle F, w \rangle,$$

for every $w \in \mathcal{V}$. The steady state, integral version of the Stokes equations is

$$\alpha \langle\langle v, w \rangle\rangle = \langle F, w \rangle,$$

for every $w \in \mathcal{V}$.

(b) Show that if $v, \bar{v} \in \mathcal{V}$ are solutions of the steady state, integral version of the Stokes equations, then $v = \bar{v}$. *Hint:* Use the fact that $\langle\langle \cdot, \cdot \rangle\rangle$ is an inner product on $\mathcal{V}$ (See Exercise 5).

(c) Apply the Galerkin method to the steady-state version of Stokes's equations to show that the constants $b_1, \ldots, b_N$ in the approximate solution

$$v^N = \sum_{n=1}^{N} b_n w_n$$

must satisfy the linear equations

$$\alpha \sum_{n=1}^{N} b_n \langle\langle w_n, w_k \rangle\rangle = \langle F, w_k \rangle,$$

for $k = 1, \ldots, N$. Show that this system of linear equations always has a unique solution. The existence of (weak) solutions to the Stokes equations will follow from this by taking limits: $v = \lim_{N \to \infty} v^N$ (with the appropriate notions from functional analysis).

7. Show that the trilinear function $B$ has the property

$$B(v, w, w) = \int_\Omega v \cdot \nabla(\frac{|w|^2}{2}) \, dV,$$

for any vector fields $v$, $w$. Use this to show that

$$B(v, w, w) = 0$$

if $v \in \mathcal{V}$ and $w$ is any vector field.

# 11 Waves in Elastic Materials

In this chapter we apply the general theory developed for continuum mechanics (See Appendix B) to the special case where the continuum $\Omega$ is an elastic material. Such a material is modeled mathematically by specifying the nature of the stress tensor $T$ as regards the way it responds to deformations of the material. This theory of elasticity has equations of motion for $\Omega$ that are nonlinear, but in certain cases the equations can be approximated by the linearized equations

$$u_{tt} = c^2 \nabla^2 u + b^2 \nabla \operatorname{div}(u) + F,$$

and this leads to a model for elastic waves in solids. In addition to the case where $\Omega$ is a 3-dimensional solid, we also study 1-dimensional and 2-dimensional elastic materials whose linearized equations are the wave equations for vibrating strings and membranes. In these cases, the linearized equations have the form

$$u_{tt} = c^2 \nabla^2 u + F,$$

which are called (non-homogeneous) *wave equations* and are, as we shall see, good approximations to the PDEs that model the propagation of waves in elastic strings and membranes. The nonlinear equations are also useful in their own right in that they serve to explain, for example, the shapes of the catenary (a hanging chain) and minimal surfaces (soap bubble surfaces).

For pedagogical reasons we first discuss one-dimensional, then two-dimensional, and finally three-dimensional elastic materials.

## 11.1.    Strings: 1-D Elastic Materials

We model the string by a curve $C_t$ in $\mathbb{R}^3$ indexed by time $t$, and as $t$ varies, these curves comprise the motion of the string. Analytically, the motion of the string is given by a map: $\alpha_t : I \to \mathbb{R}^3$, with components

$$\alpha_t(s) = (\alpha_t^1(s), \alpha_t^2(s), \alpha_t^3(s)),$$

and $C_t \equiv \alpha_t(I)$ is the position of the string at time $t$. The derivative with respect to the parameter $s$ is denoted by a prime: $\alpha_t' = d\alpha_t/ds$. Often it will be convenient to to suppress $t$ in the notation and write just $\alpha$ for $\alpha_t$.

*Note*: The parameter $s$ does *not* denote arc length. The conservation of mass and linear momentum equations for the string are (cf. Appendix B)

$$\rho \det G^{1/2} = \rho_0 \det G_0^{1/2},$$

$$\rho \frac{\partial^2 \alpha}{\partial t^2} = \rho F + \operatorname{div}_C(T).$$

Here $G$ is the $1 \times 1$ matrix $G = \alpha_t' \cdot \alpha_t'$, and so $\det G^{1/2} = |\alpha_t'|$. (This notation is to display unity with the membrane and solid cases where $G$ is $2 \times 2$ and $3 \times 3$ respectively.) The angular momentum equation leads to the requirement that $Tw$ be a tangential vector field along $C_t$ if $w$ is. Thus there is a scalar field $f$ such that

$$Tw = fw$$

for every tangential vector field $w$. It can be shown that the theory for elastic 1-dimensional materials gives a stress tensor $T$ with this property, and so the model here is for an elastic string. In essence, the string can develop stresses only in the tangential direction (tension or compression, but no bending stresses), and the stress is proportional to the strain:

$$T\alpha_t' = f\alpha_t'.$$

See Figure 11.1 As we shall see, the proportionality factor $f$, besides depending on the parameter $s$ that parametrizes the curve, depends on $\alpha_t'$ as well.

To analyze the equations of motion further, notice first that the conservation of mass equation is

$$\rho = \rho_0 \frac{|\alpha_0'|}{|\alpha_t'|}. \tag{11.1}$$

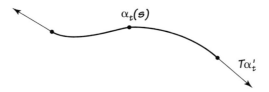

Figure 11.1.    The stresses in an elastic string are tangential.

This gives the mass density $\rho$ in terms of the initial mass density $\rho_0$ times the ratio of the change of length factors $|\alpha_0'|/|\alpha_t'|$.

From the above property of the stress tensor $T$ and from the definition of the curve divergence $\mathrm{div}_C$, we find (exercise)

$$\mathrm{div}_C(T) = \frac{\partial f}{\partial s} e^1 + f \kappa N,$$

where $\kappa$ is the curvature of the curve (at time $t$), $N$ is the unit normal vector field along the curve, and $e^1 = \alpha'/|\alpha'|^2$ is the reciprocal basis vector to the tangent basis vector $e_1 = \alpha'$. Using this and the equation (11.1) for $\rho$, we can write the linear momentum equation as

$$\frac{\partial^2 \alpha}{\partial t^2} = F + \frac{|\alpha'|}{\rho_0 |\alpha_0'|} \frac{\partial f}{\partial s} e^1 + \frac{|\alpha'|}{\rho_0 |\alpha_0'|} f \kappa N.$$

This vector equation can be resolved into three component equations as follows. The natural basis is the frame of orthogonal vectors $\{e^1, N, B\}$, consisting of the tangent, unit normal, and unit binormal. Thus the tangential, normal, and binormal components of the linear momentum equation are

$$\frac{\partial^2 \alpha}{\partial t^2} \cdot e_1 = F \cdot e_1 + \frac{|\alpha'|}{\rho_0 |\alpha_0'|} \frac{\partial f}{\partial s}, \tag{11.2}$$

$$\frac{\partial^2 \alpha}{\partial t^2} \cdot N = F \cdot N + \frac{|\alpha'|}{\rho_0 |\alpha_0'|} f \kappa, \tag{11.3}$$

$$\frac{\partial^2 \alpha}{\partial t^2} \cdot B = F \cdot B. \tag{11.4}$$

These equations allow for a quite general motion of the string. The motion is completely determined once the intitial conditions and boundary conditions are specified. The initial conditions specify the postion and velocity of each point along the curve at time zero:

**Initial Conditions:**

$$\alpha(s, 0) = p(s),$$

$$\frac{\partial \alpha}{\partial t}(s, 0) = v(s)$$

Here $p, v : I \to \mathbb{R}^3$ are two given vector-valued functions.

The boundary conditions specify what happens at the two ends of the string (if the string is closed or infinite in extent, there are no boundary conditions). One type of BC specifies the displacement of the two ends of the string: $\alpha(p_i, t) = \beta_i(t)$, $i = 1, 2$, where $p_1, p_2$ are the two endpoints of $I$ and $\beta_1, \beta_2$ are given (vector-valued) functions. To implement these BCs, take two thin metal rods, bent in the shape of the curves $\beta_1, \beta_2$, and attach the string to these rods with O-rings as shown in Figure 11.2. Then arrange a mechanism (motors of some sort) to move the O-rings along the rods so that their positions at time $t$ are $\beta_1(t)$ and $\beta_2(t)$. Another type of BC is to specify the traction at the two ends: $Tv = w$ at the two ends. Here $v = \pm\alpha'/|\alpha'|$ is the outward-directed tangent at the boundary, and $w$ is the given traction. Because of our assumption about the stress tensor, this says that $fv = w$. Often it is the case that the given traction is proportional to $\beta - \alpha$, i.e., to the difference between a specified position $\beta$ of the endpoint and its actual position $\alpha$. Then the BC has the form $mTv = h(\beta - \alpha)$.

The more general boundary condition is

**Boundary Conditions:**

$$mTv + h\alpha = \beta \tag{11.5}$$

on $\partial C_t$, $\forall t$. Here $m \geq 0, h \geq 0$ are constants. The model includes both types of BC mentioned above:

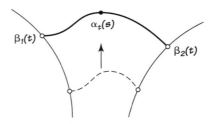

Figure 11.2.    Implementation of the specified displacement boundary condition.

**($m = 0$) Prescribed Displacements**

**($m \neq 0$) Prescribed Traction**

Figure 11.3 illustrates how to implement these BCs at the left endpoint of the string. As before, a thin metal rod is bent in the shape of the curve $\beta/h$, and the string is attached to this rod by an armature consisting of two pieces. The first piece is a sleeve that can slide freely along the rod. The second piece is also a sleeve, pinned to the first sleeve and able to rotate about the pin. There is a coil spring at the pin, which resists rotation in either direction, and a spring to which the string is attached. At any time during the motion, the tangent line to the string at the boundary is forced to line up with the sleeve there.

To simplify the discussion, from here on we assume that the motion takes place entirely in one plane, the $x$-$y$ plane, and that the curve $C_t$ is the graph of a function $u(x, t)$, for each time $t$. Thus we use $x$ as the parameter that was previously denoted by $s$, and the assumed form of the curve is

**Assumption**

$$\alpha(x, t) = (x, u(x, t), 0),$$

for $x \in [0, L]$. Thus for each time $t$, the graph of the function $u(x, t)$ over the interval $[0, L]$ represents the position of the string. Figure 11.4 shows a typical situation, where the string is fastened at its two endpoints $x = 0$ and $x = L$. Based on the above assumption, the three equations above for the components of the linear momentum simplify to (see the exercises)

$$u_{tt} u_x = F^1 + F^2 u_x + \frac{(1 + u_x^2)^{1/2}}{\rho_0 (1 + u_{0x}^2)^{1/2}} \frac{\partial f}{\partial x}, \tag{11.6}$$

$$u_{tt} = F^2 - F^1 u_x + \frac{f u_{xx}}{\rho_0 (1 + u_{0x}^2)^{1/2} (1 + u_x^2)^{1/2}}, \tag{11.7}$$

$$0 = F^3. \tag{11.8}$$

Here $u_{0x}$ stands for $u_x$ evaluated at $t = 0$. The last of these equations is to be expected: Since we wish the motion to be constrained to the $x$-$y$ plane, the component of force perpendicular to this plane must vanish.

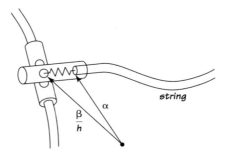

Figure 11.3.   Implementation of the BC: $mTv = h(\beta/h - \alpha)$ at the left endpoint of the string (assuming $h \neq 0$).

The first two equations simplify further if we make the assumption

**Assumption**   $F^1 = 0$.

This says that there is no component of force in the $x$ direction. Then we can substitute the second equation into the first to get

$$\frac{f u_{xx} u_x}{\rho_0 (1 + u_{0x}^2)^{1/2} (1 + u_x^2)^{1/2}} = \frac{(1 + u_x^2)^{1/2}}{\rho_0 (1 + u_{0x}^2)^{1/2}} \frac{\partial f}{\partial x},$$

or

$$\frac{u_x u_{xx}}{1 + u_x^2} = \frac{f_x}{f}.$$

This integrates to

$$\frac{1}{2} \ln(1 + u_x^2) = \ln f + c.$$

Thus $f$ is related to $u$ by

$$f = k(1 + u_x^2)^{1/2}, \tag{11.9}$$

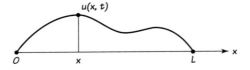

Figure 11.4.   A vibrating string with ends fixed at $x = 0$ and $x = L$.

where $k$ is a constant. Substituting this value of $f$ into equation (11.7) reduces that equation to

**A Nonlinear Wave Equation:**

$$u_{tt} = \left( \frac{k}{\rho_0 (1 + u_{0x}^2)^{1/2}} \right) u_{xx} + F^2. \tag{11.10}$$

This equation governs the motion of an elastic string, under the assumptions of planar motion and transverse external forces only. The solutions $u$ of this equation are the same as the solutions of equations (11.6)–(11.7), $f$ being specified by equation (11.9).

Equation (11.10) is nonlinear because of the expression involving $u_{0x}$. Note that $u_{0x} = u_x(\cdot, 0)$ gives the slope of the tangent line at each point along the initial displacement $u(x, 0), x \in [0, L]$. In the case where the initial displacement is zero, $u(x, 0) = 0$, or in the case where the displacement $u$ is small in magnitude, so that it may be safely assumed that $u_{0x}^2 \approx 0$, the nonlinear wave equation (11.10) reduces to

**The 1-D Wave Equation:**

$$u_{tt} = c^2 u_{xx} + F^2,$$

where, as is customary, we have introduced the constant

$$c = \sqrt{\frac{k}{\rho_0}}.$$

The above PDE is the (non-homogeneous) *1-dimensional wave equation* and, to the extent that our approximations are valid, models the motion of a vibrating string. This PDE is entirely analogous to the 1-dimensional heat equation, with the obvious exception that it is second order in the time derivative. This latter difference will hardly affect our means of analytically expressing the solutions in terms of Fourier series, but of course the physics embodied in the series solution will be totally different.

From a physical point of view, the motion will be completely determined once we know the initial conditions and boundary conditions. There are two initial conditions, since the initial displacement and initial velocity are considered given:

**Initial Conditions:**

$$u(x, 0) = \phi(x),$$

$$u_t(x, 0) = \psi(x),$$

for $0 \le x \le L$.

The general type of boundary conditions given above when applied to the present situation give the following (exercise)

**Boundary Conditions:**

$$-\kappa_1 u_x(0, t) + h_1 u(0, t) = g_1(t), \tag{11.11}$$

$$\kappa_2 u_x(L, t) + h_2 u(L, t) = g_2(t), \tag{11.12}$$

for all $t > 0$.

*Note*: These BCs are identical to those for 1-D heat problems. Because of this, we know that for homogeneous BCs ($g_1 = 0$, $g_2 = 0$), the corresponding Sturm–Liouville problem will have eigenvalues that are either negative or zero. Further, the zero eigenvalue occurs only for the BCs

$$u_x(0, t) = 0,$$

$$u_x(L, t) = 0.$$

For a heat problem, these BCs correspond to an insulated boundary. In the present circumstance of a vibrating string, these BCs mean something entirely different. The string must move so that it has a horizontal tangent line at each endpoint. The endpoints, however, are free to move in the $y$ direction within the plane of the motion. An apparatus to physically implement these BCs, as well as the general type of BCs, is shown in Figure 11.5.

## 11.1.1.  Series Solution of the Wave Equation IBVP

The solution of the general wave equation IBVP in terms of a series of Sturm-Liouville eigenfunctions is entirely analogous to that for the heat equation. For this reason we only make a few general comments on how the technique applies to the wave equation.

The first step in the solution is to transform, if necessary, to an IBVP with homogeneous BCs, i.e., to a semi-homogeneous problem. This amounts to

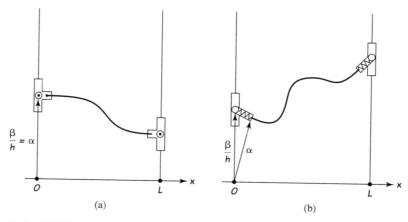

Figure 11.5.    (a) The horizontal tangent line BC and (b) the general type of BC for an elastic string.

writing $u$ as

$$u = S + U$$

and requiring $S$ to satisfy the given non-homogeneous BCs. This then leaves a semi-homogeneous problem for $U$.

The solution of a semi-homogeneous wave equation problem is the same as for the heat equation, except now the time-dependent factors must satisfy a second-order ODE. Specifically,

**Theorem 11.1 (Semi-Homogeneous Problem)**     *Consider the 1-D wave equation*

$$u_{tt} = c^2 u_{xx} + F,$$

*with homogeneous boundary conditions*

$$-\kappa_1 u_x(0, t) + h_1 u(0, t) = 0,$$

$$\kappa_2 u_x(L, t) + h_2 u(L, t) = 0,$$

*and initial conditions:*

$$u(x, 0) = \phi(x),$$

$$u_t(x, 0) = \psi(x).$$

Let $\{X_n\}_{n=0}^{\infty}$ and $\{-\lambda_n^2\}_{n=0}^{\infty}$ be the eigenfunctions and eigenvalues for the corresponding Sturm–Liouville problem:

$$X'' = -\lambda^2 X,$$

$$-\kappa_1 X'(0) + h_1 X(0) = 0,$$

$$\kappa_2 X'(L) + h_2 X(L) = 0.$$

Then the solution of the semi-homogeneous problem can be expressed analytically by

$$u(x,t) = T_0(t) + \sum_{n=1}^{\infty} T_n(t) X_n(x),$$

where the time-dependent factor $T_n(t)$ is the solution of the initial value problem

$$T_n''(t) + \lambda_n^2 c^2 T_n(t) = F_n(t), \tag{11.13}$$

$$T_n(0) = A_n, \tag{11.14}$$

$$T_n'(0) = B_n. \tag{11.15}$$

Here $F_n(t)$, $A_n$, and $B_n$ are the coefficients in the eigenfunction expansions of $F(x,t)$, $\phi(x)$, and $\psi(x)$ respectively. By convention, $\lambda_0 = 0$, and we take $F_0 = 0$, $A_0 = 0$, $B_0 = 0$, when zero is not an eigenvalue of the Sturm–Liouville BVP.

**Remark**   As with the heat equation, we can express the time-dependent factors by integral formulas. Namely, using variation of parameters to solve the DE (11.13) and then choosing the two arbitrary constants such that (11.14)–(11.15) are satisfied, we get (exercise)

$$T_0(t) = A_0 + B_0 t + \int_0^t (t-s) F_0(s)\,ds, \tag{11.16}$$

$$T_n(t) = A_n \cos a_n t + a_n^{-1} B_n \sin a_n t \tag{11.17}$$

$$+ a_n^{-1} \int_0^t \sin[a_n(t-s)] F_n(s)\,ds, \tag{11.18}$$

where for convenience, we have let

$$a_n = \lambda_n c.$$

**Corollary 11.1 (Homogeneous Problem)**   *The solution of the homogeneous wave equation IBVP is*

$$u(x, t) = A_0 + B_0 t + \sum_{n=1}^{\infty} (A_n \cos a_n t + a_n^{-1} B_n \sin a_n t) X_n(x), \qquad (11.19)$$

*where $a_n = \lambda_n c$ and $\{X_n\}_{n=0}^{\infty}$, $\{-\lambda_n^2\}_{n=0}^{\infty}$ are the eigenfunctions and eigenvalues. The numbers $A_n$ and $B_n$ in the time-dependent part are the coefficients in the eigenfunction expansion of the initial position $\phi(x)$ and velocity $\psi(x)$ respectively:*

$$\phi(x) = \sum_{n=0}^{\infty} A_n X_n(x),$$

$$\psi(x) = \sum_{n=0}^{\infty} B_n X_n(x).$$

*By convention, $A_0 = 0$, $B_0 = 0$ when zero is not an eigenvalue of the problem.*

**Example 11.1   (Fundamental Modes)**   As a first example we consider the most standard (and simplest) situation for a wave on a string. Assume that both ends of the string are held fixed, there is no external force ($F = 0$), and there is an initial displacement of string but no initial velocity ($\psi = 0$). Thus consider the wave equation IBVP

$$u_{tt} = c^2 u_{xx} \qquad \text{for } x \in [0, L],\ t > 0,$$

$$u(0, t) = 0 \qquad \text{for } t > 0,$$

$$u(L, t) = 0 \qquad \text{for } t > 0,$$

$$u(x, 0) = \phi(x) \qquad \text{for } x \in [0, L],$$

$$u_t(x, 0) = 0 \qquad \text{for } x \in [0, L].$$

It is easy to see that the eigenfunctions of the problem are $X_n(x) = \sin \frac{n\pi}{L} x$, $n = 1, 2, 3, \ldots$, and thus by the corollary, the series representation of the solution is

$$u(x, t) = \sum_{n=1}^{\infty} A_n \cos \frac{n\pi c}{L} t \sin \frac{n\pi}{L} x, \qquad (11.20)$$

where the $A_n$'s are the Fourier sine cofficients of the initial displacement $\phi$. There are some general observations about the motion of the string that can be made regardless of its initial displacement.

The series solution (11.20) that describes the motion is a superposition of the infinitely many *fundamental modes* of vibration

$$u_n(x, t) = \cos \frac{n\pi c}{L}t \ \sin \frac{n\pi}{L}x.$$

For a given $n$, the string can be made to vibrate in the $n$th fundamental mode by choosing the initial displacement to be $\phi(x) = \sin \frac{n\pi}{L}x$. The description of this mode of vibration is simple. For a fixed $t$, the graph of $u_n(x, t)$ is a sine wave with amplitude $|\cos \frac{n\pi c}{L}t|$ and period $2L/n$. The amplitude is a maximum of 1 initially, when $t = 0$, and decreases to zero at time $t = L/(2nc)$ (so the string is flat at this time). After this time, the string vibrates past the equilibrium position (the $x$-axis) to a maximum displacement of 1 at time $t = L/(nc)$, then back through the equilibrium position at $t = 3L/(2nc)$, and then returns to its initial position at $t = 2L/(nc)$. Thus the motion repeats this pattern with period

$$\frac{1}{n} \cdot \frac{2L}{c}$$

and is called a *simple harmonic motion*. See Figure 11.6. It is important to note that the $n$th fundamental mode has period (time for one complete oscillation) that is $1/n$ times the period of the first fundamental mode. Thus the ratios of the

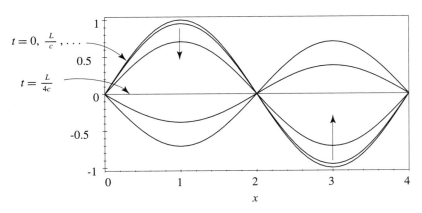

Figure 11.6.   Graph of $u_2(x, t) = \cos \frac{2\pi c}{L}t \ \sin \frac{2\pi}{L}x$, the second fundamental mode of vibration of the string.

periods are

$$1, \frac{1}{2}, \frac{1}{3}, \frac{1}{4}, \frac{1}{5}, \ldots,$$

which is known as the *harmonic sequence*. This name arose historically from the observation of the musical notes produced by stringed instruments. A string vibrating in the first fundamental mode produces a certain note (actually a vibration in the surrounding air, which is transmitted to the ear). By placing a finger at the midpoint along the length of the string, the vibration is forced into the second fundamental mode, which produces a higher note (twice the number of oscillations in the same time). This pattern, when repeated, produces higher and higher notes in the musical scale and forms the basis for musical harmony and composition. Thus while Fourier originally proposed his series in connection with the heat equation, we see that it is from the wave equation that *harmonic analysis* (as a modern area of mathematics) derives its name.

If the initial displacement $\phi$ of the string is not a multiple of one of the eigenfunctions, then the motion of the string will involve a combination of all the fundamental modes:

$$u(x, t) = \sum_{n=1}^{\infty} A_n u_n(x, t).$$

The sound produced by the vibration will be a combination of all the notes from the fundamental modes, but since $\lim_{n \to \infty} A_n = 0$, the higher-pitched notes will contribute less to the overall sound.

**Example 11.2   (A Triangular Pulse)**   As a particuar case of the last example, suppose the displacement of the string is

$$\phi(x) = 2H(x),$$

where $H$ is the triangular pulse function

$$H(x) = H_{a,p,b,L}(x) = \begin{cases} 0 & \text{for } 0 \le x < a, \\ (x - a)/(p - a) & \text{for } a \le x < p, \\ (x - b)/(p - b) & \text{for } p \le x < b, \\ 0 & \text{for } b \le x \le L. \end{cases} \qquad (11.21)$$

We let $H_n = H_{n,a,p,b,L}$ denote the $n$th Fourier sine series coefficient of $H = H_{a,p,b,L}$:

$$H(x) = \sum_{n=1}^{\infty} H_n \sin \frac{n\pi}{L} x.$$

The result of an exercise from a previous chapter gives the following formula for these coefficents:

$$H_{n,a,p,b,L} =$$

$$\frac{2L}{(p-a)(b-p)n^2\pi^2} \left[ (p-b) \sin \frac{n\pi}{L} a + (b-a) \sin \frac{n\pi}{L} p + (a-p) \sin \frac{n\pi}{L} b \right].$$

It is now a simple task to use a computer algebra system to study the motion of the string graphically. The dynamics of the motion can *only* be captured by having the computer do an animation (see the exercises), but plotting (in separate pictures) $u(x, t)$ for selected times $t$ will give an indication of the motion. This is shown in Figure 11.7 for the initial displacement $\phi = 2H_{4,5,7,20}$. As you can see, the initial triangular displacement quickly separates into two similar, but half the size, triangular displacements that move in opposite directions (initially) down the string. These are called left and right *moving waves*. Note that the left moving wave reaches the boundary first and is reflected there (exhibited by the displacement of the string in the negative direction), and after the right moving wave is similarly reflected, both waves approach one another and combine to produce a maximum negative displacement.

While a computer animation displays the motion most vividly, there are additional graphical means that enable us to understand the motion more thoroughly. One is a *spacetime* plot, which simply means a plot of $u$ as a function of two variables space $x$ and time $t$. The result, of course, is a surface, and in this case the picture is the one shown in Figure 11.8. Note that the figure contains all the information shown in Figure 11.7. Namely, the intersection of the surface with a plane perpendicular to the time axis at $t = t_0$ is the curve representing the displacement of the string at time $t_0$. Thus slicing through the surface at an appropriate sequence of times yields the pictures in Figure 11.7. The spacetime surface for the string's motion can be thought of as a total picture of the overall motion. It also reveals most clearly how the left and right moving waves on the string separate, reflect from the boundaries, and recombine.

**Example 11.3    (Triangular Pulse and Oscillating Boundary)**    Suppose we extend the situation described in the last example by now having the right end oscillate harmonically with relatively small amplitude. Everything else is the

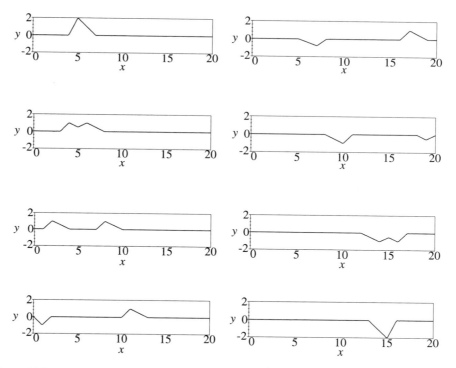

Figure 11.7.    Successive positions of the string in Example 11.2 starting from an initial triangular displacement $\phi = 2H_{4,5,7,20}$.

same. Thus the IBVP is

$$u_{tt} = c^2 u_{xx} \qquad \text{for } x \in [0, L], \ t > 0,$$

$$u(0, t) = 0 \qquad \text{for } t > 0,$$

$$u(L, t) = q \sin \omega t \qquad \text{for } t > 0,$$

$$u(x, 0) = 2H_{4,5,7,20}(x) \qquad \text{for } x \in [0, L],$$

$$u_t(x, 0) = 0 \qquad \text{for } x \in [0, L],$$

Since the BCs are non-homogeneous, we need to transform to homogeneous boundary conditions. Thus we write $u$ as

$$u = S + U,$$

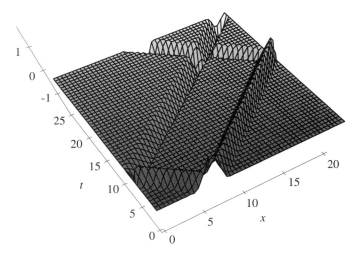

Figure 11.8.   Plot of the displacement function $u(x, t)$ in Example 11.2 as a function of two variables.

where

$$S(x, t) = \left(\frac{q}{L} \sin \omega t\right) x,$$

and $U$ is the solution to the semi-homogeneous problem

$$U_{tt} = c^2 U_{xx} + \left(\frac{q\omega^2}{L} \sin \omega t\right) x \qquad \text{for } x \in [0, L],\ t > 0,$$

$$U(0, t) = 0 \qquad\qquad\qquad \text{for } t > 0,$$

$$U(L, t) = 0 \qquad\qquad\qquad \text{for } t > 0,$$

$$U(x, 0) = 2H_{4,5,7,20}(x) \qquad\quad \text{for } x \in [0, L],$$

$$U_t(x, 0) = -\frac{q\omega}{L} x \qquad\qquad \text{for } x \in [0, L].$$

The eigenfunctions for the problem are $X_n(x) = \sin \frac{n\pi}{L}x$, and the eigenvalues are $-n^2\pi^2/L^2$, for $n = 1, 2, 3, \ldots$. The solution $U$ has the form

$$U(x, t) = \sum_{n=1}^{\infty} T_n(t) \sin \frac{n\pi}{L}x,$$

and all we have left to do is compute the $T_n$'s. First note that the forcing function in the transformed problem is

$$F(x, t) = \left( \frac{q\omega^2}{L} \sin \omega t \right) x,$$

and the coefficients in its eigenfunction expansion are

$$F_n(t) = \frac{2}{L} \int_0^L F(x, t) \sin \frac{n\pi}{L} x \, dx$$

$$= \left( \frac{q\omega^2}{L} \sin \omega t \right) \frac{2}{L} \int_0^L x \sin \frac{n\pi}{L} x \, dx$$

$$= (q\omega^2 \sin \omega t) K_n,$$

where

$$K_n = \frac{2(-1)^{n+1}}{n\pi}.$$

Further, $A_n = 2H_n = 2H_{n,4,5,7,20}$ is the coefficient of the initial displacement function $\phi(x)$ in terms of the eigenfunctions (which we computed above). Let $B_n$ be the coefficient of $-q\omega x/L$ in its eigenfunction expansion. From the above calculation of $F_n(t)$, it is easy to see that

$$B_n = -q\omega K_n.$$

For convenience, we let $a_n = n\pi c/L$. Next, we use formula (11.17) for $T_n$ and compute the integral involved to get

$$T_n(t) = A_n \cos a_n t + a_n^{-1} B_n \sin a_n t + Q_n[\omega \sin \omega t - a_n \sin a_n t], \qquad (11.22)$$

where

$$Q_n = \frac{q\omega^2 K_n}{a_n(\omega^2 - a_n^2)}.$$

In summary, we get the following solution of the wave problem in this example:

$$u(x, t) = \left( \frac{q}{L} \sin \omega t \right) x + \sum_{n=1}^{\infty} T_n(t) \sin \frac{n\pi}{L} x.$$

The series solution for the motion of the string is certainly very complicated, but with a computer algebra system, the physics embodied in the motion is easily displayed. Again, a computer animation will best display the motion. A sequence of plots of the string position at selected times is shown in Figure 11.9. The plots are for $c = 2$, $q = 0.5$, and $\omega = 2$.

Compare the motion shown here with that in Figure 11.7 of the previous example. Now the propagations down the string of the the left and right moving waves are disturbed by the waves that are generated by the oscillations at the right boundary point. This is also evident in the spacetime plot shown in Figure 11.10, which shows ripples in parts of the surface that previously (in Figure 11.8) were perfectly flat.

Note that the traveling waves set up in the string by the oscillation of the right boundary point can be separated out and plotted separately from the two traveling waves that come from the initial triangular displacement. Namely, write $u$ as

$$u = v + w,$$

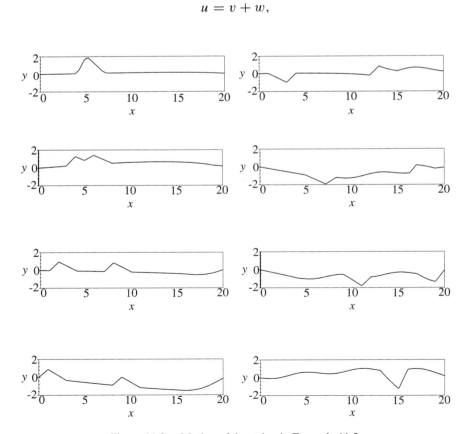

Figure 11.9.    Motion of the string in Example 11.3.

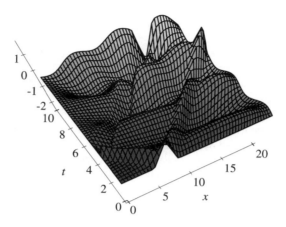

Figure 11.10.    Spacetime surface for the triangular pulse + oscillating boundary.

where $v$, $w$ are the solutions of the IBVPs

$$v_{tt} = c^2 v_{xx} \qquad \text{for } x \in [0, L], \ t > 0,$$

$$v(0, t) = 0 \qquad \text{for } t > 0,$$

$$v(L, t) = 0 \qquad \text{for } t > 0,$$

$$v(x, 0) = 2H_{4,5,7,20}(x) \qquad \text{for } x \in [0, L],$$

$$v_t(x, 0) = 0 \qquad \text{for } x \in [0, L],$$

$$w_{tt} = c^2 w_{xx} \qquad \text{for } x \in [0, L], \ t > 0,$$

$$w(0, t) = 0 \qquad \text{for } t > 0,$$

$$w(L, t) = q \sin \omega t \qquad \text{for } t > 0,$$

$$w(x, 0) = 0 \qquad \text{for } x \in [0, L],$$

$$w_t(x, 0) = 0 \qquad \text{for } x \in [0, L].$$

Of course, the IBVP for $v$ is the one we solved in the previous example. The IBVP for $w$ needs to be transformed to one with homogeneous BCs. Alternatively, note that in the above series solution for $u$ there are terms that do not involve $q$ as a factor and terms that do. Separating the series into two series with these types of terms gives $v$ and $w$ respectively.

### Exercises 11.1

1. (*The catenary*) The equations of motion formulated in the text also cover the static case where the string is at rest. Suppose the string is suspended from its two endpoints and subjected only to the force of gravity. See Figure 11.11. Instead of a string, this exercise can also be applied to a hanging rope, necklace, or chain. The curve formed by any of these is called a *catenary*, after the Latin word for chain. Galileo thought that the shape was a parabola. In this exercise you will see that the shape is actually described by the graph of a hyperbolic cosine function.

   Assume that the coordinate system is chosen such that the $x$-$y$ plane is vertical, with the positive $y$-axis directed upward. Then the density for the external force is

$$F = (0, -\rho g, 0),$$

   where $g$ is the acceleration of gravity near the earth's surface. The shape of the string is described by the graph of the displacement function $u = u(x)$, which now does not depend on time. With these assumptions, do the following:

   (a) Show that the nonlinear wave equation for $u$ reduces to the ODE

$$\frac{ku''}{(1 + u'^2)^{1/2}} = \rho g.$$

   (b) Assuming that $\rho$ = a constant, solve the above ODE for $u$ with the BCs $u(0) = 0$, $u(L) = 0$. *Hint:* multiply both sides of the ODE by $u'$ and integrate to obtain an equation involving only $u'$ and $u$. With a little algebra, this equation becomes a separable differential equation.

   (c) Explain how the constants $k$, $\rho$, $g$, and $L$ affect the shape of the graph.

2. Use the definition of $\mathrm{div}_C$ from Appendix B to show that

$$\mathrm{div}_C(T) = \frac{\partial f}{\partial s} e^1 + f \kappa N,$$

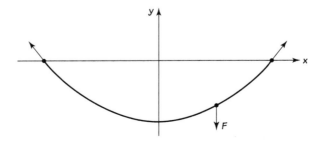

**Figure 11.11.**    A string suspended from its end points assumes the catenary shape under the force of gravity.

where $T$ has the assumed property $Tw = fw$ for any *tangential* vector field $w$ : $I \to \mathbb{R}^3$. For this, first show that since $e^1 = \alpha'/|\alpha'|^2$, it follows that

$$T^i \cdot e^1 = f \frac{\alpha'_i}{|\alpha'|^2},$$

for $i = 1, 2, 3$.

3. Show how to derive equations (11.6)–(11.8) from the general equations of motion (11.2)–(11.4).

4. (*Boundary conditions*) Derive the boundary conditions (11.11)–(11.12) from the general boundary conditions (11.5). Specifically, at the left endpoint, let $\beta(t) = (r(t), g(t))$. By the assumptions made in the text, the outward-directed normal to the boundary is

$$\nu = \frac{-\alpha'(0)}{|\alpha'(0)|} = -(1, u_x(0, t)).$$

Consequently, $mT\nu + h\alpha = \beta$ at the left endpoint gives

$$-mk = r(t), \tag{11.23}$$

$$-mku_x(0, t) + hu(x, 0) = g(t). \tag{11.24}$$

Similarly for the right endpoint. Relabeling the constants gives (11.11)–(11.12).

5. Show how to derive the integral formulas (11.16)–(11.17) for $T_0$ and $T_n$.

6. (*Parabolic pulse*) Let $K$ be the *parabolic pulse* function defined by

$$K(x) = K_{a,b,L}(x) = \begin{cases} 0 & \text{for } 0 \le x < a, \\ -4(x-a)(x-b)/(b-a)^2 & \text{for } a \le x \le b, \\ 0 & \text{for } b < x \le L. \end{cases} \tag{11.25}$$

Here $a < b$ are two numbers in the interval $[0, L]$. Let $K_n = K_{n,a,b,L}$ denote the $n$th Fourier sine coefficient of $K = K_{a,b,L}$:

$$K(x) = \sum_{n=1}^{\infty} K_n \sin \frac{n\pi}{L} x.$$

(a) Show that these coefficients are given by the formula

$$K_{n,a,b,L} =$$

$$\frac{-8L}{n^2\pi^2(b-a)^2} \left[ \frac{2L}{n\pi}(\cos \frac{n\pi b}{L} - \cos \frac{n\pi a}{L}) + (b-a)(\sin \frac{n\pi b}{L} + \sin \frac{n\pi a}{L}) \right].$$

This exercise explores the motion of a string with fixed ends and initial displacement $\phi = K_{a,b,L}$. Specifically,

$$u_{tt} = c^2 u_{xx} \qquad \text{for } x \in [0, L], \ t > 0,$$

$$u(0, t) = 0 \qquad \text{for } t > 0,$$

$$u(L, t) = 0 \qquad \text{for } t > 0,$$

$$u(x, 0) = \phi(x) \qquad \text{for } x \in [0, L],$$

$$u_t(x, 0) = 0 \qquad \text{for } x \in [0, L].$$

(b) For (i) $\phi = K_{0,1,2}$ and (ii) $\phi = K_{0,1,8}$ do an animation of the motion of the string and determine the approximate period (if any) of the motion, i.e., the first time that the string returns to its initial shape. Make sure your animation contains enough frames so that the motion does not appear jerky. In case (i), can you distinguish the left and right moving waves clearly? In case (ii), determine the first time for the right-moving wave to reach the right-hand boundary and the time when this wave, after reflecting beneath equilibrium, recombines with the original left-moving wave to produce the maximum displacement of the string below equilbrium position.

(c) For (i) $\phi = K_{1,6,8}$ and (ii) $\phi = K_{1,2,8}$, repeat part (b) of this exercise.

(d) For $\phi = K_{1,2,8} + K_{6,7,8}$, do an animation of the motion of the string and determine its period (if any). Is there a time when the string is completely flat? If so, when?

7. Do all the parts of the previous exercise, but now with $\phi$ a triangular pulse function $H_{a,p,b,L}$ (take $p = (a + b)/2$ in all cases).

8. Consider the wave problem in Exercise 6, but now with initial displacement $\phi = 0$ and initial velocities $\psi = K_{a,b,L}$. Do all the parts of Exercise 6 for this situation. Compare and contrast the differences.

9. Do all the parts of the previous exercise, but now with $\psi$ a triangular pulse function $H_{a,p,b,L}$ (take $p = (a + b)/2$ in all cases).

10. This exercise deals with the determination of steady states $S$ for the semi-homogeneous wave problem

$$u_{tt} = c^2 u_{xx} + F(x),$$

$$-\kappa_1 u_x(0, t) + h_1 u(0, t) = 0,$$

$$\kappa_2 u_x(L, t) + h_2 u(L, t) = 0,$$

$$u(x, 0) = \phi(x),$$

$$u_t(x, 0) = \psi(x).$$

*Note*: We are assuming that $F$ does not depend on $t$.

(a) Find all the steady states of this IBVP, i.e., functions $S = S(x)$ that satisfy the PDE and the BCs. Divide into cases: (i) zero is not an eigenvalue, and (ii) zero is an eigenvalue. Express your answers in terms of $F$. In case (i), show that the given IBVP has one and only one steady state $S$. In case (ii), show that steady states exist if and only if

$$\int_0^L F(x)\,dx = 0, \tag{11.26}$$

and when this conditions is met, there are infinitely many steady states, but any two of them differ by a constant.

(b) Suppose $S$ is a steady state, and let $S_n$ be it is $n$th coefficient in the eigenfunction expansion

$$S(x) = \sum_{n=0}^{\infty} S_n X_n(x).$$

Also, let $F_n$ be the $n$th coefficient of $F$ in such an eigenfunction expansion. Show that

$$S_n = a_n^{-2} F_n, \qquad \text{for } n = 1, 2, 3, \ldots.$$

Here $a_n = \lambda_n c$.

(c) Show that the series solution of of the IBVP can be written in the form

$$u(x, t) = S(x) + B_0 t + \sum_{n=1}^{\infty} \left( (A_n - S_n) \cos a_n t + a_n^{-1} B_n \sin a_n t \right) X_n(x),$$

where $S$ is a steady state. Use this to argue that in case (i), the motion of the string oscillates about the steady state $S$. In case (ii), argue that the string oscillates about $S$ if

$$v \equiv \int_0^L \psi(x)\,dx$$

is zero, and otherwise it will move off to infinity with mean speed $v$.

11. Solve the following wave problems. In each case determine the steady states and plot a spacetime surface that adequately shows the oscillation about a steady state or the movement of the string off to infinity. Also, do animations of the motion for each problem.

(a)

$$u_{tt} = u_{xx} + 1,$$

$$u(0, t) = 0,$$

$$u_x(1, t) = 0,$$

$$u(x, 0) = 0,$$

$$u_t(x, 0) = 0.$$

(b)

$$u_{tt} = u_{xx} + 1,$$

$$u(0, t) = 0,$$

$$u_x(1, t) = 0,$$

$$u(x, 0) = \sin \frac{\pi}{2} x,$$

$$u_t(x, 0) = 0.$$

(c)

$$u_{tt} = u_{xx},$$

$$u_x(0, t) = 0,$$

$$u_x(1, t) = 0,$$

$$u(x, 0) = 0,$$

$$u_t(x, 0) = \cos \frac{\pi}{4} x.$$

12. Determine the formula for the Fourier cosine coefficients $K_n^* = K_{n,a,b,L}^*$ of the parabolic pulse function $K = K_{a,b,L}$:

$$K(x) = \sum_{n=0}^{\infty} K_n^* \cos \frac{n\pi}{L} x.$$

Consider the wave problem in Exercise 6, but now with BCs

$$u_x(0, t) = 0,$$

$$u_x(L, t) = 0,$$

initial displacement $\phi = 0$, and intial velocities $\psi = K_{a,b,L}$. Do all the parts of Exercise 6 for this situation. Compare and contrast the differences.

13. Do all parts of the previous exercise, but now with $\psi = H_{a,p,b,L}$. You may take $p = (a+b)/2$.

14. (*Transforming to homogeneous BCs*) Show that the general 1-D wave problem

$$u_{tt} = c^2 u_{xx} + F, \tag{11.27}$$

$$-\kappa_1 u_x(0,t) + h_1 u(0,t) = g_1(t), \tag{11.28}$$

$$\kappa_2 u_x(L,t) + h_2 u(L,t) = g_2(t), \tag{11.29}$$

$$u(x,0) = \phi(x), \tag{11.30}$$

$$u_t(x,0) = \psi)(x), \tag{11.31}$$

can be transformed to one with homogeneous BCs by letting $u = S+U$ and choosing $S$ of the form

(a) $S(x,t) = A(t)x + B(t)$, if not both $h_1$, $h_2$ are zero.

(b) $S(x,t) = A(t)x^2 + B(t)x$, if $h_1 = 0 = h_2$ (this is the zero eigenvalue case).

15. Solve the following wave problems. Do animations and spacetime plots. Determine the period of the motion (if any). Describe how the boundary conditions and initial conditions influence the motion in each case. (Can you get a rough idea of what the motion will be before solving the problem?)

(a)
$$u_{tt} = u_{xx},$$
$$u_x(0,t) = -1,$$
$$u_x(1,t) = 1,$$
$$u(x,0) = x^2 - x,$$
$$u_t(x,0) = \sin \pi x.$$

(b)
$$u_{tt} = u_{xx},$$
$$u(0,t) = 0,$$
$$u(1,t) = 1,$$
$$u(x,0) = 2x^2 - x,$$
$$u_t(x,0) = 0.$$

(c)

$$u_{tt} = u_{xx},$$

$$u(0, t) = 0,$$

$$u(1, t) = 1,$$

$$u(x, 0) = x,$$

$$u_t(x, 0) = x - x^2.$$

(d)

$$u_{tt} = u_{xx},$$

$$-u_x(0, t) + u(0, t) = 1,$$

$$u_x(1, t) = 0,$$

$$u(x, 0) = 1,$$

$$u_t(x, 0) = \cos \pi x.$$

16. (*D'Alembert's solution*) There is an ingenious method due to the French mathematician d'Alembert for solving the 1-D wave problem:

$$u_{tt} = c^2 u_{xx},$$

$$u(0, t) = 0,$$

$$u(L, t) = 0,$$

$$u(x, 0) = \phi(x),$$

$$u_t(x, 0) = 0,$$

*without* series. The method is based on the observation that if $f : \mathbb{R} \to \mathbb{R}$ is any twice-differentiable function, then the functions defined by $v(x, t) = f(x + ct)$ and $w(x, t) = f(x - ct)$ are solutions of the homogeneous wave equation. Furthermore, $v(x, 0) = f(x) = w(x, 0)$, so the graphs of $v(\cdot, t)$ and $w(\cdot, t)$ coincide with the graph of $f$ at time zero. For later times, the graph of $v(\cdot, t)$ lies to the left of the graph of $f$, while the graph of $w(\cdot, t)$ lies to the right (verify this!). For this reason, $v$ and $w$ are called left and right moving waves. Each moves with speed $c$. Based on this, d'Alembert's solution of the above wave problem is

$$u(x, t) = \frac{1}{2}[\phi_o(x + ct) + \phi_o(x - ct)]. \tag{11.32}$$

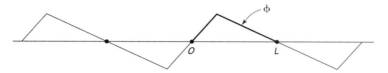

Figure 11.12.    The extension $\phi_o$ of $\phi$ to an odd, periodic function on $\mathbb{R}$.

Here $\phi_o$ denotes the extension of $\phi$ (which is defined only on the interval $[0, L]$) to an odd, periodic function on $\mathbb{R}$ of period $2L$. See Figure 11.12. Relative to this, do the following:

(a) Verify that d'Alembert's solution (11.32) actually satisfies the wave problem.

(b) For initial displacement $\phi$ = the triangular pulse in Example 11.2, show how to graph d'Alembert's solution $u(\cdot, t)$ for various times $t$ by graphing the left and right moving waves $v(x, t) = \frac{1}{2}\phi_o(x + ct)$, $w(x, t) = \frac{1}{2}\phi_o(x - ct)$ separately and then combining to obtain $u$. Plot the graph on $[0, L]$ only.

(c) D'Alembert actually gave a more general solution than (11.32), one that covers the case when there are initial velocities $u_t(x, 0) = \psi(x)$. Determine what this more general solution is. *Hint:* try adding a term of the form $g(x+ct)+h(x-ct)$ to (11.32) and determine how $g, h$ depend on $\psi$.

17. Based on what you learned in the last problem, show how to solve the wave problem

$$u_{tt} = c^2 u_{xx},$$

$$u_x(0, t) = 0,$$

$$u_x(L, t) = 0,$$

$$u(x, 0) = \phi(x),$$

$$u_t(x, 0) = 0,$$

by d'Alembert's method. Also do the following:

(a) For $\phi$ = the triangular pulse function in Example 11.2, show how to graph the solution $u$ by plotting the left and right moving waves and combining (just on $[0, L]$).

(b) Generalize the solution you obtained to allow for specified initial velocities $\psi$.

(c) Determine if d'Alembert's method can be extended to more general types of boundary conditions than the 'fixed ends' and 'horizontal tangent' BCs discussed so far.

## 11.2.    Membranes: 2-Dimensional Elastic Materials

The membrane is modeled as a surface $S_t$ in $\mathbb{R}^3$. Indexed by time $t$ and as $t$ varies, these surfaces comprise the motion of the membrane. Analytically, the motion is given by a map $\alpha_t : U \subseteq \mathbb{R}^2 \to \mathbb{R}^3$ with components

$$\alpha_t(s) = (\alpha_t^1(s), \alpha_t^3(s), \alpha_t^3(s)),$$

where $s = (s_1, s_2)$ is a point in the parameter domain $U$. For each time $t$, the surface $S_t = \alpha_t(U)$ indicates the position of the membrane. For convenience of notation, we will often suppress the $t$ from the notation and write $\alpha$ for $\alpha_t$.

The conservation of mass and linear momentum equations for the membrane are (see Appendix B)

$$\rho \det G^{1/2} = \rho_0 \det G_0^{1/2},$$

$$\rho \frac{\partial^2 \alpha}{\partial t^2} = \rho F + \text{div}_S(T).$$

Here $G = \alpha'^* \alpha'$ is the $2 \times 2$ matrix formed from the Jacobian matrix $\alpha'$ of $\alpha$. The angular momentum equation requires that $Tw$ be tangential for every tangential vector field $w$. To satisfy this and the other requirements of the angular momentum equation, we make the following assumption:

**Assumption:**    There is a scalar field $f$ such that

$$Tw = fw,$$

for every tangential vector field $w$ along $S$.

With this assumption, the calculation of the surface divergence of $T$ gives (exercise)

$$\text{div}_S(T) = \frac{\partial f}{\partial s_1} e^1 + \frac{\partial f}{\partial s_2} e^2 + \mu f n.$$

Here $e^1, e^2$ are the reciprocal basis vectors associated with $e_1 = \partial\alpha/\partial s_1$, $e_2 = \partial\alpha/\partial s_2$, $n$ is the unit normal vector field, and $\mu$ is the mean curvature of the surface (actually, twice the mean curvature in the customary usage of the term). See Appendix A for the definitions. Using this, we can write the linear momentum equation as

$$\frac{\partial^2 \alpha}{\partial t^2} = F + \frac{1}{\rho}\frac{\partial f}{\partial s_1}e^1 + \frac{1}{\rho}\frac{\partial f}{\partial s_2}e^2 + \frac{1}{\rho}\mu f n.$$

This can be expressed in component form with respect to the moving frame $\{e_1, e_2, n\}$ of basis vectors as follows. First, use the expression for $\rho$ that results from the conservation of mass equation

$$\rho = \frac{\rho_0 \det G_0^{1/2}}{\det G^{1/2}}.$$

Then the two tangential components and normal components of the equation of motion for the membrane are:

$$\frac{\partial^2 \alpha}{\partial t^2} \cdot e_1 = F \cdot e_1 + \frac{\det G^{1/2}}{\rho_0 \det G_0^{1/2}} f_{s_1}, \tag{11.33}$$

$$\frac{\partial^2 \alpha}{\partial t^2} \cdot e_2 = F \cdot e_2 + \frac{\det G^{1/2}}{\rho_0 \det G_0^{1/2}} f_{s_2}, \tag{11.34}$$

$$\frac{\partial^2 \alpha}{\partial t^2} \cdot n = F \cdot n + \frac{\det G^{1/2}}{\rho_0 \det G_0^{1/2}} \mu f. \tag{11.35}$$

These are the equations governing the motion of an elastic membrane and are quite general in form. The special case when there is no motion is also included in these equations. In particular, if there are no external forces ($F = 0$) and constant surface tension ($f = $ constant), then the first two equations are satisfied if $\alpha$ does not depend on time (so $\partial^2 \alpha / \partial t^2 = 0$), and the third equation will be satisfied if and only if the mean curvature $\mu$ of $\alpha$ vanishes. Thus the PDE

$$\mu = 0 \qquad \textbf{(Minimal Surface Equation)}$$

defines a class of surfaces, each of which can be viewed as a membrane in equilbrium under constant surface tension and no (or at least negligble) external forces (See Appendix A for the expression for $\mu$ in terms of the partial derivatives of $\alpha$). These surfaces, called *minimal surfaces*, have been well studied, and a few special examples will be considered in the exercises.

To simplify the discussion, we assume that the motion is such that at each time $t$ the surface $S_t$ is the graph of a function of two variables. Specifically,

**Assumption**

$$\alpha(x, y, t) = (x, y, u(x, y, t)),$$

for $(x, y) \in R$, where $R$ is a domain in the $x$-$y$ plane.

Here $x, y$ denote the parameters that before were denoted by $s_1, s_2$. In this situation, the membrane is viewed as a thin material stretched over the region $R$ like a drumhead and vibrating above and below the $x$-$y$ plane with $u(x, y, t)$ denoting the displacement from this plane at time $t$. With this assumption, the three components of the equation of motion reduce to

$$u_{tt}u_x = F^1 + F^3 u_x + \frac{\det G^{1/2}}{\rho_0 \det G_0^{1/2}} f_x, \tag{11.36}$$

$$u_{tt}u_y = F^2 + F^3 u_y + \frac{\det G^{1/2}}{\rho_0 \det G_0^{1/2}} f_y, \tag{11.37}$$

$$u_{tt} = -F^1 u_x - F^2 u_y + F^3 + \frac{\det G}{\rho_0 \det G_0^{1/2}} \mu f. \tag{11.38}$$

*Note*: There is *no* exponent $1/2$ on $\det G$ in equation (11.38). These equations for the motion of the elastic membrane (or its static equilibrium position when there is no motion) constitute a system of nonlinear PDEs for the unknown function $u$. Some of the nonlinearities come from the area factor

$$\det G = 1 + u_x^2 + u_y^2 \tag{11.39}$$

and the mean curvature

$$\mu = \frac{(1 + u_y^2)u_{xx} + (1 + u_x^2)u_{yy} - 2u_x u_y u_{xy}}{\det G^{3/2}}. \tag{11.40}$$

The PDEs (11.36)–(11.38) are most difficult to study, but with a few simplifying assumptions, we can reduce them to the 2-D wave equation, which will very nearly approximate the true motion of the membrane.

**Assumptions**

(1)  $F^1 = 0$, $F^2 = 0$ (external forces in the $z$ direction only).

(2)  $f$ is constant, say $f = k$.

(3)  $\rho_0$ is constant.

(4) Products involving two or more partial dervatives are zero; e.g., $u_x u_y \approx 0$.

The fourth assumption essentially requires small displacements of the membrane from equilibrium position (the $x$-$y$ plane).

   With these assumptions, equations (11.36)–(11.37) are automatically satisfied, and

$$\det G \approx 1,$$

$$\mu \approx u_{xx} + u_{yy}.$$

Thus equation (11.38) reduces to

**The 2-D Wave Equation:**

$$u_{tt} = c^2(u_{xx} + u_{yy}) + F^3,$$

where $c = \sqrt{k/\rho_0}$.

   To proceed further with the analysis of the vibrating membrane, we make the simplifying assumption that the region $R$ is a rectangle: $R = [0, L] \times [0, M]$.
   Much as in the discussion in the last section for the elastic string, it can be argued (see the exercises) that the appropriate boundary conditions are

**Boundary Conditions:**

$$-\kappa_1 u_x(0, y, t) + h_1 u(0, y, t) = g_1(y, t), \tag{11.41}$$

$$\kappa_2 u_x(L, y, t) + h_2 u(L, y, t) = g_2(y, t), \tag{11.42}$$

$$-\kappa_3 u_y(x, 0, t) + h_3 u(x, 0, t) = g_3(x, t), \tag{11.43}$$

$$\kappa_4 u_y(x, M, t) + h_4 u(x, M, t) = g_4(x, t), \tag{11.44}$$

for all $0 \leq x \leq L$ and $0 \leq y \leq M$. The physical interpretation and implementation of the BCs (11.41)–(11.44) are similar to those presented for the elastic string in the last section.

*Note*: As we have seen in the study of the heat equation, the homogeneous case of these BCs ($g_i = 0, i = 1, 2, 3, 4$) leads to two Sturm–Liouville problems with eigenvalues $-\lambda_n^2$ and $-\mu_m^2$ that are negative or zero, and so for the 2-D problem the eigenvalues $-(\lambda_n^2 + \mu_m^2)$ are zero or negative. Further, the 2-D problem has zero as an eigenvalue only in the case that $h_1 = 0 = h_2$ and $h_3 = 0 = h_4$.

The initial conditions for the membrane consist in specifying the initial displacement and initial velocty for each point on the membrane:

**Initial Conditions:**

$$u(x, y, 0) = \phi(x, y),$$

$$u_t(x, y, 0) = \psi(x, y),$$

for all $x, y$ in $R$.

Based on our experience with the 2-D heat equation and the study of the 1-D wave equation in the last section, it is easy to write out a prescription for the solution of the 2-D wave equation in terms of double series of Sturm–Liouville eigenfunctions. The details of this are left to the exercises.

**Example 11.4**   The simplest example of the 2-D wave IBVP is a homogeneous one, like

$$u_{tt} = c^2(u_{xx} + u_{yy}) \qquad \text{in } R, \text{ for } t > 0,$$

$$u = 0 \qquad \text{on } \partial R, \text{ for } t > 0,$$

$$u = \phi \qquad \text{in } R, \text{ for } t = 0,$$

$$u_t = 0 \qquad \text{in } R, \text{ for } t = 0,$$

where for simplicity, we take $R$ to be the unit square. The problem then is of a square membrane, held fixed along its boundary, and displaced from equilibrium initially by a displacement function $\phi$, which we take to be

$$\phi(x, y) = \begin{cases} 0.25 \sin 2\pi x \sin 2\pi y & \text{if } (x, y) \in [0, 1/2] \times [0, 1/2], \\ 0 & \text{otherwise.} \end{cases}$$

The solution of the wave IBVP in this case is

$$u(x, y, t) = \sum_{n=1}^{\infty} \sum_{m=1}^{\infty} A_{nm} \cos[(n^2 + m^2)^{1/2}\pi t] \sin n\pi x \sin m\pi y,$$

where the $A_{nm}$ are the Fourier coefficients of $\phi$. We have already computed these coefficients in an example on 2-D heat flow. We found that $A_{nm} = f_n f_m$, where

$$f_n = \frac{\sin[(n-2)\pi/2]}{(n-2)\pi} - \frac{\sin[(n+2)\pi/2]}{(n+2)\pi}$$

for $n \neq 2$, and

$$f_2 = 1/2$$

for $n = 2$. Figure 11.13 shows the motion of the membrane during the first second of the motion (we take $c = 1$ here). The second frame (at $t = 1/7$ sec) of the animation shows that the membrane has dropped quickly toward equilibrium position and at the same time has separated into two wave crests, one propagating parallel to the $x$-axis, and the other parallel to the $y$-axis. At $t = 2/7$ sec, part of the membrane reaches what appears to be a maximum displacement below equilibrium, and this tends to form the two propagating wave crests into a single crescent-shaped wave crest propagating symmetrically toward each of the opposite boundaries. The ends of the crescent reach the boundary first (somewhere

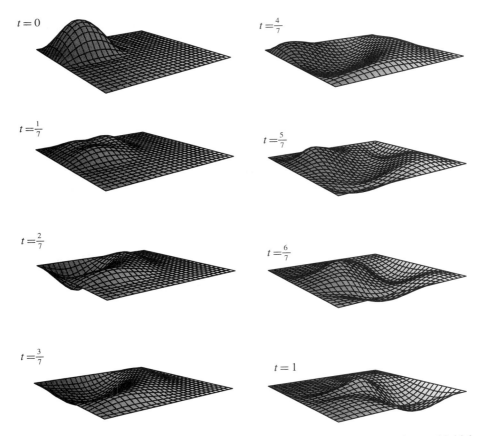

Figure 11.13.   Motion of the square membrane in Example 11.4 with fixed boundary and initial sinusoidal displacement near one corner.

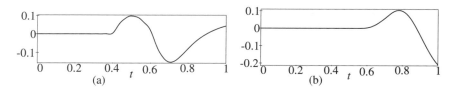

Figure 11.14.    (a) Plots of the the displacements $u(0.9, 0.1, t)$ and $u(0.1, 0.9, t)$ at the points $(0.9, 0.1)$ and $(0.1, 0.9)$. Note that these are identical. (b) Plot of the displacement at the point $(0.9, 0.9)$.

between 3/7 sec and 4/7 sec), while the middle (high point), propagating along the diagonal, appears to reach the opposite vertex at about $t = 5/7$ sec. Meanwhile, the initially displaced portion of the membrane has vibrated back across equilibrium, and a wave begins propagating again toward the opposite boundary. Eventually (by the last frame), this wave meets and combines with the reflection of the original crescent-shaped wave. An animation will give you a qualitative feel for the overall motion of the membrane, but you should probably use more than eight frames. The still pictures are best for a detailed analysis of complex motions. The comments about the times for the ends and the middle of the crescent wave to reach the opposite boundary can be clarified and made more precise. Namely, Figure 11.14 shows plots of the displacements at the points $(0.9, 0.1)$, $(0.1, 0.9)$ and $(0.9, 0.9)$ up until time $t = 1$. Note that the displacements of the first two points over time is identical, thus showing the symmetry in the wave propagation. Also, we can readily see that these two points are motionless until about time $t = 0.4$, when the propagating wave reaches these points. Similarly, the point $(0.9, 0.9)$ is motionless until about time $t = 0.6$.

**Example 11.5**    This example is similar to the last one, except that we change the BCs to those shown in Figure 11.15. Thus while holding the membrane fixed on the left and right boundaries, we allow it to move vertically over the top and bottom boundaries as long as it maintains there a horizontal tangent line (in the

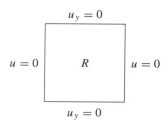

Figure 11.15.    BCs for the square membrane in Example 11.5.

$y$ direction). Additionally, we change the initial displacement to

$$\phi(x, y) = 0.5 K_{0,1/2,1}(x) K_{0,1/2,1}(y),$$

where $K_{a,b,L}$ is the parabolic pulse defined by

$$K_{a,b,L}(x) = \begin{cases} 0 & \text{for } 0 \le x < a, \\ -4(x-a)(x-b)/(b-a)^2 & \text{for } a \le x \le b, \\ 0 & \text{for } b < x \le 1. \end{cases}$$

The solution the IBVP now has the form

$$u(x, y, t) = \sum_{n=1}^{\infty} \sum_{m=0}^{\infty} A_{nm} \cos a_{nm} t \, \sin n\pi x \cos m\pi y,$$

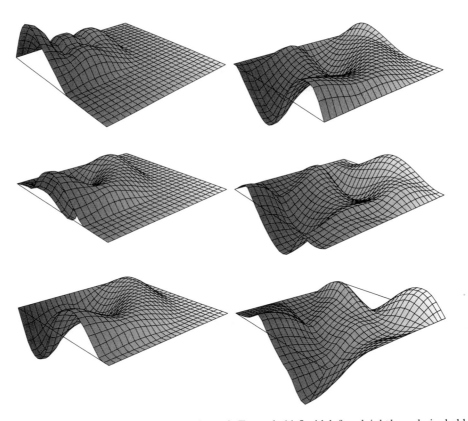

Figure 11.16.    Motion of the square membrane in Example 11.5 with left and right boundaries held fixed while the top and bottom boundaries are allowed to displace while maintaining a horizontal tangent line in the $y$ direction. The initial displacement is a parabolic pulse centered at $(1/4, 1/4)$.

where $a_{nm} = (n^2 + m^2)^{1/2}\pi$ and the $A_{nm}$ are the Fourier coefficients of $\phi$. Since $\phi$ has a special form, we find for these

$$A_{nm} = K_{n,0,1/2,1}\,K^*_{m,0,1/2,1},$$

where $K_{n,a,b,L}$ are the Fourier sine coefficients of $K_{a,b,L}(x)$, and $K^*_{m,a,b,L}$ are the Fourier cosine coefficients of the same function. The sine coefficients are given in the last exercise set, while the cosine coefficients are found to be

$$K^*_{m,a,b,L} = \tag{11.45}$$

$$\frac{8L}{m^2\pi^2(b-a)^2}\left[\frac{2L}{n\pi}\left(\sin\frac{m\pi b}{L} - \sin\frac{m\pi a}{L}\right) - (b-a)\left(\cos\frac{m\pi b}{L} + \cos\frac{m\pi a}{L}\right)\right]$$

for $m \neq 0$, and

$$K^*_{0,a,b,L} = \frac{2(b-a)}{3L}$$

for $m = 0$. For comparison with the last example, Figure 11.16 shows six frames from the motion up to time $t = 1$.

## Exercises 11.2

1. Use the definition of the surface divergence $\mathrm{div}_S$ from Appendix B to show, under the assumption made on $T$, that

$$\mathrm{div}_S(T) = \frac{\partial f}{\partial s_1}e^1 + \frac{\partial f}{\partial s_2}e^2 + \mu f n.$$

   *Hint:* use the first expression for $\mathrm{div}_S$ given in Lemma B.2 to show that

$$\mathrm{div}_S(T) = \det G^{-1/2}\left[\frac{\partial}{\partial s_1}(T(e_2 \times n)) - \frac{\partial}{\partial s_2}(T(e_1 \times n))\right].$$

   Then use this and the assumption $Tw = fw$ for tangential vector fields $w$ to get the result. (*Note:* $\det G^{1/2} = |e_1 \times e_2|$.)

2. Use the definition of the mean curvature normal $\mu$ from Appendix A to show that $\mu$ is given by equation (11.40) when $\alpha(x, y) = (x, y, u(x, y))$.

3. (*Minimal surfaces*) As mentioned in the text, static solutions (i.e., ones not depending on $t$) of the continuum equations with no external forces and constant surface tension are surfaces with vanishing mean curvature: $\mu = 0$. They are called minimal surfaces since they satisfy the extremal equations for minimizing the surface area integral. The following are some examples of minimal surfaces. For each example, (i) show

that $\mu = 0$; (ii) do several plots of the surface (especially Enneper's surface, which is difficult to visualize); (iii) compute the Gaussian curvature $\kappa$ (extra credit: produce plots in part (ii) that are "colored by Gaussian curvature.") and compute the area of the surface.

(a) (*The helicoid*):

$$\alpha(r, \theta) = (r \cos \theta, r \sin \theta, a\theta).$$

Here $a > 0$, and there is no restriction on the parameters $(r, \theta)$. For the plots and the area calculation, assume that the parameter domain is $U = [-1, 1] \times [0, 4\pi]$ and $a = 1$.

(b) (*The catenoid*):

$$\alpha(x, \theta) = (x, \cosh x \cos \theta, \cosh x \sin \theta).$$

This is the surface obtained by revolving the catenary (graph of $y = \cosh x$) about the $x$-axis. For the plots and area calculation use parameter domain $U = [-1, 1] \times [0, 2\pi]$.

(c) (*Enneper's minimal surface*):

$$\alpha(s_1, s_2) = (s_1 - s_1^3/3 + s_1 s_2^2, \ s_2 - s_2^3/3 + s_1^2 s_2, \ s_1^2 - s_2^2).$$

Here $(s_1, s_2)$ can be any point in $\mathbb{R}^2$, but for the plots and the computation of the area, use for parameter domain $U = [-1, 1] \times [-1, 1]$ and $U = [-2, 2] \times [-1, 1]$.

(d) (*Scherk's minimal surface*):

$$\alpha(x, y) = \left( x, \ y, \ \ln \frac{\cos y}{\cos x} \right),$$

where the parameters are restricted to $U = [0, \pi/2) \times [0, \pi/2)$.

4. Suppose $\gamma(t) = (f(t), g(t))$ for $t \in I$ is a curve in the $x$-$y$ plane. The corresponding surface of revolution, obtained by revloving $\gamma$ about the $x$-axis, is

$$\alpha(t, \theta) = (f(t), \ g(t) \cos \theta, \ g(t) \sin \theta),$$

for $(t, \theta) \in I \times [0, 2\pi]$ (verify this!).

(a) Compute the Gaussian and mean curvatures for such a surface of revolution.

(b) The PDE for a minimal surface reduces to an ODE when the surface is a surface of revolution. Show that the only surfaces of revloution that are minimal surfaces are planes and catenoid type surfaces (see [Kl 78]).

5. (*Boundary conditions*) The general boundary conditions for a membrane $\alpha_t : U \rightarrow \mathbb{R}^3$ are

$$mT\nu + h\alpha = \beta$$

on $\partial S$. Here $\nu$ is the outward-directed normal to the boundary of the surface: It lies in the tangent plane to $S$, is normal to the boundary curve $\partial S$, and is directed "outward" (see Definition B.3 in Appendix B for the technical definition). The term $\beta = (\beta^1, \beta^2, \beta^3)$ is a given function that can vary from point to point along the boundary and also can depend on time. Likewise, $m$ and $h$ are such functions, but for simplicity, here we will always assume that they are constants. The assumptions on the stress tensor from the text give $T\nu = f\nu$, where $f$ is a scalar function. This exercise examines a few details connected with such BCs.

(a) The general boundary conditions contain as special cases (i) specified displacement of the boundary $\alpha = \beta$, on $\partial S$, and (ii) traction proportional to the difference between specified and actual displacements $T\nu = h(\beta - \alpha)$ on $\partial S$. For each of these cases, discuss the actual physical implementation of the BCs.

(b) Under the assumptions in the text, show how to get the boundary conditions (11.41)–(11.44) from the general boundary conditions above. For this, show that from the assumptions in the text, the normal to the surface is

$$n \approx e_1 \times e_2 = (-u_x, -u_y, 1).$$

Use this and the fact that the outward-directed normals to the four parts of the boundary are $\nu = \pm\varepsilon_i \times n$, $i = 1, 2$, where $\varepsilon_1 = (1, 0, 0)$, $\varepsilon_2 = (0, 1, 0)$.

6. Describe the general series solution of the semi-homogeneous wave problem for the rectangle,

$$u_{tt} = c^2(u_{xx} + u_{yy}) + F,$$

$$-\kappa_1 u_x(0, y, t) + h_1 u(0, y, t) = 0,$$

$$\kappa_2 u_x(L, y, t) + h_2 u(L, y, t) = 0,$$

$$-\kappa_3 u_y(x, 0, t) + h_3 u(x, 0, t) = 0,$$

$$\kappa_4 u_y(x, M, t) + h_4 u(x, M, t) = 0,$$

for $x \in [0, L]$, $y \in [0, M]$, and $t > 0$. Give explicit integral formulas for the time-dependent factors $T_{nm}(t)$ in the series solution. Also, describe how to transform any nonhomogeneous problem to a semi-homogeneous one.

7. For the homogeneous wave equation $u_{tt} = u_{xx} + u_{yy}$, with the membrane fixed at the boundary ($u = 0$, on $\partial\Omega$, for $t > 0$) and with zero initial velocities ($u_t = 0$, in

$\Omega$ for $t = 0$), find the series solution that has initial displacements

$$u(x, y, 0) = K_{a,b,L}(x)K_{a',b',M}(y),$$

for $x, y \in \Omega = [0, L] \times [0, M]$. Use a computer to study the motion for the cases (i) $(a, b, L) = (0, 8, 8)$, $(a', b', M) = (0, 2, 4)$; (ii) $(a, b, L) = (0, 4, 8)$, $(a', b', M) = (0, 2, 4)$; and (iii) $(a, b, L) = (0, 4, 12)$, $(a', b', M) = (0, 2, 4)$. Determine the times it takes the waves to reach the opposite boundaries. Examine the motion of the point of original maximal displacement. Compare and contrast the type of motion in the three cases. In going from (ii) to (iii), the initial disturbance has the same geometry, but the membrane increases in length in the $x$ direction. Investigate the effect of further such increases.

8. Rework Exercise 7, except now assume that the membrane is fixed on only threee sides of the rectangle and free on the remaining side (i.e., horizontal tangent line in the direction perpendicular to that side). There are four choices for the free side, but you may do just the two cases when it is (1) $x = L$ or (2) $y = M$.

9. Rework Exercise 7, except now assume that the membrane is fixed on only two sides of the rectangle and free on the remaining two sides (i.e., horizontal tangent line in the direction perpendicular to those sides). There are four choices for the two free sides, but you may do just the two cases when they are (1) $x = L, y = M$ or (2) $y = 0, y = M$.

10. Rework Exercise 7, except now assume that the membrane is fixed on only one side of the rectangle and free on the remaining three sides (i.e., horizontal tangent line in the direction perpendicular to those sides). There are four choices for the three free sides, but you may do just the two cases when they are (1) $x = 0, x = L, y = M$ or (2) $x = L, y = 0, y = M$.

11. Find the solution of the homogeneous wave equation $u_{tt} = u_{xx} + u_{yy}$, subject to the BCs

$u = Ax/L$

$$\boxed{\begin{array}{c} R \end{array}}$$

$u = 0$ (left)     $R$     $u = Ay/M$ (right)

$u = 0$ (bottom)

and the ICs

$$u(x, y, 0) = \frac{A}{LM}xy + \frac{1}{10}\sin\frac{\pi}{L}x \sin\frac{\pi}{M}y,$$

$$u_t(x, y, 0) = 0,$$

for $x \in [0, L]$, $y \in [0, M]$. *Hint:* You will *not* need a series for this. Use a computer to analyze the motion for $A = 0$ and $A = 1$ and your choice of $L$ and $M$.

In Exercises 12–15, find the series solution of the homogeneous wave equation $u_{tt} = u_{xx} + u_{yy}$ that satisfies the given BCs and ICs. The function $K_{a,b,L}$ is the parabolic pulse function (see text). Use a computer to analyze the motion for the following two choices for initial velocities: (i) $(a, b, L) = (1, 2, 4)$, $(a', b', M) = (1, 2, 4)$ and (ii) $(a, b, L) = (1, 2, 4)$, $(a', b', M) = (1, 2, 8)$. Determine the times it takes the waves to first reach the boundaries. Compare and contrast the motions for the various cases. Identify and plot the steady state (time-independent solution of the PDE + BCs).

12. Boundary conditions:

$$u = Ax/L$$

$$u = 0 \qquad R \qquad u = Ay/M$$

$$u = 0$$

Initial conditions:

$$u(x, y, 0) = \frac{A}{LM} xy,$$

$$u_t(x, y, 0) = K_{a,b,L}(x) K_{a',b',M}(y),$$

for $x \in [0, L]$, $y \in [0, M]$. Do the analysis for $A = 0$ and $A = 1$.

13. Boundary conditions:

$$u = \sin \tfrac{\pi}{L} x$$

$$u = 0 \qquad R \qquad u = 0$$

$$u = 0$$

Initial conditions:

$$u(x, y, 0) = S(x, y),$$

$$u_t(x, y, 0) = K_{a,b,L}(x) K_{a',b',M}(y),$$

for $x \in [0, L]$, $y \in [0, M]$. Here $S$ is a solution of Laplace's equation $S_{xx} + S_{yy} = 0$ that satisfies the same BCs as $u$.

14. Boundary conditions:

$$u = \sin \tfrac{\pi}{L}x$$

$u = 0$    $R$    $u = 0$

$$u = 0$$

Initial conditions:

$$u(x, y, 0) = y \sin \frac{\pi}{L}x,$$

$$u_t(x, y, 0) = K_{a,b,L}(x)K_{a',b',M}(y),$$

for $x \in [0, L]$, $y \in [0, M]$.

15. Boundary conditions:

$$u = \sin \tfrac{\pi}{L}x$$

$u_x = 0$    $R$    $u_x = 0$

$$u_y = 0$$

Initial conditions:

$$u(x, y, 0) = S(x, y),$$

$$u_t(x, y, 0) = K_{a,b,L}(x)K_{a',b',M}(y),$$

for $x \in [0, L]$, $y \in [0, M]$. Here $S$ is a solution of Laplace's equation $S_{xx} + S_{yy} = 0$ that satisfies the same BCs as $u$.

16. (*Flag waving*) Find the series solution of the homogeneous wave equation: $u_{tt} = u_{xx} + u_{yy}$ subject to the BCs

$$u = g(x, t)$$

$u_x = 0$    $R$    $u_x = 0$

$$u = 0$$

and ICs $u = 0$, $u_t = 0$ at time zero. Do a computer analysis of the motion for the two cases (i) $g(x, t) = A \sin \omega t$ and (ii) $g(x, t) = A \sin \omega t \, \sin \frac{2\pi}{L} x$. Use $L = 1$, $M = 4$ and experiment with various choices of $\omega$ and $A > 0$ to produce few and many waves traveling toward the boundary $y = 0$.

17. Find the series solution of the semi-homogeneous problem

$$u_{tt} = u_{xx} + u_{yy} + A \sin \omega t \, K_{a,b,L}(x) K_{a',b',M}(y),$$

with $u = 0$ on the boundary and no initial displacements or velocities imparted to the membrane. Analyze the resulting motion on a computer for $L = 4$, $M = 8$ and various choices of $a$, $b$, $a'$, $b'$ (i.e., location of the parabolic disturbance) and $A$, $\omega$ (amplitude and frequency of the disturbance).

## 11.3.  Solids: 3-D Elastic Materials

We come finally to the the discussion of the motion of a three-dimensional, elastic solid that, from the Lagrangian viewpoint, is modeled by a region $\Omega_t$ in $\mathbb{R}^3$, indexed by the time $t$, and that for various times constitutes the motion of the solid. Analytically, the motion is given by a map $\alpha_t : \Omega \subseteq \mathbb{R}^3 \to \mathbb{R}^3$ with components

$$\alpha_t(s) = (\alpha_t^1(s), \alpha_t^2(s), \alpha_t^3(s)),$$

where $s = (s_1, s_2, s_3)$ is a point in the parameter domain $\Omega$. At time 0, the initial (perhaps stressed) configuration of the solid is $\Omega_0 = \alpha_0(\Omega)$, and over the time interval $t$, the solid moves and deforms into the configuration $\Omega_t = \alpha_t(\Omega)$. For convenience of notation we will often suppress the $t$ from the notation and just write $\alpha$ for $\alpha_t$.

The conservation of mass and linear momentum equations are (see Appendix B)

$$\rho \det G^{1/2} = \rho_0 \det G_0^{1/2}, \tag{11.46}$$

$$\rho \frac{\partial^2 \alpha}{\partial t^2} = \rho F + \mathrm{div}_\Omega(T). \tag{11.47}$$

In these equations, $\rho$, $T$, and $F$ are respectively the *Lagrangian* mass density, stress tensor, and external force density. The matrix $G$ is the $3 \times 3$ matrix $G = \alpha'^* \alpha'$, where $*$ denotes matrix transpose. The solid divergence $\mathrm{div}_\Omega$ is (see Appendix B)

$$\mathrm{div}_\Omega(T) = \det G^{-1/2} \frac{\partial}{\partial s_j} (\det G^{1/2} \, T e^j), \tag{11.48}$$

where we are using implied summation on the repeated index $j$. Here $e^1, e^2, e^3$ are the reciprocal basis vector fields associated with $e_j = \partial\alpha/\partial s_j$, $j = 1, 2, 3$.

The conservation of mass and linear momentum equations are the two principal equations for modeling the motion of a continuum from the Lagrangian point of view. We now want to specialize the discussion to elastic materials (technically, elastic, isotropic, and materially uniform materials), and for this we must specify how the stress tensor depends on the deformations $\alpha_t$. Deformations of $\Omega$ are also called strains, and each strain produces stresses in material comprising the solid. The general theory that describes the relation between stresses and strains, based on the assumed constitution of the solid, is called the *theory of constitutive equations* (see [MH 83], [Tru 77], [Mal 69], [Fu 65]). From this theory, we are led to the following definition of elastic materials:

**Definition 11.1 (Hooke's Law)**    The *strain tensor* (sometimes called the large strain tensor) is the second rank tensor field

$$E = \frac{1}{2}[G - I],$$

where $I$ is the $3 \times 3$ identity matrix. The continuum (solid) is called an *elastic material* (or more precisely, an elastic, isotropic, materially uniform material) if its stress tensor has the form

$$T = \lambda \operatorname{tr}(E)I + 2\mu E \qquad \text{(Hooke's law)}, \qquad (11.49)$$

where $\lambda$ and $\mu$ are positive constants (called the Lamé moduli). Equation (11.49), Hooke's law, basically postulates a particular type of linear relation between the stresses $T$ and strains $E$.

We see from Hooke's law that $T$ depends in a nonlinear way on $\alpha$, and thus the equation of motion (linear momentum equation) is a nonlinear PDE for $\alpha$. However, in linear elasticity theory, certain natural assumptions are made so that the equation of motion reduces to a linear PDE.

**Assumption 1:**    For convenience, assume that $\alpha_t : \Omega \to \mathbb{R}^3$ has the form

$$\alpha(x, y, z, t) = (x, y, z) + u(x, y, z, t),$$

with $u(x, y, z, t) = (u^1(x, y, z, t), u^2(x, y, z, t), u^3(x, y, z, t))$ representing the displacement from the unstressed position $(x, y, z)$. Here $x, y, z$ denote the parameters that before were denoted by $s_1, s_2, s_3$.

To linearize, we make the following assumption, in the equations of motion (11.46)–(11.47):

**Assumption 2:**    All products involving $u$ and its partial derivatives with two or more factors are zero; e.g.,

$$\frac{\partial u^k}{\partial x}\frac{\partial u^p}{\partial y} \approx 0, \;\; u^k\frac{\partial u^p}{\partial z} \approx 0, \ldots,$$

for all $k, p$.

From the first assumption, we get $\alpha' = I + u'$, and then the second assumption leads to the following approximation:

$$E = \frac{1}{2}[G - I] = \frac{1}{2}[\alpha'^*\alpha' - I]$$

$$= \frac{1}{2}[(I + u'^*)(I + u') - I]$$

$$= \frac{1}{2}[u' + u'^* + u'^*u']$$

$$\approx \frac{1}{2}[u' + u'^*].$$

The expression in the last line above is called the *small strain tensor*, and from here on we will assume that $E$ is equal to this. In terms of this, the stress tensor (via Hooke's law) is

$$T = \lambda \operatorname{div}(u)I + \mu(u' + u'^*).$$

Next, note that because of the second assumption, we have

$$\det G^{1/2} = \det(\alpha') = \det(I + u') \approx 1 + \operatorname{div}(u),$$

and from this and the form of the stress tensor we have

$$\det G^{1/2}Te^j \approx T^j$$

(exercise). Thus in the linear momentum equation, we have

$$\operatorname{div}_\Omega(T) = \det G^{-1/2}\operatorname{div}(T).$$

An easy calculation (as in Chapter 10 if you studied that chapter) gives

$$\operatorname{div}(T) = \mu \nabla^2 u + (\lambda + \mu) \nabla \operatorname{div}(u)$$

(exercise). Thus if we use the conservation of mass equation, then $\det G^{1/2} \rho = \rho_0 (1 + \operatorname{div}(u_0))$. The linear momentum equation can be written as

$$\rho_0 (1 + \operatorname{div}(u_0)) \frac{\partial^2 u}{\partial t^2} = \mu \nabla^2 u + (\lambda + \mu) \nabla \operatorname{div}(u) + \rho_0 (1 + \operatorname{div}(u_0)) F.$$

To simplify this further, we make the assumption that the initial displacement from equilibrium position has zero divergence (approximately):

**Assumption 3:**   $\operatorname{div}(u_0) \approx 0$.

Thus we arrive at a linear PDE for the displacement $u$ of the elastic body from its equilibrium (unstressed) position:

**Linear Elasticity Equation of Motion:**

$$\rho_0 \frac{\partial^2 u}{\partial t^2} = \mu \nabla^2 u + (\lambda + \mu) \nabla \operatorname{div}(u) + \rho_0 F. \tag{11.50}$$

This is also called the *Navier equation*. It is the principal equation for the motion, since the mass density $\rho = \rho_0 (1 + \operatorname{div}(u))^{-1}$ is determined once $u$ is known. *Note*: We will limit the discussion to the case where $\rho_0$ is constant.

The theory of linear elasticity is based upon the linear PDE (11.50) for $u$ supplemented with appropriate boundary conditions and initial conditions. This initial–boundary value problem is called the *Navier IBVP*. The boundary conditions are of three types:

**(1) Specified Displacement on the Boundary:**

$$u = g \qquad \text{on } \partial\Omega.$$

**(2) Specified Traction (Stress) on the Boundary:**

$$Tn = g \qquad \text{on } \partial\Omega.$$

**(3) A Mixture of these BCs:**   with displacement being specified on part of the boundary and traction on the remaining part.

Here $g$ is a given, time-dependent vector field on $\partial\Omega$. For the traction boundary condition there is a solvability condition that $g$ must satisfy in order for the Navier IBVP to have a solution. This is discussed in the exercises.

The initial conditions specifiy just the initial displacements and initial velocities at each of the points of $\Omega$:

**Initial Conditions:**

$$u(x, y, z, 0) = \phi(x, y, z),$$

$$u_t(x, y, z, 0) = \psi(x, y, z),$$

for all $(x, y, z) \in \Omega$. Here $\phi$ and $\psi$ are given vector fields.

**Example 11.6   (Axial and Transverse Waves)**   As a simple but somewhat idealized example of the propagation of waves in elastic solids, consider the case where $\Omega = [0, L] \times \mathbb{R} \times \mathbb{R}$ is the slab bounded between the planes $x = 0$ and $x = L$. We consider the homogeneous problem $F = 0$ and BC $u = 0$ on $\partial\Omega$ and look for solutions that depend only on $x$ and $t$:

$$u(x, t) = (u^1(x, t), u^2(x, t), u^3(x, t)).$$

Such displacements $u$ describe special types of vibrations of $\Omega$, ones for which at each time $t$ and on any plane $x = a$ in $\Omega$, the displacment $u(a, t)$ is the same at all points in this plane. The initial conditions have to depend on $x$ only in order to allow for such vibrations. We assume no initial velocities and initial displacements $\phi(x) = (\phi^1(x), \phi^2(x), \phi^3(x))$.

For $u$ of the assumed form, we find that $\operatorname{div}(u) = u_x^1$ and $\nabla \operatorname{div}(u) = (u_{xx}^1, 0, 0)$. Consequently, the Navier system of PDEs reduces to three separate (1-dimensional) wave equations:

$$u_{tt}^1 = c^2 u_{xx}^1,$$

$$u_{tt}^2 = k^2 u_{xx}^2,$$

$$u_{tt}^3 = k^2 u_{xx}^3,$$

where $c = \sqrt{(\lambda + 2\mu)/\rho_0}$ and $k = \sqrt{\mu/\rho_0}$. These latter constants represent the speeds of the respective waves, and since $c > k$, the wave $u^1$ propagates more quickly along the $x$-axis than do the the two transverse waves $u^2, u^3$ in planes perpendicular to the $x$-axis. Solving the three separate wave equations gives the following expression for $u$:

$$u(x,t) = \sum_{n=1}^{\infty} \left[ \left( \cos \frac{n\pi c}{L} t \right) (A_n^1, 0, 0) + \left( \cos \frac{n\pi k}{L} t \right) (0, A_n^2, A_n^3) \right] \sin \frac{n\pi}{L} x.$$

(11.51)

Here $A_n^1$, $A_n^2$, $A_n^3$ are the Fourier sine coefficients of $\phi^1$, $\phi^2$, $\phi^3$ respectively.

To illustrate the effect of the different speeds $c$, $k$ of propagation on the overall motion, we consider the very simple case where the initial displacement $\phi$ has components $\phi^1 = \phi^3 = 0.1 \sin \frac{\pi}{L} x$ and $\phi^2 = 0$. Since these are fundamental modes, the displacement reduces from a series to a single term:

$$u(x,t) = \left( 0.1 \cos \frac{\pi c}{L} t \, \sin \frac{\pi}{L} x, \; 0, \; 0.1 \cos \frac{\pi k}{L} t \, \sin \frac{\pi}{L} x \right).$$

To analyze the motion graphically, view the slab $\Omega$ as composed of fibers or line segments, $F_{y_0, z_0} \equiv \{(x, y_0, z_0) \mid 0 \leq x \leq L\}$, parallel to the $x$-axis and running from the plane $x = 0$ to the plane $x = L$. Each fiber undergoes the same motion, so we need to examine the motion of only one of them, say $F_{0,0}$, which is just the interval $[0, L]$ along the $x$-axis. Since the $y$-component of the displacement is zero, the motion of this fiber is entirely in the $x$-$z$ plane. The position of the fiber at each time $t$ is given by

$$\alpha(x, 0, 0, t) = (x, 0, 0) + u(x, t).$$

To plot the successive positions of the fiber, we choose $L = 1$, $\lambda = 1$, $\mu = 1$, so that the speeds are $c = \sqrt{3}$, $k = 1$. Figure 11.17 shows two sets of curves. One set is the plot of $\alpha(x, 0, 0, t)$ for times $t = 0, 0.1, 0.2, \ldots, 0.9, 1$ and displays the initial sinusoidal position of the fiber (top curve), the subsequent vibration down to the flat equilibrium position (at $t = 0.5$), and the final ($t = 1$) position below equilibrium. The other set of curves shows the paths traced out by nine points on the fiber during the overall motion. Figure 11.18 displays similar features of the motion but over the longer time interval $t = 0, \ldots, 2$. The motion of the fiber may at first appear periodic, but Figure 11.18 shows that the nine selected points have not returned to their initial positions after 2 seconds. The question of periodicity is whether the paths traced out will eventually form closed curves. This will be the case only if the speeds $c$, $k$ have a rational ratio. For our choice of constants, the ratio $c/k = \sqrt{3}$ is irrational. Each of the paths is not only not closed, but it is also a *space-filling* curve. This is indicated in Figure 11.19. In the figure, each path appears to be confined to lying inside a certain rectangle and after a long enough time period will appear to pass through every point in this rectangle. Theoretically, this will take infinitely long.

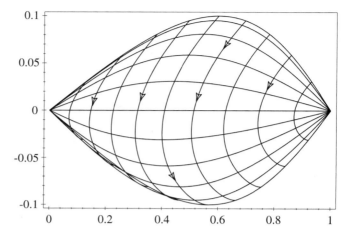

Figure 11.17.    Plots of the fiber $F_{0,0}$ at times $t = 0, 0.1, 0.2, \ldots, 1$ as well as the paths traced out by nine points on the fiber.

Even though the homogeneous Navier equation

$$\rho_0 \frac{\partial^2 u}{\partial t^2} = \mu \nabla^2 u + (\lambda + \mu) \nabla \operatorname{div}(u)$$

is a linear PDE — actually a system of three scalar, linear PDEs — it is *not* in general susceptible to analysis by the separation of variables technique (exercise).

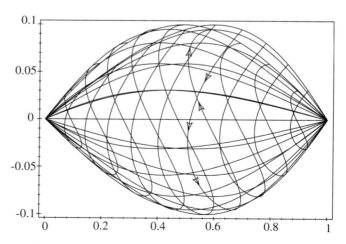

Figure 11.18.    Position of the fiber $F_{0,0}$ and the paths traced out by nine of its points over the time interval $t = 0, \ldots, 2$.

This is primarily because of the term $\nabla \operatorname{div}(u)$, which couples the system of three scalar PDEs together. In addition, there are very few special types of solutions of the Navier IBVP (as in the last example), primarily because of the boundary conditions; and for the most part these special solutions are not three-dimensional in character (see the exercises). If you studied Chapter 10, you will note that the situation here for elastic waves is similar to that for fluid flow there. As we did for fluid flow, when analytic representation of the solution to the IBVP is lacking, we turn to theoretical and numerical descriptions of the solution.

### 11.3.1.   Weak Solutions of the Navier IBVP

The theoretical and numerical descriptions of the solution to the Navier problem are based on rephrasing the problem in terms of its corresponding *integral (or weak) version*, the solutions of which are known as *weak solutions*. We describe this in an elementary way here. See Chapter 10 for a similar discussion on the Navier–Stokes equations and Chapter 13 for a more in-depth discussion of the method as it applies to the heat equation. To keep things simple and give just an overview of the subject, we omit details about differentiability and integrability of the functions involved (as well as other topics from functional analysis).

To motivate how the integral version of the Navier equation arises, we reason as follows. Suppose $u$ is a (suitably differentiable) solution of the Navier

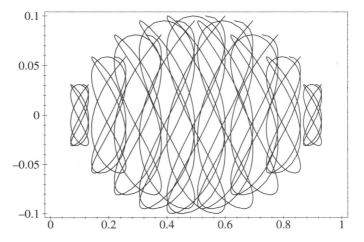

Figure 11.19.   The paths traced out by nine points on the fiber $F_{0,0}$ over the time interval $t = 0, \ldots, 6$.

equation:

$$\rho_0 \frac{\partial^2 u}{\partial t^2} = \mu \nabla^2 u + (\lambda + \mu) \nabla \operatorname{div}(u) + \rho_0 F.$$

If we take the dot product of this equation with a vector field $w$ on $\Omega$ and then integrate over $\Omega$, we find that $u$ satisfies the integral equation

$$\rho_0 \frac{d^2}{dt^2} \int_\Omega u \cdot w \, dV = \int_\Omega [\mu \nabla^2 u + (\lambda + \mu) \nabla \operatorname{div}(u)] \cdot w \, dV + \rho_0 \int_\Omega F \cdot w \, dV.$$
(11.52)

On the left side here, we have interchanged the second-order time derivative with the integral over $\Omega$. We can get a better version of this integral equation if we use integration by parts to reduce the order of the spatial derivatives from second-order to first-order. This integration by parts is interesting enough to record as a separate result:

**Theorem 11.2**    *Suppose $u$ and $w$ are vector fields on $\Omega$ with continuous derivatives to the second order. Let $T$ be the second-rank tensor field*

$$T = \lambda \operatorname{div}(u) I + \mu (u' + u'^*).$$

*Then the following identity holds:*

$$\int_\Omega [\mu \nabla^2 u + (\lambda + \mu) \nabla \operatorname{div}(u)] \cdot w \, dV \qquad\qquad (11.53)$$

$$= \int_{\partial\Omega} Tn \cdot w \, dA - \int_\Omega \left[ \mu (\nabla u^i \cdot \nabla w^i + \frac{\partial u}{\partial x_i} \cdot \nabla w^i) + \lambda \operatorname{div}(u) \operatorname{div}(w) \right] dV.$$

*For convenience of notation here, we are using $(x_1, x_2, x_3) = (x, y, z)$ as the notation for points in $\Omega$. Also, as usual, in terms of components, $u = (u^1, u^2, u^3)$, $w = (w^1, w^2, w^3)$, and there is implied summation over the repeated index $i$ in (11.53).*

**Proof**    This will be easiest to prove if we convert the integrand on the left side back to a divergence; i.e.,

$$\mu \nabla^2 u + (\lambda + \mu) \nabla \operatorname{div}(u) = \operatorname{div}(T).$$

Also note that the $i$th row of $T$ is the vector field

$$T^i = \lambda \operatorname{div}(u)\varepsilon_i + \mu(\nabla u^i + \frac{\partial u}{\partial x_i}),$$

where $\varepsilon_i$ is the $i$th standard basis vector for $\mathbb{R}^3$. Then we simply use the identity $f \operatorname{div}(E) = \operatorname{div}(fE) - \nabla f \cdot E$, which holds for any scalar field $f$ and vector field $E$, and Gauss's divergence theorem to get

$$\int_\Omega \operatorname{div}(T) \cdot w \, dV = \int_\Omega \operatorname{div}(T^i)w^i dV = \int_\Omega [\operatorname{div}(T^i w^i) - T^i \cdot \nabla w^i] dV$$

$$= \int_{\partial\Omega} (T^i \cdot n)w^i \, dA - \int_\Omega \left[ \mu(\nabla u^i \cdot \nabla w^i + \frac{\partial u}{\partial x_i} \cdot \nabla w^i) + \lambda \operatorname{div}(u) \operatorname{div}(w) \right] dV.$$

This completes the proof.

We now rewrite equation (11.52) in a more convenient form. For this we introduce some notation:

**Definition 11.2**   Suppose $u$, $w$ are vector fields on $\Omega$, and let

$$\langle u, w \rangle \equiv \int_\Omega u \cdot w \, dV,$$

$$D(u, w) \equiv - \int_\Omega \left[ \mu(\nabla u^i \cdot \nabla w^i + \frac{\partial u}{\partial x_i} \cdot \nabla w^i) + \lambda \operatorname{div}(u) \operatorname{div}(w) \right] dV.$$

These expressions define functions $\langle \cdot, \cdot \rangle$ and $D$ of two variables (the variables are $u$, $w$), and it is important to note that these functions are *bilinear* functions. That is,

$$\langle au + b\tilde{u}, w \rangle = a\langle u, w \rangle + b\langle \tilde{u}, w \rangle,$$

$$\langle u, aw + b\tilde{w} \rangle = a\langle u, w \rangle + b\langle u, \tilde{w} \rangle,$$

for all vector fields $u$, $\tilde{u}$, $w$, $\tilde{w}$ and constants $a$, $b$. Similarly for $D$. The bilinear form $D$ is called the *Dirichlet form* for the Navier IBVP.

Using this notation, we can write the integral equation (11.52) in the form

$$\rho_0 \frac{d^2}{dt^2} \langle u, w \rangle = D(u, w) + \rho_0 \langle F, w \rangle + \int_{\partial\Omega} Tn \cdot w \, dA. \qquad (11.54)$$

Bear in mind that $u$ is a time-dependent vector field on $\Omega$, and as is customary, its dependence on $t$ is not explicity indicated in writing equations like (11.52) and (11.54). It will be convenient to let $u(t)$ be the vector field $u(t)(x, y, z) \equiv u(x, y, z, t)$.

Our reasoning to this point is that if $u$ is a classical (i.e., suitably differentiable) solution to the Navier equation, then it also satisfies the integral equation (11.54) for every vector field $w$. If we now impose the boundary conditions and initial conditions, we get the following weak versions of the Navier IBVP:

**(1) Weak Version with Specified Boundary Displacement:**

For convenience, we assume that the BC is $u = 0$ on $\Omega$. The general case $u = g$ can always be transformed to one with homogeneous BCs. We let $\mathcal{V}$ be the set of vector fields on $\Omega$ that vanish on the boundary:

$$\mathcal{V} = \{ w \mid w = 0 \text{ on } \partial\Omega \}.$$

Note that if we restrict the integral equation (11.54) to just those $w \in \mathcal{V}$, then the boundary integral there is zero. Thus the weak version of the IBVP can be formulated as, Determine a time-dependent vector field $u$ with $u(t) \in \mathcal{V}$, for all $t$, that satisfies

$$\rho_0 \frac{d^2}{dt^2} \langle u, w \rangle = D(u, w) + \rho_0 \langle F, w \rangle, \tag{11.55}$$

for all $w \in \mathcal{V}$, and that also satisfies the initial conditions

$$u(0) = \phi, \quad u_t(0) = \psi.$$

**(2) Weak Version with Specified Boundary Traction:**

The boundary condition here is $T n = g$ on $\partial\Omega$, and this allows us to rewrite the boundary integral in equation (11.54) with the data $g$ alone. In particular, in the homogeneous BC case where $g = 0$, the boundary integral is zero. Thus for the traction boundary condition, we can here take $\mathcal{V}$ to be the larger set of all vector fields on $\Omega$. The weak version of this IBVP can be formulated as, Determine a vector field $u$ that satisfies

$$\rho_0 \frac{d^2}{dt^2} \langle u, w \rangle = D(u, w) + \rho_0 \langle F, w \rangle + \int_{\partial\Omega} g \cdot w \, dA, \tag{11.56}$$

for all $w$, and that also satisfies the initial conditions

$$u(0) = \phi, \quad u_t(0) = \psi.$$

Having reformulated the Navier IBVP in terms of its weak version — either equation (11.55) or equation (11.56) — we need to have a method for finding solutions of these equations. A solution of the weak version, called a weak solution since it need not be a differentiable function, is determined first by calculating approximations to it and then taking the limit of these approximations to obtain, in theory, a weak solution. This method is known variously as the *Galerkin method* or the *Ritz–Galerkin method*, depending on the specific setting.

The calculation of approximate solutions amounts to selecting a set $\{w_n\}_{n=1}^N$ of linearly independent vector fields in $\mathcal{V}$ and restricting the search for solutions to the subspace $\mathcal{V}_N \subset \mathcal{V}$ that is the span of these vector fields. Thus we look for solutions of the form

$$ u^N(t) = \sum_{n=1}^N b_n(t) w_n, $$

with $b_1, \ldots, b_N$ unknown scalar functions of $t$. These unknowns are determined by substituting the above expression for $u^N$ in the weak equation, equation (11.55) or equation (11.56), and restricting the $w$ there to lie in $\mathcal{V}_N$. This leads to a second-order system of ODE's for the $b_n$'s, which together with the initial conditions will determine these functions and thus give the approximate solution $u^N$. The details of this are left to the exercises. This is also discussed in Chapter 10 for the Navier–Stokes equations and in the exercises in Chapter 13 for the heat equation.

The first part of the Ritz–Galerkin method, which calculates the approximate solutions $u^N$, is known as the *finite element method* when special choices for the basis $\{w_n\}_{n=1}^N$ are made, and it forms an important and practical numerical scheme for PDEs in general. The second part of the Ritz–Galerkin method involves taking limits of the approximate solutions to obtain true solutions to the weak problem. In addition, the question of uniqueness of the weak solution must be addressed as well as whether the weak solution is also a strong solution (i.e., is differentiable and satisfies the original Navier IBVP). We do not discuss these latter aspects here.

## Exercises 11.3

1. This exercise motivates why $E = \frac{1}{2}(G - I)$ is the strain tensor for the solid. Intuitively, strain is $\Delta L/L$, i.e., the change in length divided by the total length. Thus if you stretch a 10 inch metal rod to a length of 11 inches, you have placed a strain of 1/10 on it. This strain produces stresses in the bar, and the strain and stress tensors $E, T$ are introduced to model, at the macroscopic level, the complex process that the metal rod is experiencing at the atomic and molecular levels.

First, assume that $\alpha_0 =$ the identity map, so that $\Omega_0 = \Omega$ is the initial configuration of the solid. Initially, the distance between two points $s$, $s_0$ in $\Omega$ is $|s - s_0|$, and after undergoing deformation over the time interval $t$, their distance apart is $|\alpha_t(s) - \alpha_t(s_0)|$. To get an approximation to the difference of these two distances (lengths), verify first that the Taylor series expansions of $\alpha_t$ about $s_0$ and of the square root function about 1 are

$$\alpha_t(s) = \alpha_t(s_0) + \alpha_t'(s_0)(s - s_0) + \cdots,$$

$$r^{1/2} = 1 + \frac{1}{2}(r - 1) + \cdots.$$

Then use this to show that

$$|\alpha_t(s) - \alpha_t(s_0)| - |s - s_0| \approx |\alpha_t'(s_0)(s - s_0)| - |s - s_0|$$

$$\approx \frac{1}{2}(|\alpha_t'(s_0)(s - s_0)|^2 - |s - s_0|^2)$$

$$= [E(s_0, t)(s - s_0)] \cdot (s - s_0).$$

In the general case, where $\alpha_t = \phi_t \circ \alpha_0$ and $\alpha_0$ is not the identity map, apply the above argument to flow $\phi_t$ and reason why $E = \frac{1}{2}(G - I)$ is an appropriate expression for the strain tensor.

2. Use Assumptions 1 and 2 to show that

   (a) $\text{div}(T) = \mu \nabla^2 u + (\lambda + \mu) \nabla \, \text{div}(u)$.

   (b) $\det(I + u') \approx 1 + \text{div}(u)$.

   (c) $(I + u')^{-1} \approx I - u'$.

   (d) $\det G^{1/2} T e^j \approx T^j$, for $j = 1, 2, 3$. First, show that the rows of the matrix $(\alpha')^{-1}$ are $e^1$, $e^2$, $e^3$. Then observe that the rows of $(\alpha')^{-1} T$ are $T e^1$, $T e^2$, $T e^3$, and thus these latter are approximately the same as $T^1$, $T^1$, $T^3$. Finally, use part (b) to get the end result.

3. This exercise deals with some properties of the Dirichlet form

$$D(u, w) \equiv -\int_{\Omega} \left[ \mu (\nabla u^i \cdot \nabla w^i + \frac{\partial u}{\partial x_i} \cdot \nabla w^i) + \lambda \, \text{div}(u) \, \text{div}(w) \right] dV.$$

   (a) Show that

$$\frac{\partial u}{\partial x_i} \cdot \nabla w^i = \frac{\partial w}{\partial x_i} \cdot \nabla u^i,$$

   for any two vector fields $u$, $w$.

*Note:* There is implied summation throughout this exercise. Use this to show that $D$ is *symmetric:* $D(u, w) = D(w, u)$.

(b)  Show that

$$\nabla u^i \cdot \nabla u^i + \frac{\partial u}{\partial x_i} \cdot \nabla u^i = \sum_{i,j=1}^{3} \frac{1}{2} \left( \frac{\partial u^i}{\partial x_j} + \frac{\partial u^j}{\partial x_i} \right)^2 .$$

(c)  Suppose $u$ is a solution of the static, homogeneous Navier equation

$$\mu \nabla^2 u + (\lambda + \mu) \nabla \operatorname{div}(u) = 0,$$

with homogeneous boundary condition of either type, i.e., either $u = 0$ on $\partial \Omega$ or $Tn = 0$ on $\partial \Omega$. Show that $u$ also satisfies the system of PDEs

$$\frac{\partial u^i}{\partial x_j} + \frac{\partial u^j}{\partial x_i} = 0 \qquad \text{(for } i, j = 1, 2, 3\text{)}, \tag{11.57}$$

$$\operatorname{div}(u) = 0. \tag{11.58}$$

*Hint:* Use equation (11.53) and the above. Show that the equation $\operatorname{div}(u) = 0$ is a redundant part of the system (11.57)–(11.58). Show that for any skew symmetric matrix $A$ and any vector $b \in \mathbb{R}^3$, the vector field $u$ defined by

$$u(x) = Ax + b$$

satisfies the system (11.57) of PDEs. Here we are using $x = (x_1, x_2, x_3)$ for the points (vectors) in $\mathbb{R}^3$. One can show, although it is considerably more difficult, that conversely, any solution of the system (11.57) must necessarily have the form $u(x) = Ax + b$, with $A$ skew symmetric. Vector fields $u$ of this form are known as *infinitesimal Euclidean symmetries*, and the result in this part of the exercise shows us that they are the only types of solutions possible for the homogeneous Navier equation with homogeneous boundary conditions.

(d)  *(Solvability condition)* Just like the Neumann problem, the static Navier equation with traction boundary condition

$$\mu \nabla^2 u + (\lambda + \mu) \nabla \operatorname{div}(u) + \rho_0 F = 0 \qquad \text{in } \Omega, \tag{11.59}$$

$$Tn = g \qquad \text{on } \partial \Omega, \tag{11.60}$$

has a condition on the data $F, g$ that must necessarily be satisfied if the problem is to have a solution. Specifically, show that if the BVP (11.59)–(11.60) has a solution, then necessarily $F$ and $g$ must satisfy

$$\rho_0 \int_{\Omega} F \cdot w \, dV = \int_{\partial \Omega} g \cdot w \, dA,$$

for every infinitesimal Euclidean symmetry $w(x) = Ax + b$. Also show that there is not a unique solution to the BVP (11.59)–(11.60).

4. (*Eigenvalues and eigenvector fields*) Consider the homogeneous problem, i.e, a homogeneous Navier equation

$$\rho_0 u_{tt} = \mu \nabla^2 u + (\lambda + \mu) \nabla \operatorname{div}(u)$$

and homogeneous BCs. Suppose we look for special solutions of the form

$$u(x, y, z, t) = v(x, y, z)\tau(t),$$

where $v = (v^1, v^2, v^3)$ is a vector field on $\Omega$, and $\tau$ is a scalar function of $t$ only.

(a) Show that if for some constant $k$, $v$ satisfies

$$\mu \nabla^2 v + (\lambda + \mu) \nabla \operatorname{div}(v) = kv \qquad (11.61)$$

and the BCs, and if $\tau$ satisfies

$$\tau'' = \rho_0^{-1} k\tau,$$

then $u = v\tau$ is a special solution satisfying the PDE and BCs. The number $k$ is called an *eigenvalue* for the *Navier Operator*

$$N \equiv \mu \nabla^2 + (\lambda + \mu) \nabla \operatorname{div},$$

and $v$ is a corresponding *eigenvector field*. In linear algebra terminology, we are trying to solve $Nv = kv$ for nontrivial ($v \neq 0$) vectors $v \in \mathcal{V}$. Here $\mathcal{V}$ denotes the set of vector fields on $\Omega$ that satisfy the BCs (either $v = 0$ on $\partial\Omega$ or $Tn = 0$ on $\partial\Omega$).

(b) Show that for zero displacement BC ($u = 0$ on $\partial\Omega$), the eigenvalues $k$ of the Navier operator are negative: $k < 0$. For the zero traction BC ($Tn = 0$ on $\partial\Omega$), show that the eigenvalues are negative or zero, $k \leq 0$, and that in the zero eigenvalue case, $k = 0$, the corresponding eigenvector fields are infinitesimal Euclidean symmetries. You may use the facts from the last problem for this second assertion (also remember how useful equation (11.53) is).

5. (*The Ritz–Galerkin method*) For the weak version (11.55) of the Navier equation with zero displacement boundary condition, complete the discussion of how to determine approximate solutions $u^N(t) = \sum_{n=1}^{N} b_n(t)w_n$. In particular:

(a) Write down a second-order system of ODEs that the $b_n$'s must satisfy in order for $u^N$ to satisfy (11.55). (*Hint:* Restrict the $w$'s in (11.55) to just the basis elements $w_k$, $k = 1, \ldots, N$.) Show that this system can be rewritten as $b_n''(t) =$

$H_n(b_1(t), \ldots, b_N(t), t)$, for $n = 1, \ldots, N$. Then rewrite this as a first-order system.

(b) Choose some appropriate initial conditions for the system in part (a) (cf. Exercise 5(iv) in Exercises 10.5) and discuss how the existence and uniqueness theorem from the Introduction applies.

(c) Discuss the static case where $F$ does not depend on $t$. Show that time-independent approximations of the form $u^N = \sum_{n=1}^{N} b_n w_n$ can be found by solving a certain linear system of algebraic equations for the constants $b_1, \ldots, b_N$. Show that this system always has a unique solution.

6. (*Separation of the spatial variables*) Consider now only the case of zero boundary displacement: $u = 0$ on $\partial\Omega$. Investigate the possibility of finding eigenvector fields $v$ of the Navier operator ($Nv = kv$ in $\Omega$ and $v = 0$ on $\partial\Omega$) by the separation of variables method. Assuming that the boundary of $\Omega$ consists of planes (parallel to the coordinate planes), it is natural to look for eigenvector fields $v$ of the form

$$v = (X_1 Y_1 Z_1, X_2 Y_2 Z_2, X_3 Y_3 Z_3), \tag{11.62}$$

where the $X_i$'s are functions of $x$ only, the $Y_i$'s are functions of $y$ only, and the $Z_i$'s are functions of $z$ only.

(a) Write out explicitly the system of ODEs for the $X$'s, $Y$'s, and $Z$'s that arises from substituting the $v$ of the form (11.62) into the eigenvalue equation $Nv = kv$ (where $N \equiv \mu\nabla^2 + (\lambda + \mu)\nabla \operatorname{div}$ is the Navier operator). This is the *Sturm–Liouville system*, and it consists of three nonlinear ODEs for nine unknown functions. Observe that it is a coupled system, and therefore you cannot separate variables as you have done before.

(b) Suppose $\Omega = [0, L] \times [0, M] \times [0, N]$ is a box bounded by the planes $x = 0$, $x = L$, $y = 0$, $y = M$, $z = 0$, and $z = N$. Show that the only separated eigenvector field $v$ (i.e., of the form (11.62)) is the trivial one: $v = 0$. For this, note that the Sturm–Liouville BCs are $X_i(0) = 0$, $Y_i(0) = 0$, $Z_i(0) = 0$ and $X_i(L) = 0$, $Y_i(M) = 0$, $Z_i(N) = 0$, for $i = 1, 2, 3$, and you may use the fact that imposition of additional BCs (like $Y_2'(0) = 0$) is not possible (it leads only to the trivial solution). *Hint:* look at what the Sturm–Liouville system reduces to for $x = 0$, $y = 0$, and $z \in [0, N]$ arbitrary.

(c) Suppose $\Omega = [0, L] \times [0, M] \times \mathbb{R}$ is a square conduit, bounded by planes $x = 0$, $x = L$, $y = 0$, $y = M$, and unbounded in the $z$ direction. Show that the only separated eigenvector fields $v$ are the ones of the form

$$v = (0, 0, X_3 Y_3).$$

Use this to find the general series solution of the Navier equation $\rho_0 u_{tt} = Nu$, with BCs $u = 0$ on $\partial\Omega$ and ICs

$$u(x, y, 0) = (0, 0, \phi_3(x, y)), \quad u_t(x, y, 0) = (0, 0, \psi_3(x, y)),$$

for $(x, y) \in [0, L] \times [0, M]$. Observe that $u$ does not depend on $z$, and $\operatorname{div}(u) = 0$ in $\Omega$.

(d) Suppose $\Omega = [0, L] \times \mathbb{R} \times \mathbb{R}$ is an infinite slab bounded between the two planes $x = 0, x = L$. What can you say about the separated eigenvector fields of the form (11.62)?

7. Consider the situation described in Example 11.6, where the general solution with zero initial velocities is given by equation (11.51). Assume now that $L = 4$, while the other constants are the same. Let $K_{a,b,L}$ be the parabolic pulse function described in Exercises 11.1, Exercise 6, and assume that the initial displacement $\phi$ has components $\phi^1(x) = \phi^3(x) = K_{1,2,4}(x)$ and $\phi^2(x) = 0$. Analyze the resulting motion of the solid by doing the following:

(a) As in Example 11.6, all the fibers in $\Omega$ parallel to the $x$-axis undergo the same type of motion, so it suffices to follow the motion of one fiber, say $F_{0,0} = \{(x, 0, 0) | 0 \le x \le L\}$. Plot its position for various times $t$ between $t = 0$ and $t = 1$. In the same plot include plots of the paths traced out by the seven points $x = 1/2, 1, 3/2, 2, 5/2, 3, 7/2$. Do the same for times between 1 and $a$, and times between $a$ and $b$, for appropriately chosen times $a, b$, that help you see the motion most clearly. Use more time intervals if you wish. Describe what is going on in your plots.

(b) For the same seven points in part (a), plot the paths they trace out over the three time intervals $t = 0...a$, $t = 0...b$, and $t = 0...c$, with $a, b, c$ appropriately chosen times (separate pictures for each). Do these paths appear to be space filling?

(c) Find the times (if any) at which each of the seven particle paths first pass back through the initial position (equilibrium position).

8. Rework the last exercise, except this time assume that there is no initial displacement ($\phi = 0$) and the initial velocity distribution $\psi$ has components $\psi^1(x) = \psi^3(x) = K_{1,2,4}(x)$ and $\psi^2(x) = 0$. Compare and contrast the differences in these two exercises.

# 12 The Heat IBVP in Polar Coordinates

Recall that the general form of the 2-D heat flow IBVP (which does *not* include convection or other physical situations) is

$$u_t = \alpha^2 \nabla^2 u + F \quad \text{in } D, \text{ for } t > 0; \qquad (12.1)$$

$$\kappa \nabla u \cdot n + hu = g \qquad \text{on } \partial D, \text{ for } t > 0; \qquad (12.2)$$

$$u = \phi \qquad \text{in } D, \text{ for } t = 0. \qquad (12.3)$$

Here $D \subset \mathbb{R}^2$ is a planar region.

We have seen that when the BCs (12.3) are inhomogeneous, then the solution $u$ of this heat flow IBVP is found by transforming the problem to homogeneous BCs $u = S + U$. In addition to giving a homogeneous BC problem of the same type for $U$, this transformation technique introduces a BVP for $S$:

$$\nabla^2 S = 0 \qquad \text{in } D, \qquad (12.4)$$

$$\kappa \nabla S \cdot n + hu = g \qquad \text{on } \partial D,. \qquad (12.5)$$

which includes as special cases the Dirichlet and Neumann BVPs.

So far, we have only discussed solutions of the IBVP (12.1)–(12.3) and the BVP (12.4)–(12.5) for the particular case where the domain $D$ is a rectangle. Here in this chapter we discuss how to solve these types of problems when the geometry of the domain $D$ is circular in nature, i.e., when $D$ is the image of a rectangle $E$ under the polar coordinate map

$$p(r, \theta) = (r \cos \theta, r \sin \theta).$$

Thus we assume that $D = p(E)$ throughout the chapter. The primary examples of such regions $D$ are circular disks and annuli. These regions, when viewed in terms of polar coordinates, are rectangles, and so by transforming the heat equation into a PDE expressed in terms of polar coordinate variables $r, \theta$, we can use the separation of variables method, which generally requires a rectilinear geometry in order to get Sturm–Liouville BCs.

## 12.1.   The Polar Coordinate Map

The polar coordinate map $p : \mathbb{R}^2 \to \mathbb{R}^2$ transforms subsets of $\mathbb{R}^2$ (the $r$-$\theta$ plane) into subsets of $\mathbb{R}^2$ (the $x$-$y$ plane) in a very particular way. For example, $p$ transforms each vertical line $r = a$ into a circle $x^2 + y^2 = a^2$ (the action of $p$ is often viewed as wrapping the vertical line around the corresponding circle), and $p$ transforms each horizontal line $\theta = b$ into a line through the origin making an angle $b$ with the $x$-axis. This is illustrated in Figure 12.1. Based on this, it is not hard to see that $p$ transforms the rectangle $E = [r_1, r_2] \times [0, 2\pi]$ into the annulus $D_{r_1 r_2}$ with radii $0 \le r_1 < r_2$. This is shown in Figure 12.2. In the special case where $r_1 = 0$ and $r_2 = a$, the annulus becomes the disk $D_a$ of a radius $a$ centered at the origin.

Now it is easy to see that the polar coordinate map $p$ is not 1 to 1 (for example, all the points $(0, \theta)$ get mapped to the same point $(0, 0)$ by $p$), and this aspect of $p$ will influence the Sturm–Liouville problems that arise in polar coordinates. Before we get to this, and in preparation for it, let's make the following definition:

**Definition 12.1**   Suppose $D = p(E)$ is the image of some set $E$ in the $r$-$\theta$ plane under the polar coordinate map $p$. We say that two points $(r, \theta), (r', \theta') \in E$ are

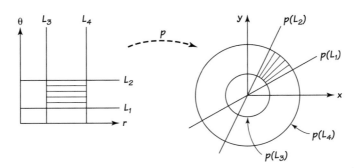

Figure 12.1.   Action for the polar coordinated map: $p$.

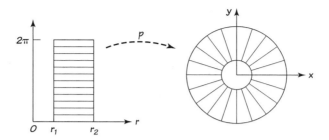

Figure 12.2.   $p$ transforms the rectangle $E = [r_1, r_2] \times [0, 2\pi]$ into the annulus $D_{r_1 r_2}$.

*equivalent* if

$$p(r, \theta) = p(r', \theta').$$

We use the notation

$$(r, \theta) \sim (r', \theta')$$

for equivalent points.

The two main examples illustrating this concept of equivalent points are the following:

(1) **Annulus:** For the annulus $D_{r_1, r_2} = p(E)$, where $E = [r_1, r_2] \times [0, 2\pi]$, we see that the equivalent points in $E$ are

$$(r, 0) \sim (r, 2\pi),$$

for any $r \in [r_1, r_2]$. *Note*: We are assuming here that $r_1 > 0$, so that the annulus does not degenerate to a disk. If we identify equivalent points in the rectangle $E$, that is, mentally think of equivalent points as the same point, then the rectangle $E$ turns into an annulus. This is illustrated in Figure 12.3.

(2) **Disk:** For the disk $D_a = p(E)$, where $E = [0, a] \times [0, 2\pi]$, we see that equivalent points in $E$ are

$$(0, \theta) \sim (0, \theta'),$$

$$(r, 0) \sim (r, 2\pi),$$

for any $\theta, \theta' \in [0, 2\pi]$ and any $r \in [0, a]$.

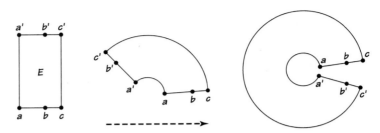

Figure 12.3. Identifying equivalent points on the boundary of the rectangle $E$ makes it (topologically) an annulus.

The notion of equivalent points in $E$ is actually an equivalence relation on $E$, and the set $E/\sim$ of equivalence classes is a space that topologists call the *identification space* for the map $p$. From general principles, the spaces $E/\sim$ and $D$ are topologically the same (homeomorphic).

## 12.2. The Laplacian in Polar Coordinates

Suppose now we have a region $D$ in the $x$-$y$, plane and consider a solution $u$ of the heat equation

$$u_t = \alpha^2(u_{xx} + u_{yy}) \quad \text{on } D \times [0, \infty].$$

At the same time, also consider a solution $S$ of Laplace's equation

$$S_{xx} + S_{yy} = 0 \quad \text{on } D.$$

Assuming that $D = p(E)$ is the image of a rectangle $E$ under the polar coordinate map $p$, we can transform the functions $u$ and $S$ over to functions $w$ and $K$ on $E$ by composition with $p$:

$$w(r, \theta, t) \equiv u(r \cos \theta, r \sin \theta, t), \tag{12.6}$$

$$K(r, \theta) \equiv S(r \cos \theta, r \sin \theta). \tag{12.7}$$

The relationships among these functions and their domains are illustrated in Figure 12.4

With this way of transforming $u$, $S$ into functions $w$, $K$ in polar coordinates, we can ask how this transformation extends to the respective partial derivatives.

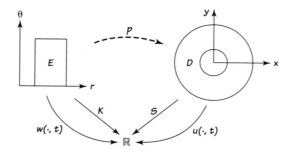

Figure 12.4. The functions $u$, $S$ and their polar coordinate representatives $w$, $K$ related via the polar coordinate map $p$.

Using the chain rule (and suppressing the dependence on $p$), we easily find

$$w_r = u_x \cos\theta + u_y \sin\theta,$$

$$w_\theta = -u_x r \sin\theta + u_y r \cos\theta,$$

$$w_{rr} = u_{xx} \cos^2\theta + u_{yy} \sin^2\theta, +2u_{xy} \sin\theta \cos\theta.$$

$$w_{\theta\theta} = u_{xx} r^2 \sin^2\theta + u_{yy} r^2 \cos^2\theta - 2u_{xy} r^2 \sin\theta \cos\theta$$
$$- u_x r \cos\theta - u_y r \sin\theta$$

From this, we see that

$$w_{rr} + \frac{1}{r^2} w_{\theta\theta} + \frac{1}{r} w_r = u_{xx} + u_{yy}.$$

Rearranging the terms on the left-hand side and putting in all the variables, we get

**The Laplacian in Polar Coordinates:**

$$\left( w_{rr} + \frac{1}{r} w_r + \frac{1}{r^2} w_{\theta\theta} \right)\Bigg|_{(r,\theta)} = \left( u_{xx} + u_{yy} \right)\Bigg|_{\substack{x=r\cos\theta \\ y=r\sin\theta}}. \qquad (12.8)$$

Thus if $w$ is regarded as the polar coordinate expression of $u$, then the left-hand side of equation (12.8) represents the Laplacian of $w$ in polar coordinates. From this we get

**Theorem 12.1**   *With the above assumptions on u and S, the polar coodinate expressions w and K for these functions satisfy the following:*

**The Heat Equation in Polar Coordinates:**

$$w_t = \alpha^2 \left( w_{rr} + \frac{1}{r} w_r + \frac{1}{r^2} w_{\theta\theta} \right),$$

*on $E \times [0, \infty]$;*

**Laplace's Equation in Polar Coordinates:**

$$K_{rr} + \frac{1}{r} K_r + \frac{1}{r^2} K_{\theta\theta} = 0,$$

*on $E$;*

**The Equivalence Conditions:**   *If $(r, \theta) \sim (r', \theta')$ are equivalent points in $E$, then*

$$w(r, \theta, t) = w(r', \theta', t),$$

$$K(r, \theta) = K(r', \theta').$$

The equivalence conditions on $w$ and $K$ are easily seen to follow from their definitions (12.6)–(12.7) in terms of compositions of $u$ and $S$ with the polar coordinate map $p$. The equivalence conditions are very important for two reasons:

(1) They allow us to prove the converse of the assertion in the theorem. Namely, if $w$ and $K$ are solutions of the heat equation and Laplace's equation in polar coordinates respectively, and if they satisfy the equivalence conditions, then there are functions $u$ and $S$ that satisfy the heat equation and Laplace's equation in Cartesian coordinates respectively and are related to $w$ and $K$ by (12.6)–(12.7).

(2) They give us the the extra parts to the Sturm–Liouville boundary conditions that are essential in determining $w$ and $K$ by the separation of variables technique.

The above discussion shows how to transform the heat equation and Laplace's equation into corresponding PDEs in polar coordinates and also illustrates how the equivalence conditions arise in the transformed problem. The other parts to the IBVP or BVP, namely the boundary conditions and initial condition (for the IBVP only), must also be transformed to polar coordinates.

## 12.3.   The BCs and IC in Polar Coordinates

To transform the BCs into polar coordinates, we first need the expression for the gradient in polar coordinates. Here is how to get this. In the above calculations we found that

$$w_r = u_x \cos\theta + u_y \sin\theta;$$

$$w_\theta = -u_x r \sin\theta + u_y r \cos\theta.$$

Solving this system of equations for $u_x$, $u_y$ gives

$$u_x = w_r \cos\theta - w_\theta \frac{1}{r} \sin\theta,$$

$$u_y = w_r \sin\theta + w_\theta \frac{1}{r} \cos\theta.$$

Thus we get

**The Gradient in Polar Coordinates:**

$$\nabla u = w_r e_1 + \frac{1}{r} w_\theta e_2,$$

where $e_1$ and $e_2$ are unit vectors in the radial and angular directions respectively:

$$e_1 = (\cos\theta, \sin\theta),$$

$$e_2 = (-\sin\theta, \cos\theta).$$

We can now use this to rewrite the boundary condition

$$\kappa \nabla u \cdot u + h u = g$$

in terms of polar coordinates. For the given function $g$, we let $G$ be its polar coordinate expression defined by

$$G(r, \theta, t) = g(r \cos\theta, r \sin\theta, t).$$

For the two regions of interest to us here, the disk $D_a$ and annulus $D_{r_1 r_2}$, the outward-directed unit normal $n$ to the boundary is easy to express in polar coordinates.

**For the disk** $D_a$: $n = e_1$.

**For the annulus** $D_{r_1 r_2}$: $n = \pm e_1$,

where the plus sign $(+)$ is used on the boundary circle of radius $r_2$ and the minus sign $(-)$ on the circle of radius $r_1$. From this we see that

$$\kappa \nabla u \cdot n = \kappa w_r$$

for the disk, and

$$\kappa \nabla u \cdot n = \pm \kappa w_r$$

for the annulus. Thus the boundary conditions for the disk and annulus are

$$\kappa w_r(a, \theta, t) + h w(a, \theta, t) = G(a, \theta, t)$$

and

$$-\kappa w_r(r_1, \theta, t) + h w(r_1, \theta, t) = G(r_1, \theta, t),$$

$$\kappa w_r(r_2, \theta, t) + h w(r_2, \theta, t) = G(r_2, \theta, t),$$

respectively. These are to hold for all $\theta \in [0, 2\pi]$ and $t > 0$. For convenience in the sequel we will just write $G(\theta, t)$ for $G(a, \theta, t)$ and $G_1(\theta, t), G_2(\theta, t)$ for $G(r_1, \theta, t), G(r_2, \theta, t)$.

The initial condition can be rewritten in polar coordinates by

$$w(r, \theta, 0) = \psi(r, \theta),$$

where $\psi(r, \theta) \equiv \phi(r \cos \theta, r \sin \theta)$. Note: Because of the form of the initial condition function $\psi$, it must also satisfy the equivalence conditions $\psi(r, 0) = \psi(r, 2\pi)$, for all $r \in [0, a]$, and $\psi(0, \theta) = \psi(0, \theta')$, for all $\theta, \theta' \in [0, 2\pi]$. Thus in particular, if $\psi$ does not depend on $r$, then it must be a constant function.

## 12.4.   The Heat IBVP for a Disk

Based on the above discussion, we can now write the heat IBVP for the disk $D_a = p(E)$, where $E = [0, a] \times [0, 2\pi]$, in polar coordinates as follows. For this we let $H(r, \theta, t) = F(r \cos \theta, r \sin \theta, t)$ be the expression for the internal heat source/sink function in polar coordinates.

**The Heat Equation:**

$$w_t = \alpha^2 (w_{rr} + \frac{1}{r} w_r + \frac{1}{r^2} w_{\theta\theta}) + H \qquad \text{on } E \times [0, \infty].$$

**The Boundary Condition:**

$$\kappa w_r(a, \theta, t) + hw(a, \theta, t) = G(\theta, t) \qquad \text{for } \theta \in [0, 2\pi], t > 0$$

**The Equivalence Conditions:**

$$w(r, 0, t) = w(r, 2\pi, t) \qquad \text{for } r \in [0, a], t > 0, \qquad (12.9)$$

$$w(0, \theta, t) = w(0, \theta', t) \qquad \text{for } \theta, \theta' \in [0, 2\pi], t > 0. \qquad (12.10)$$

**The Initial Condition:**

$$w(r, \theta, 0) = \psi(r, \theta) \qquad \text{for all } (r, \theta) \in E.$$

## 12.5. Solution of the Homogeneous Problem

The separation of variables technique applied to the polar coordinate version of the *homogeneous* heat equation leads to some new Sturm–Liouville problems. As usual we look for special solutions of the form

$$w(r, \theta, t) = R(r)\Theta(\theta)T(t).$$

Substituting this in the homogeneous PDE gives

$$R\Theta T' = \alpha^2 \left( R''\Theta T + \frac{1}{r} R'\Theta T + \frac{1}{r^2} R\Theta'' T \right).$$

Rearranging this gives

$$\frac{T'}{\alpha^2 T} = \frac{1}{R} \left( R'' + \frac{1}{r} R' \right) + \frac{1}{r^2} \frac{\Theta''}{\Theta}.$$

Since the left-hand side of this equation involves only $t$, while the right-hand side does not, each side is a constant, say $k$. This constant $k$, being an eigenvalue of the Laplacian with heat equation boundary conditions, must be negative or zero, say $k = -\lambda^2$ with $\lambda \geq 0$ (this is proven in the exercises in Chapter 13). Thus we get that $R$, $\Theta$, $T$ must satisfy

$$\frac{T'}{\alpha^2 T} = -\lambda^2,$$

$$\frac{1}{R}\left(R'' + \frac{1}{r}R'\right) + \frac{1}{r^2}\frac{\Theta''}{\Theta} = -\lambda^2.$$

The second equation above can be further separated as follows. The ratio $\Theta''/\Theta$ must be constant (since we can rearrange the last equation with this ratio on one side and an expression involving only functions of $r$ on the other). Call this constant $\mu$. Then the three separated equations for $R$, $\Theta$, and $T$ are

$$T' = -\lambda^2\alpha^2 T,$$

$$\Theta'' = \mu\Theta,$$

$$R'' + \frac{1}{r}R' + \left(\lambda^2 + \frac{\mu}{r^2}\right)R = 0$$

The homogeneous BC and the two equivalence conditions lead directly to the following conditions on $R$ and $\Theta$:

$$\kappa R'(a) + hR(a) = 0,$$

$$\Theta(0) = \Theta(2\pi),$$

$$R(0)\Theta(\theta) = R(0)\Theta(\theta') \text{ for all } \theta, \theta'.$$

The last equivalence condition says that $R$ is defined at 0 and that either $R(0) = 0$ or $\Theta$ is a constant function. Thus we arrive at two Sturm–Liouville problems:

**The Angular Sturm–Liouville Problem:**

$$\Theta'' = \mu\Theta, \tag{12.11}$$

$$\Theta(0) = \Theta(2\pi). \tag{12.12}$$

**The Radial Sturm–Liouville Problem:**

$$R'' + \frac{1}{r}R' + \left(\lambda^2 + \frac{\mu}{r^2}\right)R = 0, \tag{12.13}$$

$$\kappa R'(a) + hR(a) = 0. \tag{12.14}$$

In addition, we have the requirement that either $R(0) = 0$ or the corresponding $\Theta$ used in $w = R\Theta T$ must be a constant function.

*Note*: Since the radial SL problem involves the eigenvalues $\mu$ of the angular SL problem, we solve the angular problem first.

The angular SL problem has BCs that are called periodic. It is not to hard to see that the eigenvalues must have the form

$$\mu = -m^2, \qquad m = 0, 1, 2, 3, \ldots$$

(with $m \geq 0$), and the eigenfunctions are

$$\Theta_0(\theta) = 1,$$

$$\Theta_m^1(\theta) = \cos m\theta,$$

$$\Theta_m^2(\theta) = \sin m\theta.$$

Using the values $\mu = -m^2$ in the radial SL problem gives a second order DE:

**Bessel's Equation:**

$$R'' + \frac{1}{r}R' + \left(\lambda^2 - \frac{m^2}{r^2}\right)R = 0. \tag{12.15}$$

Power series solutions of this type of equation are usually discussed in undergraduate differential equations courses. The discussion of the general solution depends on the value of $\lambda$, so we divide into cases. The first case does not require series for its solution.

(1) ($\lambda = 0$) In this case, Bessel's equation is

$$R'' + \frac{1}{r}R' - \frac{m^2}{r^2}R = 0,$$

which upon multiplication by $r^2$ is recognized as a Cauchy–Euler equation

$$r^2 R'' + r R' - m^2 R = 0.$$

To solve this, we divide into two cases:

(a) ($m = 0$) In this case the general solution is given by

$$R(r) = A + B \ln r,$$

and to satisfy the equivalence condition that $R$ be defined at zero, we must take $B = 0$ to get rid of the $\ln r$ term. Thus $R(r) = A$ is a constant function, and of course we want $A \neq 0$ so that we have something other than the trivial function. Now, since in this case $R(0) \neq 0$, the other part of the equivalence condition (i.e., that the corresponding $\Theta$ be a constant function) holds because we are assuming that $m = 0$, and we have already found that $\Theta(\theta) = 1$ when $\mu = -m^2 = 0$. Finally, the BC $\kappa R'(a) + hR(a) = 0$ reduces in this case to $hA = 0$. So in problems where $h = 0$ (i.e., the boundary is insulated) $R(r) = A$ is an eigenfunction. Otherwise it is not.

(b) ($m \neq 0$) In this case the general solution of the Cauchy–Euler equation is

$$R(r) = Ar^m + Br^{-m},$$

and as above, we must take $B = 0$ in order for $R$ to be defined at zero. However, the boundary condition applied to this $R$ is

$$A(\kappa ma^{m-1} + ha^m) = 0.$$

This gives $\kappa = -h/(ma)$, which is contrary to our assumption that $\kappa$ is always nonnegative. Thus there are no nontrivial eigenfunctions in this case.

(2) ($\lambda > 0$) In this case it is customary to change the scale and transform Bessel's equation as follows. We look for a solution $R$ of the form

$$R(r) = Q(\lambda r).$$

We denote the independent variable for $Q$ by $x$. The change of variable transformation is $x = \lambda r$, and since $R' = \lambda Q'$ and $R'' = \lambda^2 Q''$, we get from Bessel's equation for $R$ the following equation for $Q$:

$$Q'' + \frac{1}{x}Q' + \left(1 - \frac{m^2}{x^2}\right)Q = 0. \tag{12.16}$$

This is just Bessel's equation with $\lambda = 1$. The solutions of this DE depend on $m$, and so we build that into the notation. One solution of this Bessel equation is

**The Bessel Function (First Kind) of Order $m$:**

$$J_m(x) = \sum_{k=0}^{\infty} \frac{(-1)^k x^{2k+m}}{2^{2k+m} k! (m+k)!}$$

$$= \frac{x^m}{2^m m!} \left( 1 + \sum_{k=1}^{\infty} \frac{(-1)^k x^{2k}}{2^{2k} k! (m+1) \cdots (m+k)} \right).$$

The graphs of $J_0$, $J_1$, $J_2$, $J_3$, and $J_4$ are shown in Figure 12.5. The Bessel functions are very interesting and important functions, and they have many special properties and relations (some of which will be explored in the exercises).

With $J_m$ as one solution of the Bessel DE (12.16), we need to find another independent solution $Y_m$ in order to write down the general solution of this second-order DE. There are various methods for doing this. As an exercise, verify that the expression for $Y_m$ in [ON 91, p. 384] can be written as follows:

**Bessel Function (Second Kind) of Order $m$:**

$$Y_m(x) = \frac{2}{\pi} J_m(x) \left[ \ln\left(\frac{x}{2}\right) + \gamma \right] + \frac{2^m}{\pi x^m} \sum_{k=0}^{\infty} \frac{\beta_{mk}}{2^{2k} k!} x^{2k}, \qquad (12.17)$$

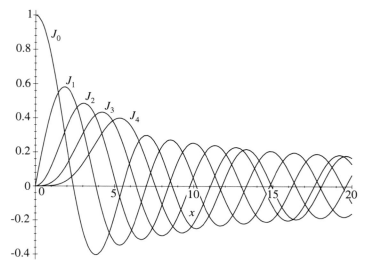

Figure 12.5.   The graphs of the Bessel functions: $J_0$, $J_1$, $J_2$, $J_3$, and $J_4$.

where $\gamma$ is Euler's constant [ON 91] and

$$\beta_{mk} = \begin{cases} -(m-1-k)! & \text{if } k \le m-1, \\ (-1)^{k-m-1}(h_{k-m}+h_k)/(k-m)! & \text{if } k > m-1. \end{cases} \qquad (12.18)$$

In the above expression for $\beta_{mk}$, we are using

$$h_p = \sum_{i=1}^{p} \frac{1}{i}$$

for convenience of notation. The graphs of $Y_0$, $Y_1$, $Y_2$, $Y_3$, and $Y_4$ are shown in Figure 12.6. Note that these Bessel functions of the second kind are not defined for $x = 0$.

With this we get that the general solution of the DE (12.16) is $H_m(x) = AJ_m(x) + BY_m(x)$. Returning to the original Bessel equation (12.15) involving $\lambda$, we get that its general solution is given by

$$R_m(r) = AJ_m(\lambda r) + BY_m(\lambda r).$$

Now, the equivalence condition that $R$ be defined at 0 requires us to take $B = 0$ to get rid of the term $Y_m$, which is not defined at zero. Thus

$$R_m(r) = J_m(\lambda r),$$

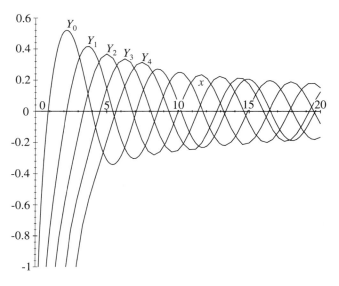

Figure 12.6.   The graphs of the Bessel functions: $Y_0$, $Y_1$, $Y_2$, $Y_3$, and $Y_4$.

where we take $A = 1$ for convenience. Note that $R_m(0) = 0$ when $m \neq 0$ and $R_0(0) \neq 0$, but in this latter case the corresponding $\Theta$ function is $\Theta_0 = 1$. Thus the other equivalence condition is satisfied regardless of the value of $m$.

What remains is to satisfy the boundary condition on $R_m$. This is where the determination of the eigenvalues $-\lambda^2$ is finally accomplished. The boundary condition gives the equation

$$\kappa \lambda J_m'(\lambda a) + h J_m(\lambda a) = 0,$$

which the $\lambda$'s must satisfy. One can show that there is an infinite sequence

$$\lambda_{m1}, \lambda_{m2}, \lambda_{m3}, \ldots.$$

of solutions of this equation. *Note*: in the equation, $m$ is fixed, and for each $m$ we are denoting the sequence of solutions by $\{\lambda_{mn}\}_{n=1}^{\infty}$. It can also be shown that

$$\lim_{n \to \infty} \lambda_{mn} = \infty.$$

Both of these facts are visually evident from looking at the graph of $\kappa \lambda J_m'(\lambda a) + h J_m(\lambda a)$ as a function of $\lambda$. For example, in the case when $\kappa = 1, h = 2$ (so that the flux of heat across the boundary is equal to the twice the temperature there), the graph of $\lambda J_m'(\lambda a) + 2 J_m(\lambda a)$ for $m = 1$ and $a = 1$ is shown in Figure 12.7. As another example, consider the case of unit radius $a = 1$ where the temperature is held fixed at at zero along the boundary (i.e., $\kappa = 0, h = 1$). Then the $\lambda$'s are the zeros of $J_m$; that is; $J_m(\lambda) = 0$. Glancing back at Figure 12.5, you can get a picture of the location of some of the zeros for $J_0, J_1, J_2, J_4$, and also confirm that these tend to $\infty$ as $n \to \infty$.

Thus we see that in general, for each $m$ we get

**Radial Eigenfunctions:**    $\{R_{mn}(r)\}_{n=1}^{\infty}$ given by

$$R_{mn}(r) = J_m(\lambda_{mn} r).$$

The separation of variables technique presented above gives us basic solutions of the form $R(r)\Theta(\theta)T(t)$ that also satisfy the boundary and equivalence conditions. By the superposition principle, sums of these will also be solutions satisfying the BC and ICs. Thus we get

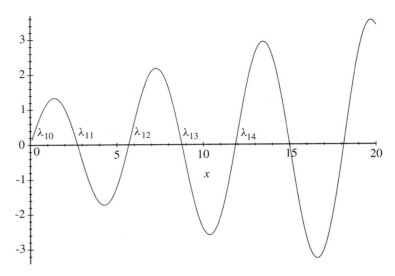

Figure 12.7. Graph of $\lambda J_1'(\lambda) + 2J_1(\lambda)$, showing the zeros $\{\lambda_{1n}\}_{n=1}^{\infty}$ of this function.

### The General Solution of the Homogeneous Heat Problem for the Disk $D_a$:

$$w(r, \theta, t) = A_0 + \sum_{m=0}^{\infty} \sum_{n=1}^{\infty} e^{-\lambda_{mn}^2 \alpha^2 t} J_m(\lambda_{mn} r)[A_{mn} \cos m\theta + B_{mn} \sin m\theta]$$

$$= A_0 + \sum_{n=1}^{\infty} A_{0n} e^{-\lambda_{0n}^2 \alpha^2 t} J_0(\lambda_{0n} r)$$

$$+ \sum_{m=1}^{\infty} \left[ \sum_{n=1}^{\infty} A_{mn} e^{-\lambda_{mn}^2 \alpha^2 t} J_m(\lambda_{mn} r) \right] \cos m\theta \qquad (12.19)$$

$$+ \sum_{m=1}^{\infty} \left[ \sum_{n=1}^{\infty} B_{mn} e^{-\lambda_{mn}^2 \alpha^2 t} J_m(\lambda_{mn} r) \right] \sin m\theta.$$

The eigenvalues $\{\lambda_{mn}\}_{n=1, m=0}^{\infty \ \infty}$ are the solutions of the

### Eigenvalue Equation:

$$\kappa \lambda J_m'(\lambda a) + h J_m(\lambda a) = 0.$$

Special cases of the eigenvalue equation are

(a) Boundary held at 0 degrees: $J_m(\lambda a) = 0$.

(b) Insulated boundary: $J_m'(\lambda a) = 0$.

*Note*: $\lambda = 0$ is not included in the solution set of the eigenvalue equation. The case of a zero eigenvalue occurs only when the boundary is insulated ($h = 0$, $\kappa \neq 0$) and the constant term $A_0$ in the series solution is a corresponding eigenfunction. By convention, we take $A_0 = 0$ when the boundary is *not* insulated.

We now specialize the general solution above, which involves the arbitrary constants $A_0$, $A_{mn}$, $B_{mn}$, so that $w$ satisfies the initial condition. As usual, we expect that the arbitrary constants will be chosen to be the Fourier coefficients of $\psi$ in its expansion in terms of the eigenfunctions of the problem. This is so, since taking $t = 0$ in equation (12.19) gives

**Fourier–Bessel Series for $\psi$:**

$$\psi(r, \theta) = A_0 + \sum_{n=1}^{\infty} A_{0n} J_0(\lambda_{0n} r)$$

$$+ \sum_{m=1}^{\infty} \left[ \sum_{n=1}^{\infty} A_{mn} J_m(\lambda_{mn} r) \right] \cos m\theta \qquad (12.20)$$

$$+ \sum_{m=1}^{\infty} \left[ \sum_{n=1}^{\infty} B_{mn} J_m(\lambda_{mn} r) \right] \sin m\theta.$$

Notice how the series expansion for $\psi$ consists of three distinct parts. The first part is a series that does not involve $\theta$, and so if the initial temperature distribution $\psi = \psi(r)$ is independent of $\theta$ (angularly symmetric), then this is the only part needed in the expansion of $\psi$ in terms of the eigenfunctions. In this case, the other $A_{nm}$'s and all the $B_{mn}$'s are zero. However, $\psi$ will generally depend on $\theta$ as well, in which case the latter two series involving cosines and sines of $\theta$ are needed too.

Based on the general Sturm–Louville theory, one can derive the following integral formulas for the Fourier coefficients in the above series expansion (see the exercises):

**Fourier–Bessel Coefficients for $\psi$:**

$$A_{0n} = \frac{1}{2\pi L_{0n}} \int_0^{2\pi} \int_0^a \psi(r, \theta) J_0(\lambda_{0n} r) r \, dr \, d\theta, \qquad (12.21)$$

$$A_{mn} = \frac{1}{\pi L_{mn}} \int_0^{2\pi} \int_0^a \psi(r, \theta) J_m(\lambda_{mn} r) \cos m\theta \, r \, dr \, d\theta, \qquad (12.22)$$

$$B_{mn} = \frac{1}{\pi L_{mn}} \int_0^{2\pi} \int_0^a \psi(r, \theta) J_m(\lambda_{mn} r) \sin m\theta \, r \, dr \, d\theta, \qquad (12.23)$$

and $A_0 = 0$ unless the boundary condition is an insulated boundary, in which case

$$A_0 = \frac{1}{\pi a^2} \int_0^{2\pi} \int_0^a \psi(r, \theta) r dr \, d\theta. \tag{12.24}$$

In these formulas the normalizing constants are given by

$$L_{mn} = \int_0^a J_m(\lambda_{mn}r)^2 r dr, \tag{12.25}$$

for $m = 0, 1, 2, \ldots$. *Note*: In these integral formulas there is a factor $r$ that occurs with the differential $dr$. This is a *weight factor* $\omega(r) = r$, which comes from the general Sturm–Liouville theory (see Theorem 4.1).

The general solution (12.19) of the homogeneous heat problem for a disk has the features we would expect in any homogeneous heat problem

$$\lim_{t \to \infty} w(r, \theta, t) = A_0,$$

and thus the temperature decays to $A_0 = 0$ when the boundary is not insulated. However, when the boundary is insulated, the eventual temperature distribution is a uniform constant temperature $A_0$, which is given by (12.24). This is the average value of the initial temperature distribution on the disk $D_a$ (exercise).

**Example 12.1 (Fixed Temperature Zero on the Boundary)** Suppose the disk has radius $a = 1$, and the boundary is held at a fixed temperature

$$w(1, \theta, t) = 0 \qquad \text{for all } \theta, t.$$

Thus the radial eigenfunctions are $R_{mn}(r) = J_m(\lambda_{mn}r)$, where the eigenvalues $\{\lambda_{mn}\}_{n=1}^\infty$ are the roots of

$$J_m(\lambda) = 0,$$

for $m = 0, 1, \ldots$. Consider heating the disk up initially with temperature distribution:

$$\psi(r) = \begin{cases} 1 - 4r^2 & \text{if } 0 \le r \le 1/2, \\ 0 & \text{if } 1/2 < r \le 1. \end{cases}$$

The graph of $\psi$ is shown in Figure 12.8. Computer plots of a function $f(r, \theta)$ in polar coordinates can be done in *Maple* using the `cylinderplot` command

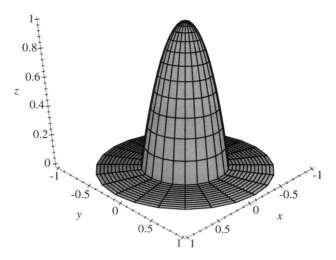

Figure 12.8.    The graph of the initial temperature distribution $\psi$ in Example 12.1.

```
with(plots):
cylinderplot([r,p,f(r,p)],r=0..a,p=0..2*Pi);
```

Here $p$ stands for the angular variable $\theta$. In this example $\psi$ does not depend on $\theta$, and so its graph is a surface of revolution as shown.

Since $\psi$ depends only on $r$ and the boundary is not insulated (so $A_0 = 0$), we see that the Fourier–Bessel expansion (12.20) reduces to

$$\psi(r) = \sum_{n=1}^{\infty} A_{0n} J_0(\lambda_{0n}r).$$

This type of problem uses only the eigenvalues $\{\lambda_{0n}\}_{n=1}^{\infty}$, and the solution of the heat equation has the form

$$w(r, t) = \sum_{n=1}^{\infty} A_{0n} e^{-\lambda_{0n}^2 \alpha^2 t} J_0(\lambda_{0n}r).$$

Thus all we need are the coefficients $A_{0n}$ in the series representation of $\psi$ in terms of Bessel functions (the $J_0(\lambda_{0n}r)$). The fact that this representation is possible is interesting in itself. Figure 12.9 shows the graphs of the $J_0(\lambda_{0n}r)$'s for $n = 1, 2, 3, 4$, and the theory guarantees us that an appropriate linear combination of enough of these functions will closely approximate $\psi$. The coefficients $A_{n0}$ in this linear combination are given by the above integral formula (12.21). Using

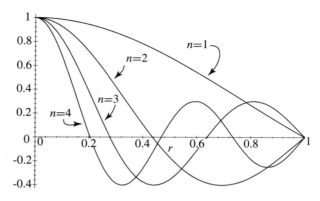

Figure 12.9.   The graphs of the $J_0(\lambda_{0n}r)$ for $n = 1, 2, 3, 4$.

these formulas, the calculations, either by hand or a CAS (see the exercises) yield

$$L_{0n} = \int_0^1 J_0(\lambda_{0n}r)^2 \, r \, dr = \frac{J_1(\lambda_{0n})^2}{2}, \tag{12.26}$$

$$A_{0n} = \frac{2}{J_1(\lambda_{0n})^2} \int_0^{1/2} (1 - 4r^2) J_0(\lambda_{0n}r) \, r \, dr$$

$$= \frac{4 J_2(\lambda_{0n}/2)}{\lambda_{0n}^2 J_1(\lambda_{0n})^2}. \tag{12.27}$$

Next, we need to calculate the values of the eigenvalues $\lambda_{0n}$. These are the roots of the equation

$$J_0(\lambda) = 0,$$

i.e., the zeros of the Bessel function $J_0$. Some of these zeros are shown graphically in Figure 12.10. Since there is no exact formula for these zeros, they must be computed numerically. The following *Maple* code computes the first twenty zeros and stores the results in an array called c0.

```
alias(J=BesselJ); c0:=array(1..20); n:=1;
for k to 62 do
   z := fsolve(J(0,x)=0,x=k..k+1);
   if type(z,float) then c0[n]:=z; n:=n+1; fi od;
```

The above algorithm is rather crude, but it is necessitated by the fact that *Maple* must be told where to look for the roots of a nonpolynomial equation. From Figure 12.10 we see that the first twenty roots lie in the interval $[1, 60]$, and to

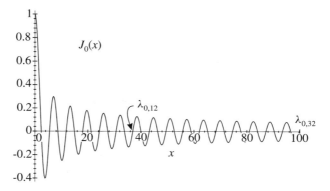

Figure 12.10.   The eigenvalues $\lambda_{0,1}, \ldots, \lambda_{0,32}$ for the heat problem of fixed zero temperature on the boundary of the unit disk are precisely the first 32 zeros of the Bessel function $J_0$.

make sure that all the roots are found, this interval is divided into small enough subintervals so that each subinterval contains *at most* one root (in this case, subintervals of length 1 will do). Since *Maple* returns no value (in floating point decimal representation) when it finds no root, we have used a `type` statement to avoid putting any nonnumerical data types in the array c0. The results of this calculation are shown in Table 12.1. Since the calculations take some time to complete (on present-day computers) and you might want to use them in other worksheets, it is a good idea to save the results in a file by using a command like `save c0,'J0z.m'`. This writes the contents of the entire array c0 to a file named J0z.m. Note the back quotes around the file name and the m as the file extension. Having saved the zeros of $J_0$, you can now load them into any worksheet by using the command `read 'J0z.m'`. This automatically establishes an array named c0 in the worksheet and stores the previously saved values in it.

With all the calculations complete, we turn to plotting the approximate solution

$$w(r, t, N) = \sum_{n=1}^{N} A_{0n} e^{-\lambda_{0n}^2 t} J_0(\lambda_{0n} r)$$

Table 12.1.   The first twenty zeros of the Bessel function $J_0$.

| | | | |
|---|---|---|---|
| 2.404825558 | 18.07106397 | 33.77582021 | 49.48260990 |
| 5.520078110 | 21.21163663 | 36.91709835 | 52.62405184 |
| 8.653727913 | 24.35247153 | 40.05842576 | 55.76551076 |
| 11.79153444 | 27.49347913 | 43.19979171 | 58.90698393 |
| 14.93091771 | 30.63460647 | 46.34118837 | 62.04846919 |

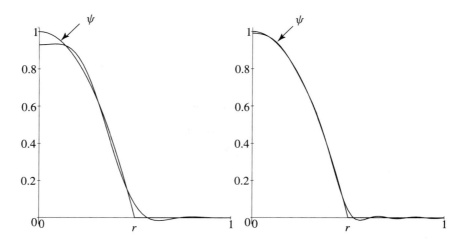

Figure 12.11.   Plots of $\psi(r)$ and its Bessel series approximations $w(r, 0, N)$ for $N = 5$ and $N = 10$, for Example 12.1.

of this heat flow problem (we have taken $\alpha = 1$). To get a feel for the number $N$ of terms to use in the plots, we first plot the approximations

$$w(r, 0, N) = \sum_{n=1}^{N} A_{0n} J_0(\lambda_{0n} r) \approx \psi(r),$$

to $\psi$ for various values of $N$. This will also illustrate how functions like $\psi$ can

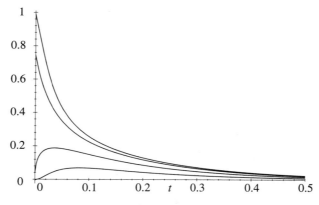

Figure 12.12.   Plot of the evolution of temperatures at the points $(0, 0)$, $(0.25, 0)$, $(0.5, 0)$ and $(0.75, 0)$ in the disk for Example 12.1.

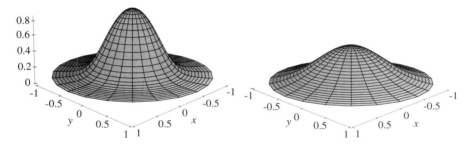

Figure 12.13.    The approximate temperature distribution $w(r, t, N)$ at times $t = 0.01, 0.05$, for Example 12.1.

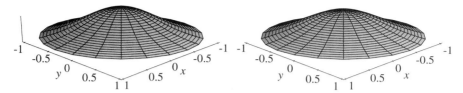

Figure 12.14.    The approximate temperature distribution $w(r, t, N)$ at times $t = 0.07, 0.09$.

be represented by Bessel series. The plots of $w(r, 0, N)$ for $N = 5$ and $N = 10$ are shown in Figure 12.11.

Now we know that the temperature decays to zero throughout the entire disk

$$\lim_{t \to \infty} w(r, t) = 0,$$

and to get a practical estimate of how long this takes, we plot the evolution of temperatures at several points in the disk. This is shown in Figure 12.12. Using this information, we select intermediate times $t = 0.01, 0.05, 0.07, 0.09$, and plot the approximate temperature distributions $w(r, \theta, N)$ at these times. This gives the results shown in Figures 12.13–12.14.

Example 12.1 is characteristic of what occurs when the initial temperature distribution $\psi$ is angularly symmetric (independent of $\theta$) and the BCs and heat source/sink $H = 0$ are likewise angularly symmetric. Then each temperature distribution $w(r, t)$ is a surface of revolution. In this case it would suffice simply to plot the 2-D profiles of these surfaces in the $\theta = 0$ plane. The ensuing exercises will explore the differences that arise when some of the data depends on $\theta$.

### Exercises 12.5

(I). **Properties of Bessel functions.** This group of exercises will make you more familiar with the Bessel functions $J_m$. All you know about these functions at this point is their definition in terms of the power series

$$J_m(x) = \sum_{k=0}^{\infty} \frac{(-1)^k x^{2k+m}}{2^{2k+m} k! (k+m)!}. \tag{12.28}$$

The series definition leads directly to all of the following results:

1. Use the ratio test to show that the series (12.28) converges absolutely for all values of $x$.

2. The following is a fundamental identity, expressing the derivative of the $m$th Bessel function in terms of the $(m-1)$st and $(m+1)$st Bessel functions:

$$J_m'(x) = \frac{1}{2}[J_{m-1}(x) - J_{m+1}(x)], \tag{12.29}$$

for $m = 1, 2, 3, \ldots$. For the case $m = 0$, we have the identity

$$J_0'(x) = -J_1(x). \tag{12.30}$$

(a) Prove the special case (12.30) by differentiating the series expression for $J_0(x)$ term by term, factoring out $-1$ from the result and reindexing the summation in the series.

(b) By differentiating the series (12.28) for $J_m(x)$ term by term, separating the result into two parts, and reindexing the summation in one of these parts, show that identity (12.29) holds.

(c) Most computer algebra systems (CAS's) have the Bessel functions built in and are programmed to recognize the basic identities for Bessel functions. Verify this for the above identities. In *Maple*, the Bessel functions $J_m(x)$ are denoted by BesselJ(m,x). This is easy to remember but cumbersome to work with. Thus you might want to alias this name to J. The following *Maple* code,

```
alias(J=BesselJ);
diff(J(m,x),x);
```

does this and then calculates the derivative of $J(m, x)$. *Note*: Your CAS may give different-looking results, but various identities (like those in Exercise 4 below) will show that they are the same as above.

(d) The identity

$$J_0'(x) = -J_1(x)$$

says (among other things) that the places where $J_1(x)$ is zero are the same as those where $J_0(x)$ has a horizontal tangent line. Use a CAS to graph

$J_0$ and $J_1$ and verify this geometrically. Do the same for the identity

$$J_1'(x) = \frac{1}{2}[J_0(x) - J_2(x)].$$

3. By differentiating the series for $x^{m+1} J_{m+1}(x)$ term by term, prove the identity

$$\frac{d}{dx}[x^{m+1} J_{m+1}(x)] = x^{m+1} J_m(x), \tag{12.31}$$

for $m = 0, 1, 2, \ldots$. Use this identity to show that

$$\int r^{m+1} J_m(\lambda r)\, dr = \lambda^{-1} r^{m+1} J_{m+1}(\lambda r).$$

This integral version of the identity (12.31) is useful in computing certain Bessel coefficients. Verify that the CAS you are using gets the same result for this integral.

4. Another basic identity for Bessel functions is

$$J_m(x) = \frac{x}{2m}[J_{m+1}(x) + J_{m-1}(x)]. \tag{12.32}$$

Prove identity (12.32) by manipulating the series representations for $J_{m+1}(x)$ and $J_{m-1}(x)$.

5. Verify that your CAS gets the following results

$$\int J_m(\lambda r)^2\, r\, dr = \frac{r^2}{2}[J_m(\lambda r)^2 - J_{m-1}(\lambda r)J_{m+1}(\lambda r)], \tag{12.33}$$

for $m = 1, 2, 3, \ldots$, and

$$\int J_0(\lambda r)^2\, r\, dr = \frac{r^2}{2}[J_0(\lambda r)^2 + J_1(\lambda r)^2] \tag{12.34}$$

for $m = 0$:

6. Use a CAS to calculate the first twenty roots of the following equations. Store these values in files, because you will need them to work some of the ensuing problems:

(a) $J_0(\lambda) = 0$.

(b) $J_0'(\lambda) = 0$.

(c) $J_1(\lambda) = 0$. *Hint*: Use part (b).

(d) $J_1'(\lambda) = 0$.

7. Use the results in Exercise 5 to derive the following expressions for some of the normalizing constants:

(a) If $\lambda_{0n}$ are the roots of $J_0(\lambda) = 0$, then

$$L_{0n} = J_1(\lambda_{0n})^2/2.$$

(b) If $\lambda_{0n}$ are the roots of $J_0'(\lambda) = 0$, then

$$L_{0n} = J_0(\lambda_{0n})^2/2.$$

(c) If $\lambda_{1n}$ are the roots of $J_1(\lambda) = 0$, then

$$L_{1n} = -J_0(\lambda_{1n})J_2(\lambda_{1n}).$$

Does this seem plausible? Recall that *by definition*,

$$L_{1n} = \int_0^1 J_1(\lambda_{1n}r)^2 \, r \, dr,$$

and so $L_{1n}$ is always positive. To make the above result more convincing, plot the graphs of $J_0(\lambda)$ and $J_2(\lambda)$ in the same picture and visually check that these functions have values at $\lambda_{1n}$, $n = 1, \ldots, 20$, that are of opposite sign (use the values of $\lambda_{1n}$ calculated in Exercise 6 above).

**(II). Bessel series.** As you have seen in the reading, the solution of heat problems by series requires the expansion of the initial temperature distribution $\psi(r, \theta)$ in a double summation series of Bessel functions and sine/cosine functions. Before doing exercises on this, it will be instructive to look at the simpler task of expanding a function $f(r)$ in a single series of Bessel functions. Thus for a given $m = 1, 2, \ldots$ we wish to express $f(r)$ as

$$f(r) = \sum_{n=1}^{\infty} A_{mn} J_m(\lambda_{mn}r),$$

where the $\lambda_{mn}$s are the eigenvalues for one of the types of Sturm–Liouville problems. For simplicity, we *assume that $f$ is defined on the unit interval* $[0, 1]$. Thus the cooresponding heat problem is for a disk of radius $a = 1$, and the $\lambda_{mn}$s are the roots of

$$\kappa \lambda J_m'(\lambda) + h J_m(\lambda) = 0.$$

The Bessel coefficients are given by the integral formulas

$$A_{mn} = \frac{1}{L_{mn}} \int_0^1 f(r) J_m(\lambda_{mn}r) r \, dr.$$

The Bessel series in the case $m = 0$ is exceptional, it has the form

$$f(r) = A_0 + \sum_{n=1}^{\infty} A_{0n} J_0(\lambda_{0n} r),$$

where $A_0 = 0$ if the boundary is not insulated, and if the boundary is insulated, it is given by

$$A_0 = 2 \int_0^1 f(r) r \, dr.$$

Some of the calculations of the Bessel coefficients can be done exactly (in closed form), but in most cases the calculations will have to be done numerically with a computer.

8. One exact result is that $f(r) = r^m$, for any value of $m = 0, 1, 2, \ldots$, has a nice Bessel expansion:

$$r^m = \sum_{n=1}^{\infty} \frac{2 J_m(\lambda_{mn} r)}{\lambda_{mn} J_{m+1}(\lambda_{mn})},$$

where the $\lambda_{mn}$'s are the roots of

$$J_m(\lambda) = 0.$$

One particular example of this is

$$1 = \sum_{n=1}^{\infty} \frac{2 J_0(\lambda_{0n} r)}{\lambda_{0n} J_1(\lambda_{0n})}.$$

For this exercise do the following:

(a) Prove that the stated expansion is correct.

(b) For $m = 0$ and $m = 1$ graph the approximate partial sums

$$\sum_{n=1}^{N} \frac{2 J_m(\lambda_{mn} r)}{\lambda_{mn} J_{m+1}(\lambda_{mn})},$$

for various values of $n$.

9. Suppose $\lambda_{0n}$ are the roots of $J_0(\lambda) = 0$, and consider the Bessel expansions

$$f(r) = \sum_{n=1}^{\infty} A_{0n} J_0(\lambda_{0n} r),$$

for the following functions:

(a) "Witches Hat":

$$f(r) = \begin{cases} 1 - 2r & \text{if } 0 \leq r \leq 1/2, \\ 0 & \text{if } 1/2 < r \leq 1. \end{cases}$$

(b) "Parabolic Pulse": For constants $a, b$, with $0 \leq a < b \leq 1$:

$$K_{a,b}(r) = \begin{cases} 0 & \text{if } 0 \leq r < a, \\ -4(r-a)(r-b)/(b-a)^2 & \text{if } a \leq r \leq b, \\ 0 & \text{if } b < r \leq 1. \end{cases}$$

For each function, write down the explicit integral formula for the coefficients $A_{0n}$ and use a CAS to plot the approximating partial sums

$$\sum_{n=1}^{N} A_{0n} J_0(\lambda_{0n} r),$$

for various values of $N$. Do (b) for $a = 0, b = 1$, and $a = 1/4, b = 1/2$.

10. Suppose $\lambda_{0n}$ are the roots of $J_0'(\lambda) = 0$, and consider the Bessel expansions

$$f(r) = A_0 + \sum_{n=1}^{\infty} A_{0n} J_0(\lambda_{0n} r)$$

for the following functions:

(a) "Cowboy Hat":

$$f(r) = \begin{cases} 1 - 4r^2 & \text{if } 0 \leq r \leq 1/2, \\ 0 & \text{if } 1/2 < r \leq 1. \end{cases}$$

(b) "Witches Hat" (see previous problem).

(c) "Parabolic Pulse" (see previous problem).

For each function write down the explicit integral formula for the coefficients $A_{0n}$ and use a CAS to plot the approximating partial sums

$$A_0 + \sum_{n=1}^{N} A_{0n} J_0(\lambda_{0n} r),$$

for various values of $N$. Do (c) for $a = 0, b = 1$, and $a = 1/4, b = 1/2$.

11. Consider the Bessel expansions

$$f(r) = \sum_{n=1}^{\infty} A_{1n} J_1(\lambda_{1n} r),$$

with the $\lambda_{1n}$s the roots of $J_1(\lambda) = 0$. Redo Exercise 9 for this case. Also, find these expansions in the case that the $\lambda_{1n}$s are the roots of $J_1'(\lambda) = 0$ and $f$ is as in Exercise 10.

12. If $\psi = \psi(r)$ is a function of $r$ only, show that the integral formulas (12.22)–(12.23) for its Fourier–Bessel coefficients give $A_{mn} = 0$ and $B_{mn} = 0$ for all $m > 0$.

**(III).** **Fourier–Bessel series.** We now expand on the last group of exercises and consider the case where the function has some angular dependence. Thus for a given function $\psi(r, \theta)$ that *satisfies the equivalence conditions*, consider expanding $\psi$ in the series of sines/cosines and Bessel functions as in equation (12.20). This, in general, can get rather tedious, so we just consider a few special cases. To clarify the strategy, let $\mathcal{A}_m(r)$, $\mathcal{B}_m(r)$ stand for the Bessel series that occur in Eq.(12.20) (the parts that depend on $r$ in the large square brackets). Thus we rewrite equation (12.20) as

$$\psi(r, \theta) = \mathcal{A}_0(r) + \sum_{m=1}^{\infty} \mathcal{A}_m(r) \cos m\theta + \sum_{m=1}^{\infty} \mathcal{B}_m(r) \sin m\theta. \tag{12.35}$$

14. Find the Fourier–Bessel expansions for the following functions. The Bessel functions are $J_m(\lambda_{mn} r)$, where the $\lambda_{mn}$'s are the roots of $J_m(\lambda) = 0$.

(a) $\psi(r, \theta) = K_{a,b}(r) \sin \theta$, where $K_{a,b}$ is the parabolic pulse function.

(b)

$$\psi(r, \theta) = \begin{cases} r \sin 2\theta & \text{for } 0 \le \theta \le \pi, \\ 0 & \text{for } \pi < \theta \le 2\pi. \end{cases}$$

Here $r \in [0, 1]$.

*Hint*: In part (a), the series (12.35) reduces to one term: $\mathcal{B}_1(r) \sin \theta$. Thus all you need to do is to expand $K_{a,b}(r)$ in terms of the Bessel series $\mathcal{B}_1(r)$. In part (b), note that $\psi(r, \theta) = r\, g(\theta)$ for an appropriate function $g$. First expand $g$ in its Fourier cosine/sine series and then expand $r$ in the appropriate Bessel series.

15. Rework Exercise 12, but now with Bessel functions that are $J_m(\lambda_{mn} r)$, where the $\lambda_{mn}$'s are the roots of $J_m'(\lambda) = 0$.

**(IV). Homogeneous heat problems for the disk.** Consider the homogeneous heat equation in polar coordinates

$$w_t = \alpha^2 \left( w_{rr} + \frac{1}{r} w_r + \frac{1}{r^2} w_{\theta\theta} \right),$$

for the unit disk $a = 1$. In the following problems, $K_{b,c}$ is the parabolic pulse function. Find the series solution for each problem and analyze the heat flow in the disk by plotting the temperature distributions at appropriate times $t$. Determine the approximate time at which the disk reaches its steady state. Also plot the flow of the heat flux vector field

$$-\nabla u = -(w_r e_1 + \frac{1}{r} w_\theta e_2).$$

*Note*: in the case where $w$ does not depend on $\theta$, the heat flux is radial, and so you need not plot the flow lines in this case.

16. Boundary condition: fixed temperature 0. Initial condition: $\psi(r) = K_{1/4,1/2}(r)$.
17. Boundary condition: Insulated. Initial condition: $\psi(r) = K_{1/4,1/2}(r)$.
18. Boundary condition: fixed temperature 0. Initial condition: $\psi(r) = K_{1/4,1/2}(r)\sin\theta$.
19. Boundary condition: Insulated. Initial condition: $\psi(r,\theta) = K_{1/4,1/2}(r)\sin\theta$.
20. Insulated. Initial condition: $\psi(r,\theta) =$ the function in Exercise 14(b) or 15(b) above, depending on which of these you worked. Boundary condition: fixed temperature 0 or insulated.
21. In the case of an insulated boundary, prove that in general, the Bessel coefficient

$$A_0 = \frac{1}{\pi a^2} \int_0^{2\pi} \int_0^a \psi(r,\theta) r\, dr\, d\theta$$

is the average value of the initial temperature distribution over the disk $D_a$. *Hint*: Use the change of variable formula, Theorem A.1 in Appendix A.

## 12.6. The Heat Equation for an Annulus

The heat flow problem for an annulus $D_{r_1 r_2} = p(E)$, where $E = [r_1, r_2] \times [0, 2\pi]$, is very similar to that for a disk, but it has some important differences. It obviously has an additional boundary condition and only one equivalence condition. Thus for the annulus, the IBVP problem consists of

**The Heat Equation:**

$$w_t = \alpha^2 \left( w_{rr} + \frac{1}{r} w_r + \frac{1}{r^2} w_{\theta\theta} \right) + H \qquad \text{on } E \times (0, \infty);$$

**The Boundary Conditions:**

$$-\kappa_1 w_r(r_1, \theta, t) + h_1 w(r_1, \theta, t) = G_1(\theta, t),$$

$$\kappa_1 w_r(r_2, \theta, t) + h_1 w(r_2, \theta, t) = G_2(\theta, t) \qquad \text{for } \theta \in [0, 2\pi], t > 0;$$

**The Equivalence Condition:**

$$w(r, 0, t) = w(r, 2\pi, t) \qquad \text{for } r \in [r_1, r_2], t > 0;$$

**The Initial Condition:**

$$w(r, \theta, 0) = \psi(r, \theta) \qquad \text{for all } (r, \theta) \in E.$$

The work we did above in solving the homogeneous problem for a disk carries over with minor changes to the homogeneous problem for an annulus.

## 12.7.   The Homogeneous Problem for an Annulus

We consider the homogeneous heat equation and homogeneous boundary conditions for the annulus $D_{r_1 r_2}$. Again we look for solutions of the heat equation that have the form

$$w(r, \theta, t) = R(r)\Theta(\theta)T(t)$$

and that also satisfy the BCs and the equivalence condition. We get the same Sturm–Liouville problem for $\Theta$ and, in the notation from above, find that the eigenvalues are

$$\mu = -m^2, \qquad m = 0, 1, 2, 3, \ldots,$$

and the eigenfunctions are

$$1, \quad \cos m\theta, \quad \sin m\theta.$$

The Sturm–Liouville problem for $R$ is still a Bessel equation, but with two BCs and no equivalence condition.

**The Radial Sturm–Liouville Problem:**

$$R'' + \frac{1}{r}R' + \left(\lambda^2 - \frac{m^2}{r^2}\right)R = 0, \tag{12.36}$$

$$-\kappa_1 R'(r_1) + h_1 R(r_1) = 0, \tag{12.37}$$

$$\kappa_2 R'(r_2) + h_2 R(r_2) = 0, \tag{12.38}$$

As above, we now divide into cases depending on whether $\lambda$ is zero or not:

(1) ($\lambda = 0$) In this case Bessel's equation reduces to the Cauchy–Euler equation

$$r^2 R'' + rR' - m^2 R = 0.$$

(a) ($m = 0$) The general solution of the Cauchy–Euler DE in this case is

$$R(r) = A + B \ln r,$$

and so

$$R'(r) = \frac{B}{r}.$$

Substituting these in the two boundary conditions and rearranging slightly gives

$$h_1 A + (h_1 \ln r_1 - \kappa_1/r_1)B = 0,$$

$$h_2 A + (h_2 \ln r_2 + \kappa_2/r_2)B = 0.$$

Of course, this system of equations with unknowns $A$, $B$ has a nontrivial solution if and only if the determinant of its coefficient matrix is zero, i.e.,

$$\begin{vmatrix} h_1 & h_1 \ln r_1 - \kappa_1/r_1 \\ h_2 & h_2 \ln r_2 + \kappa_2/r_2 \end{vmatrix} = h_1 h_2 \ln(r_2/r_1) + h_1\kappa_2/r_2 + h_2\kappa_1/r_1$$

$$= 0.$$

Now, since $r_2 > r_1$, it follows that $\ln(r_2/r_1) > 0$. Then a little thought will convince you that the only way the above equation can be zero is for both $h_1 = 0$ and $h_2 = 0$. In this case the boundary conditions on $R$ are

$$R'(r_1) = 0,$$
$$R'(r_2) = 0;$$

i.e., each boundary of the annulus is insulated. In this case we see that $B = 0$, and the eigenfunction is

$$R(r) = 1.$$

(b) $(m \neq 0)$ In this case there are no nontrivial eigenfunctions (see the exercises).

(2) $(\lambda > 0)$ In this case (as we saw above) the general solution of Bessel's equation is

$$R(r) = A J_m(\lambda r) + B Y_m(\lambda r).$$

Substituting this in the two BCs gives the following two equations:

$$\left[ h_1 J_m(\lambda r_1) - \kappa_1 \lambda J'_m(\lambda r_1) \right] A + \left[ h_1 Y_m(\lambda r_1) - \kappa_1 \lambda Y'_m(\lambda r_1) \right] B = 0,$$

$$\left[ h_2 J_m(\lambda r_2) + \kappa_2 \lambda J'_m(\lambda r_2) \right] A + \left[ h_2 Y_m(\lambda r_2) + \kappa_2 \lambda Y'_m(\lambda r_2) \right] B = 0.$$

In order for this system to have nontrivial solutions, the determinant of its coefficient matrix must be zero. This determinant is

$$f_m(\lambda) = h_1 h_2 [J_m(\lambda r_1) Y_m(\lambda r_2) - J_m(\lambda r_2) Y_m(\lambda r_1)] \qquad (12.39)$$
$$+ \lambda h_1 \kappa_2 [J_m(\lambda r_1) Y'_m(\lambda r_2) - J'_m(\lambda r_2) Y_m(\lambda r_1)]$$
$$+ \lambda h_2 \kappa_1 [J_m(\lambda r_2) Y'_m(\lambda r_1) - J'_m(\lambda r_1) Y_m(\lambda r_2)]$$
$$+ \lambda^2 \kappa_1 \kappa_2 [J'_m(\lambda r_2) Y'_m(\lambda r_1) - J'_m(\lambda r_1) Y'_m(\lambda r_2)].$$

The eigenvalues $\lambda$ are the roots of the equation

$$f_m(\lambda) = 0,$$

and the general Sturm–Liouville theory guarantees us the existence (for each $m$) of an increasing sequence

$$\{\lambda_{mn}\}_{n=1}^{\infty}$$

of eigenvalues and that $\lim_{n\to\infty} \lambda_{mn} = \infty$. For each eigenvalue $\lambda_{mn}$, we take the corresponding

**Radial Eigenfunctions:**

$$R_{mn}(r) = a_{mn} J_m(\lambda_{mn} r) + b_{mn} Y_m(\lambda_{mn} r), \qquad (12.40)$$

where the coefficients are given by

$$a_{mn} = \kappa_1 \lambda_{mn} Y_m'(\lambda_{mn} r_1) - h_1 Y_m(\lambda_{mn} r_1),$$

$$b_{mn} = h_1 J_m(\lambda_{mn} r_1) - \kappa_1 \lambda_{mn} J_m'(\lambda_{mn} r_1).$$

*Note*: these latter formulas for the $a$'s and $b$'s are not unique. Many other choices for these constants will do. For example, when the boundary is totally insulated ($h_1 = 0 = h_2$ and $\kappa_1 > 0$, $\kappa_2 > 0$), we can take $a_{mn} = Y_m'(\lambda_{mn} r_1)$ and $b_{mn} = -J_m'(\lambda_{mn} r_1)$.

This completes the determination of the eigenfunctions and eigenvalues for the homogeneous annulus problem, and we arrive at its general solution:

**The General Solution of the Homogeneous Problem for the Annulus $D_{r_1 r_2}$:**

$$w(r, \theta, t) = A_0 + \sum_{m=0}^{\infty} \sum_{n=1}^{\infty} A_{mn} e^{-\lambda_{mn}^2 \alpha^2 t} R_{mn}(r) \Theta_{mn}(\theta)$$

$$= A_0 + \sum_{n=1}^{\infty} A_{0n} e^{-\lambda_{0n}^2 \alpha^2 t} R_{0n}(r)$$

$$+ \sum_{m=1}^{\infty} \left[ \sum_{n=1}^{\infty} A_{mn} e^{-\lambda_{mn}^2 \alpha^2 t} R_{mn}(r) \right] \cos m\theta \qquad (12.41)$$

$$+ \sum_{m=1}^{\infty} \left[ \sum_{n=1}^{\infty} B_{mn} e^{-\lambda_{mn}^2 \alpha^2 t} R_{mn}(r) \right] \sin m\theta.$$

Here, for convenience of notation, we are using

$$\Theta_{mn}(\theta) = A_{mn} \cos m\theta + B_{mn} \sin m\theta$$

and $R_{mn}$ for the eigenfunction given in equation (12.40).

*Note*: By convention, $A_0 = 0$ in the formula (12.41) for the general solution for any set of boundary conditions except for the set of BCs where *both* parts of the boundary are insulated.

The solution here for the annulus is entirely similar to that for the disk, except now the radial eigenfunctions are more complicated, and the equation for the eigenvalues $\lambda_{mn}$ is rather lengthy. Nevertheless, on a computer, the solution for the annulus can be programmed as easily as that for the disk.

**Example 12.2**   Consider the annulus with $r_1 = 1$ and $r_2 = 2$. Suppose that both parts of the boundary are insulated. Then the expression (12.39) reduces to just the last term (since $h_1 = 0 = h_2$), and the equations for the eigenvalues $\lambda$ are

$$J'_m(2\lambda)Y'_m(\lambda) - J'_m(\lambda)Y'_m(2\lambda) = 0,$$

for $m = 0, 1, \ldots$. With $\{\lambda_{mn}\}_{n=1}^{\infty}$ denoting the roots of the $m$th equation, the radial eigenfunctions are

$$R_{mn}(r) = Y'_m(\lambda_{mn})J_m(\lambda_{mn}r) - J'_m(\lambda_{mn})Y_m(\lambda_{mn}r).$$

For a general initial temperature distribution $\psi$, we would need all of this data. However, to keep things simple, suppose we take

$$\psi(r, \theta) = K_{1,2}(r)\sin\theta,$$

where $K_{a,b}$ is the parabolic pulse function from the last exercise set. Because of the special form of $\psi$, it is easy to see that the general solution (12.41) reduces in this case to

$$w(r, \theta, t) = A_0 + \left[\sum_{n=1}^{\infty} B_{1n}e^{-\lambda_{1n}^2 t}R_{1n}(r)\right]\sin\theta.$$

As usual, we take $\alpha = 1$. The constants $B_{1n}$ are the Bessel coefficients for the expansion of $K_{1,2}$ in terms of the eigenfunctions $R_{1n}$. Thus

$$B_{1n} = L_{1n}^{-1}\int_1^2 K_{1,2}(r)R_{1n}(r)\, r\, dr,$$

where $L_{1n} = \int_1^2 R_{1n}(r)^2\, r\, dr$ is the normalizing constant. Neither of these integrals is computable in closed form, so they will have to be calculated numerically on a computer. The constant $A_0$ is the average value of the initial temperature distribution over the annulus. It is easy to see, by graphing $\psi$, that

$$A_0 = 0,$$

and thus $\lim_{t \to \infty} w(r, \theta, t) = 0$ is the eventual temperature distribution in the annulus. However, there is some interesting intermediate behavior as the temperature approaches zero. To investigate this, we use the following *Maple* code. The code employs the identities

$$J_1'(x) = J_0(x) - \frac{J_1(x)}{x},$$

$$Y_1'(x) = Y_0(x) - \frac{Y_1(x)}{x},$$

and assumes that the eigenvalues $\{\lambda_{1n}\}_{n=1}^{10}$ have been previously calculated, stored in an array named c, and saved in a file named annz.m.

```
with(plots): alias(J=BesselJ,Y=BesselY); read 'annz.m';
F:=x->J(0,x)-J(1,x)/x; G:=x->Y(0,x)-Y(1,x)/x;
R:=(n,r)->G(c[n])*J(1,c[n]*r)-F(c[n])*Y(1,c[n]*r);
L:=array(1..10); B:=array(1..10);
for n to 10 do
  L[n]:=evalf(Int(R(n,r)^2*r,r=1..2)):
  B[n]:=evalf(Int(R(n,r)*4*(r-1)*(2-r)*r,r=1..2)/L[n]):
od;
M:=proc(r,t,N)s:=0;for n to N do
   s:=s+B[n]*exp(-c[n]^2*t)*R(n,r) od end;
w:=(r,p,t,N)->M(r,t,N)*sin(p);
```

The following figures show some of the analysis of the heat flow. As usual, we first plot the temperature evolution at several points in the plate in order to gauge how quickly the temperature goes to zero. This is shown in Figure 12.15. Note that for early times, $t < 0.1$, some portions of the plate are cooling down and others heating up as the heat diffuses toward the boundaries. However, as indicated in the figure, all points (at least the three shown) along the radius $\theta = \pi/2$ rapidly approach the same temperature and then decay uniformly together to zero. This is exhibited more clearly in Figure 12.16. Note that the temperature function has the form

$$w(r, \theta, t) = M(r, t) \sin \theta,$$

and so the temperature profiles along any radius $\theta = \theta_0$ will be similar to those in Figure 12.16. Thus for $t > 0.1$, the function $M$ does not depend on $r$

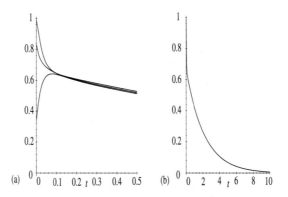

Figure 12.15. (a) Plot of the temperature evolution at the points with polar coordinates $(1.1, \pi/2)$, $(1.3, \pi/2)$, $(1.5, \pi/2)$ in the plate for Example 12.2. (b) temperature evolution at the point with polar coordinates $(1.5, \pi/2)$

(approximately), $M(r, t) \approx M(t)$, and consequently, $w(r, \theta, t) \approx M(t) \sin \theta$, for large $t$. Transforming this back to Cartesian coordinates gives a temperature function

$$u(x, y, t) \approx \frac{M(t)y}{(x^2 + y^2)^{1/2}},$$

for $t > 0.1$. This is a circular band that eventually flattens out into the annulus $D_{1,2}$, since $\lim_{t \to \infty} M(t) = 0$. This is shown clearly in Figure 12.17, which also shows the interesting short-term behavior of the temperature distribution in the annulus before it flattens out into a band.

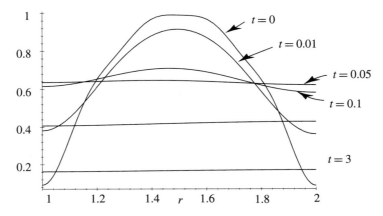

Figure 12.16. Plot of some temperature distributions $w(r, \pi/2, t) = M(r, t)$ along the radius $\theta = \pi/2$ in the plate for Example 12.2. The distributions shown are for times $t = 0, 0.01, 0.05, 0.1, 1, 3$.

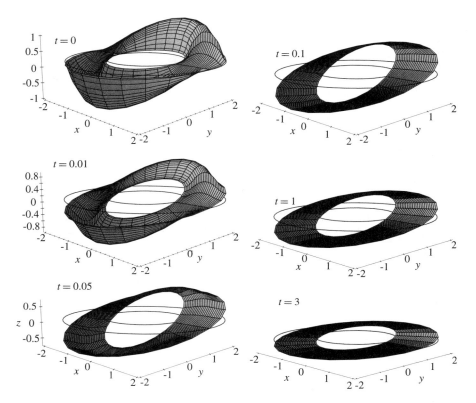

Figure 12.17.  The temperature distribution in the annulus of Example 12.2 at times $t = 0, 0.01, 0.05, 0.1, 1, 3$.

## Exercises 12.7

1. Show that in the radial Sturm–Liouville problem for the annulus, the case where $\lambda = 0$ and $m \neq 0$ has no nontrivial eigenfunctions.

2. Determine the integral formulas for the Fourier–Bessel coefficients $A_0, A_{mn}, B_{mn}$ in equation (12.41).

3. The following are homogeneous heat IBVPs for the annulus $r_1 = 1, r_2 = 2$. In each case determine the series solution and use a computer to analyze the heat flow in the annulus. *Note*: some of these problems can take awhile, the time depending particularly on the number of eigenvalues and numerical integrations computed. So try to optimize your code and otherwise have something else to do while the computer is churning away.

   (a) Boundary conditions: fixed temperature zero on both parts of the boundary. Initial condition: $\psi(r, \theta) = K_{1,2}(r) \sin \theta$.

   (b) Boundary conditions: insulated on both parts of the boundary. Initial condition: $\psi(r, \theta) = K_{1,2}(r)$.

(c) Boundary conditions: insulated on both parts of the boundary. Initial condition:

$$\psi(r, \theta) = \begin{cases} K_{1,2}(r) \sin \theta & \text{for } \theta \in [0, \pi], \\ 0 & \text{for } \theta \in (\pi, 2\pi). \end{cases}$$

(d) Boundary conditions: insulated on the circle $r = 1$ and fixed temperature zero on the circle $r = 2$. Initial condition: $\psi(r, \theta) = K_{1,2}(r) \sin 2\theta$.

(e) Boundary conditions: insulated on both parts of the boundary. Initial condition:

$$\psi(r, \theta) = \begin{cases} c & \text{for } \theta \in [\theta_1, \theta_2], \\ 0 & \text{otherwise.} \end{cases}$$

Here $c$ is a positive constant and the IC represents the idealized situation of initially heating the sector $\theta_2 \le \theta \le \theta_2$ of the annulus to a constant temperature $c$.

4. [*The semi-homogeneous IBVP (disk and annulus)*] Show how to solve the semi-homogeneous heat problems for the disk and the annulus. You may combine the two cases in your discussion because of the similar structure of the solutions. Specifically, note that the solution to the homogeneous problems for both the disk and the annulus can be written in the form

$$w(r, \theta, t) = T_0(t) + \sum_{m=0}^{\infty} \sum_{n=1}^{\infty} R_{mn}(r) \Theta_{mn}(\theta, t), \tag{12.42}$$

where

$$\Theta_{mn}(\theta, t) = T_{mn}^1(t) \cos m\theta + T_{mn}^2(t) \sin m\theta. \tag{12.43}$$

For the homogeneous problem: $T_0(t) = $ a constant, and $T_{mn}^1(t)$, $T_{mn}^2(t)$ are constant multiples of $\exp(-\lambda_{mn}^2 \alpha^2 t)$. To solve the semi-homogeneous problem, take $T_0$, $T_{mn}^1$, and $T_{mn}^2$ as arbitrary functions and see what conditions the semi-homogeneous problem dictates for them. Note that since $R_{mn}$ and $\Theta_{mn}(\cdot, t)$ (for each $t$) satisfy their respective boundary conditions, the function $w$ satisfies the homogeneous BCs for the heat problem, regardless of the choice of $T_0$, $T_{mn}^1$, and $T_{mn}^2$. Substitute the above series expression (12.42) for $w$ into the heat equation and expand the heat source/sink function in terms of the eigenfunctions of the problem:

$$H(r, \theta, t) = H_0(t) + \sum_{m=0}^{\infty} \sum_{n=1}^{\infty} R_{mn}(r) \Gamma_{mn}(\theta, t), \tag{12.44}$$

where

$$\Gamma_{mn}(\theta, t) = H_{mn}^1(t) \cos m\theta + H_{mn}^2(t) \sin m\theta. \tag{12.45}$$

From this find the initial value problems (ODE+IC) that determine all the $T$'s.

## 12.8.   BVPs in Polar Coordinates

The final element in the description of how to solve heat problems in polar coordinates (specifically just for disks and annuli) is the technique for transforming the general problem into a semi-homogeneous one. This amounts to splitting the general heat problem into two parts by writing

$$w = K + W$$

and requiring that $K$ satisfy Laplace's equation in polar coordinates

$$K_{rr} + \frac{1}{r} K_r + \frac{1}{r^2} K_{\theta\theta} = 0,$$

together with the non-homogeneous boundary conditions originally imposed on $w$ and the equivalence conditions as well. Once we determine the $K$ that satisfies the BVP, then the remaining IBVP for $W$ will be semi-homogeneous. Thus in this section we concentrate on solving boundary value problems in polar coordinates. For this we restrict ourselves to regions that are either a disk or an annulus.

As usual, we use the separation of variables technique to find solutions of Laplace's equation in polar coordinates that have the form

$$K(r, \theta) = R(r)\Theta(\theta).$$

It is also required that such a $K$ satisfy the equivalence conditions, which for the regions we consider are

**For the Disk $D_a$:**

(1)  $\Theta(0) = \Theta(2\pi)$.

(2)  $R$ is defined at $r = 0$, and either $R(0) = 0$ or $\Theta$ is a constant function.

**For the Annulus $D_{r_1 r_2}$:**   $\Theta(0) = \Theta(2\pi)$.

It is easy to see that separation of variables leads to

$$\frac{r^2 R'' + r R'}{-R} = \frac{\Theta''}{\Theta} = k,$$

and the equivalence condition on $\Theta$ gives us a complete Sturm–Liouville problem

$$\Theta'' = k\Theta$$

$$\Theta(0) = \Theta(2\pi),$$

whose eigenvalues are

$$k = -m^2, \qquad m = 0, 1, 2, \ldots,$$

and whose eigenfunctions are

$$1, \quad \cos m\theta, \quad \sin m\theta.$$

The Sturm–Liouville problem for $R$ is incomplete in that it only requires that $R$ satisfy the Cauchy–Euler equation

$$r^2 R'' + r R' - m^2 R = 0.$$

As we have seen, two linearly independent solutions of this are

for $m = 0$,  $1$,  $\ln r$;

for $m > 0$,  $r^m$,  $r^{-m}$.

Thus the functions of the form $R(r)\Theta(\theta)$ that satisfy the heat equation in polar coordinates and the equivalence condition are

$$1, \quad \ln r, \quad r^{\pm m} \cos m\theta, \quad r^{\pm m} \sin m\theta.$$

By the superposition principle, the same is true of the more general series expression formed by taking linear combinations of these functions. Since the annulus problem has only one equivalence condition, we have thus arrived at the

**General Solution of Laplace's Equation for an Annulus:**

$$K(r, \theta) = A_0 + C_0 \ln r$$

$$+ \sum_{m=1}^{\infty} r^m (A_m \cos m\theta + B_m \sin m\theta)$$

$$+ \sum_{m=1}^{\infty} r^{-m} (C_m \cos m\theta + D_m \sin m\theta). \qquad (12.46)$$

The arbitrary constants $A_m$, $B_m$, $C_m$, $D_m$ are chosen such that (12.46) satisfies the two boundary conditions in the annulus problem.

For the disk problem we must take into account the other equivalence condition, which requires that $K$ be defined at $r = 0$ and that $K(0, \theta) = K(0, \theta')$ for every $\theta, \theta'$. This condition can be met by the series expression (12.46) if we delete the $\ln r$ term and all the terms with $r^{-m}$. Doing so, we arrive at the

**General Solution of Laplace's Equation for a Disk:**

$$K(r, \theta) = A_0 + \sum_{m=1}^{\infty} r^m (A_m \cos m\theta + B_m \sin m\theta). \tag{12.47}$$

The arbitrary constants $A_m$, $B_m$, again are chosen such that the series expression (12.47) satisfies the boundary condition for the disk problem.

We turn now to the solution of specific problems, i.e., the determination of the arbitrary constants in the series solutions for the disk and annulus. For this we limit the discussion to just Dirichlet-type boundary conditions.

## 12.8.1.   The Dirichlet Problem for a Disk

For the disk $D_a$, the Dirichlet boundary condition on $K$ is

$$K(a, \theta) = G(\theta).$$

In terms of the series expression (12.47) for $K$, this condition is

$$A_0 + \sum_{m=1}^{\infty} a^m (A_m \cos m\theta + B_m \sin m\theta) = G(\theta). \tag{12.48}$$

Thus we see that $A_0$, $a^m A_m$, and $a^m B_m$ are the Fourier cosine and sine coefficients of the given function $G$. Hence these are given by

$$A_0 = \frac{1}{2\pi} \int_0^{2\pi} G(\theta) d\theta, \tag{12.49}$$

$$A_m = \frac{1}{\pi a^m} \int_0^{2\pi} G(\theta) \cos m\theta \, d\theta, \tag{12.50}$$

$$B_m = \frac{1}{\pi a^m} \int_0^{2\pi} G(\theta) \sin m\theta \, d\theta, \tag{12.51}$$

for $m = 1, 2, \ldots$.

**Example 12.3**   Suppose the given boundary function is

$$G(\theta) = \begin{cases} \sin 2\theta, & 0 \le \theta \le \pi, \\ 0, & \pi < \theta \le 2\pi. \end{cases}$$

The calculation of the Fourier coefficients for $G$ is straightforward. *Note*: the integrals reduce to integrals from 0 to $\pi$ only, since $G = 0$ in the interval from $\pi$ to $2\pi$.

$$A_0 = \frac{1}{2\pi} \int_0^\pi \sin 2\theta \, d\theta = 0,$$

$$A_m = \frac{1}{\pi a^m} \int_0^\pi \sin 2\theta \, \cos m\theta \, d\theta$$

$$= \frac{1}{\pi a^m} \int_0^\pi \frac{1}{2}[\sin(m+2)\theta - \sin(m-2)\theta]d\theta$$

$$= \frac{1}{2\pi a^m} \left[ \frac{-\cos(m+2)\theta}{m+2} + \frac{\cos(m-2)\theta}{m-2} \right]_0^\pi$$

$$= \begin{cases} 0 & \text{if } m = 2k, \\ \frac{-4}{(2k+1)(2k-3)\pi a^m} & \text{if } m = 2k-1, \end{cases}$$

$$B_m = \frac{1}{\pi a^m} \int_0^\pi \sin 2\theta \, \sin m\theta \, d\theta$$

$$= \frac{1}{\pi a^m} \int_0^\pi \frac{1}{2}[\cos(m-2)\theta - \cos(m+2)\theta]d\theta$$

$$= \begin{cases} 1/(2a^2) & \text{if } m = 2, \\ 0 & \text{if } m \neq 2. \end{cases}$$

Putting all this together, we get that the solution of this problem is

$$K(r,\theta) = \frac{1}{2a^2}r^2 \sin 2\theta - \frac{4}{\pi} \sum_{k=1}^{\infty} \frac{r^{2k-1}}{(2k+1)(2k-3)\pi a^{2k-1}} \cos(2k-1)\theta.$$

To graph an approximation to this solution, we take $a = 1$ for the radius and use the following *Maple* code:

```
with(plots):
A:=k->-4/((2*k+1)*(2*k-3)*Pi);
K:=proc(r,p,N) s:=0;
   for k to N do s:=s+a(k)*r^(2*k-1)*cos((2*k-1)*p) od;
   s:=s+r^2*sin(2*p)/2 end;
p1:=cylinderplot([r,p,K(r,p,10)],r=0..1,p=0..2*Pi):
p2:=spacecurve([cos(t),sin(t),0],t=0..2*Pi,color=black):
```

```
display3d({p1,p2});
```

Note that we have chosen to create two plot structures, p1, p2, the first the surface that is the plot of $K$, and the second the circular boundary of the unit disk. The resulting plot is shown in Figure 12.18.

**Example 12.4**   This example illustrates that it is often not necessary to use the integral formulas for the coefficients $A_m$, $B_m$. For simplicity we assume that the radius is $a = 1$ in this example. Here we suppose the boundary data is specified by

$$G(\theta) = \cos 3\theta.$$

The Fourier cosine/sine series for this function obviously consists of a single term, i.e, $A_3 = 1$, and all the other $A_m$'s and $B_m$'s are zero. Thus the series solution reduces to

$$K(r, \theta) = r^3 \cos 3\theta.$$

The plot of this function is shown in Figure 12.19. This solution of the Dirichlet problem, given in polar coordinates, can be easily transformed back into Cartesian coordinates as follows. Using the angle addition trigonometric

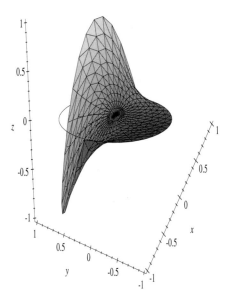

Figure 12.18.   Plot of the approximate solution $K(r, \theta, 10)$ to the solution $K(r, \theta)$ in Example 12.3

formulas, we rewrite $K$ as follows:

$$K(r, \theta) = r^3 \cos 3\theta = r^3 \cos(2\theta + \theta)$$

$$= r^3 [\cos 2\theta \cos \theta - \sin 2\theta \sin \theta]$$

$$= r^3 [(\cos^2 \theta - \sin^2 \theta) \cos \theta - 2 \sin^2 \theta \cos \theta]$$

$$= r^3 \cos^3 \theta - 3r^3 \sin^2 \theta \cos \theta.$$

Thus we see that $K(r, \theta) = S(r \cos \theta, r \sin \theta)$, where

$$S(x, y) = x^3 - 3xy^2.$$

## 12.8.2.  The Dirichlet Problem for an Annulus

The two boundary conditions in the Dirichlet problem for the annulus $D_{r_1 r_2}$ are

$$K(r_1, \theta) = G_1(\theta),$$

$$K(r_2, \theta) = G_2(\theta),$$

for $0 \le \theta \le 2\pi$. Here $G_1, G_2$ are given functions on the interval $[0, 2\pi]$. Thus substituting the series expression (12.46) for the general solution of Laplace's

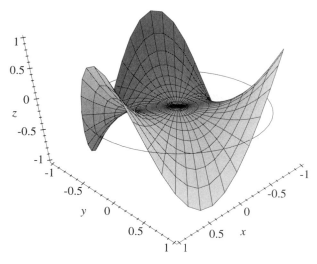

Figure 12.19.    Plot of $K(r, \theta) = r^3 \cos 3\theta$ in Example 12.4

equation for the annulus into the above boundary conditions, we get

$$G_1(\theta) = A_0 + C_0 \ln r_1$$

$$+ \sum_{m=1}^{\infty} (A_m r_1^m + C_m r_1^{-m}) \cos m\theta$$

$$+ \sum_{m=1}^{\infty} (B_m r_1^m + D_n r_1^{-m}) \sin m\theta, \qquad (12.52)$$

$$G_2(\theta) = A_0 + C_0 \ln r_2$$

$$+ \sum_{m=1}^{\infty} (A_m r_2^m + C_m r_2^{-m}) \cos m\theta$$

$$+ \sum_{m=1}^{\infty} (B_m r_2^m + D_m r_2^{-m}) \sin m\theta. \qquad (12.53)$$

These two equations suffice to determine completely all the coefficients $A_m$, $B_m$, $C_m$, $D_m$. Regrettably, this determination is not all that straightforward. In fact, we are led to some systems of equations to solve for these coefficients. To see this, we expand $G_1$ and $G_2$ into their respective Fourier cosine–sine series

$$G_1(\theta) = a_0^1 + \sum_{m=1}^{\infty} (a_m^1 \cos m\theta + b_m^1 \sin m\theta), \qquad (12.54)$$

$$G_2(\theta) = a_0^2 + \sum_{m=1}^{\infty} (a_m^2 \cos m\theta + b_m^2 \sin m\theta). \qquad (12.55)$$

If we now equate the cosine and sine coefficients in these two series with the corresponding coefficients in the series (12.52)–(12.53), then we get the following systems of two equations in two unknowns:

**(1) The constant coefficients:**

$$A_0 + (\ln r_1)C_0 = a_0^1,$$

$$A_0 + (\ln r_2)C_0 = a_0^2, \qquad (12.56)$$

**(2) The cosine coefficients:**

$$r_1^m A_m + r_1^{-m} C_m = a_m^1,$$

$$r_2^m A_m + r_2^{-m} C_m = a_m^2; \qquad (12.57)$$

**(3) The sine coefficients:**

$$r_1^m B_m + r_1^{-m} D_m = b_m^1,$$

$$r_2^m B_m + r_2^{-m} D_m = b_m^2; \tag{12.58}$$

for $n, m = 1, 2, \ldots.$ It is an easy exercise to show that since $r_1 \neq r_2$, each of these systems has a unique solution. In principle, (12.56)–(12.58) constitute infinitely many systems of equations to solve. However, it is often the case that only finitely many of the Fourier sine–cosine coefficients of $G_1$ and $G_2$ are nonzero, in which case there are only finitely many systems of equations to solve.

**Example 12.5**   Consider the case where the annulus has radii $r_1 = 1$ and $r_2 = 2$, and the boundary data are:

$$G_1(\theta) = 0,$$

$$G_2(\theta) = 1 + \cos\theta.$$

We see that all the Fourier coefficients for $G_1$ are zero and that the only nonzero Fourier coefficients for $G_2$ are $a_0^2 = 1$ and $a_1^2 = 1$. Also, note that since $r_1 = 1$, we have $\ln r_1 = \ln 1 = 0$. Now we consider each of the systems of equations in order.
   $m = 0$ :

$$A_0 = 0,$$

$$A_0 + \ln 2 C_0 = 1.$$

Thus $A_0 = 0$ and $C_0 = 1/\ln 2$.
   $m = 1$ :

$$A_1 + C_1 = 0,$$

$$2A_1 + \frac{1}{2}C_1 = 1.$$

Thus $A_1 = 2/3$ and $C_1 = -2/3$.
   $m > 1$ :

$$A_m + C_m = 0,$$

$$2^m A_m + 2^{-m} C_m = 0.$$

Thus $A_m = 0$ and $C_m = 0$ for all $m > 1$. Similarly, since the systems for the sine coefficients are homogeneous, we get $B_m = 0$ and $D_m = 0$ for all $m > 0$.

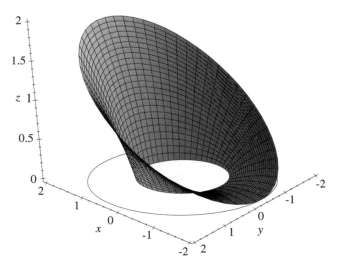

Figure 12.20.    Plot of the solution $K(r, \theta)$ in Example 12.5

Consequently, the solution to the problem is

$$K(r, \theta) = \frac{\ln r}{\ln 2} + \frac{2}{3} \left( r - \frac{1}{r} \right) \cos \theta.$$

The plot of this solution is shown in Figure 12.20. It is easy to see that the solution to this problem transformed back to Cartesian coordinates is

$$S(x, y) = \frac{\ln(x^2 + y^2)}{\ln 2} + \frac{2}{3} x \left( 1 - \frac{1}{x^2 + y^2} \right).$$

**Exercises 12.8**

The following exercises deal with the Dirichlet problem for the disk and the annulus.

I. **Disk of radius** $a = 1$: Consider the Dirichlet BVP for the unit disk in polar coordinates:

$$K_{rr} + \frac{1}{r} K_r + \frac{1}{r^2} K_{\theta\theta} = 0,$$

$$K(1, \theta) = G(\theta),$$

where $K$ is assumed to satisfy the equivalence conditions for the disk. For each of the following boundary data $G(\theta)$:

(a) Write down explicitly the solution $K$ to the BVP.

(b) If the solution, which generally will be a series, reduces to a finite sum of terms, transform $K$ back into the solution $S$ in Cartesian coordinates: $K(r, \theta) = S(r \cos \theta, r \sin \theta)$.

(c) Graph the surface $z = S(x, y)$, which is the solution of the BVP in Cartesian coordinates. For clarity, also graph the boundary circle of the disk in the same picture.

  1. $G(\theta) = \cos 2\theta$.

  2. $G(\theta) = \sin 2\theta + \cos 3\theta$.

  3. $G(\theta) = \cos 5\theta$.

  4. $G(\theta) = \sin^2 \theta$. (use trigonometric identities to write $G(\theta)$ first as a sum of trigonometric functions.)

  5.

$$G(\theta) = \begin{cases} \sin \theta & \text{if } 0 \leq \theta \leq \pi, \\ 0 & \text{if } \pi < \theta \leq 2\pi. \end{cases}$$

**II.  Annulus with radii $r_1 = 1$ and $r_2 = 2$:** Consider the Dirichlet BVP for the annulus with radii 1 and 2 in polar coordinates:

$$K_{rr} + \frac{1}{r} K_r + \frac{1}{r^2} K_{\theta\theta} = 0,$$

$$K(1, \theta) = G_1(\theta),$$

$$K(2, \theta) = G_2(\theta),$$

where $K$ is assumed to satisfy the equivalence condition for the annulus. For each of the following boundary data $G_1(\theta), G_2(\theta)$:

(a) Write down explicitly the solution $K$ to the BVP.

(b) If the solution, which generally will be a series, reduces to a finite sum of terms, transform $K$ back into the solution $S$ in Cartesian coordinates: $K(r, \theta) = S(r \cos \theta, r \sin \theta)$.

(c) Graph the surface $z = S(x, y)$, which is the solution of the BVP in Cartesian coordinates. For clarity, also graph the two boundary circles of the annulus in the same picture.

  6. $G_1(\theta) = 3, G_2(\theta) = 5$.

  7. $G_1(\theta) = 0, G_2(\theta) = \sin^2 \theta$.

  8. $G_1(\theta) = 1 + \cos \theta, G_2(\theta) = 1 + \sin \theta$.

9. $G_1(\theta) = 1/2$ and

$$G_2(\theta) = \begin{cases} \sin\theta & \text{if } 0 \leq \theta \leq \pi, \\ 0 & \text{if } \pi < \theta \leq 2\pi. \end{cases}$$

(You may use the work you did in Exercise 5 above.)

# 13 Solution of the Heat IBVP in General

In this chapter we discuss the solution of the heat IBVP on a *general* domain $\Omega$ in $\mathbb{R}^3$ (technically, $\Omega$ is a bounded, open, connected set with boundary $\partial\Omega$ having a suitable degree of smoothness). The *solution* is given in theoretical terms using techniques from functional analysis that merely predict the existence of the solution and its uniqueness. The same techniques lead to schemes for calculating the solution numerically. The discussion here is similar to that we gave for the Navier–Stokes equations in Chapter 10 and the Navier equation in Chapter 11, and is mainly heuristic, giving just a rough outline of the theory, which can only be more fully understood with a background in functional analysis and differential geometry.

The need to have a theory for general domains $\Omega$ should be clear. We know how to solve the heat IBVP when $\Omega$ is the unit interval, unit square, or unit cube. For arbitrary intervals, rectangles, and boxes, we can transform the problem by a translation, rotation, and scaling to arrive at an IBVP similar to the ones we know how to solve (the Laplacian is not invariant under scaling transformations, so there are minor changes in the techniques). Similarly, we saw in the last chapter how to solve 2-D problems for the disk and annulus by transforming to polar coordinates. The transformation technique extends to many other situations. For example, when $\Omega$ is a cylinder, sphere, cone, or torus, we can transform the heat IBVP to one in cylindrical, spherical, conical, or toroidal coordinates. The analytical description of the solution in terms of series and the calculation of the coefficients from integral formulas can get rather involved (as you might guess from the work in the last chapter) and will not be discussed here. However, for "real" situations where the object $\Omega$ being heated is, say, an apple, a pear, or a flounder fillet, the corresponding heat problem is too complex to be solved explicitly by any of the above idealized situations. Nevertheless, the existence of solutions for such complex domains $\Omega$ and their expansions in terms of eigenfunctions can at least be exhibited *theoretically*, and this chapter discusses the details for this.

## 13.1.   Weak Solutions of Poisson's Equation

Before discussing the heat IBVP, we consider the BVP

$$\nabla^2 u = f \qquad \text{in } \Omega \tag{13.1}$$

$$a\nabla u \cdot n + bu = g \qquad \text{on } \partial\Omega \tag{13.2},$$

on a general domain $\Omega$ in $\mathbb{R}^3$ (with the restriction in the first paragraph). Here $f$, $g$ are given scalar fields on $\Omega$ and $\partial\Omega$ respectively. The PDE (13.1) is Poisson's equation, and the boundary condition (13.2) involves two constants $a, b \geq 0$ (more generally, $a, b$ could be scalar fields defined on $\partial\Omega$, but for simplicity, we assume that they are constants). The case $a = 0$ is the Dirichlet problem, and the case $b = 0$ is the Neumann problem. The case $a > 0, b > 0$ is often called the *Robin problem*. A solution of this BVP can be considered as a steady-state temperature distribution for a heat IBVP.

The contemporary method for proving existence and uniqueness of solutions to such boundary value problems involves the concept of *weak* solutions. By generalizing the notion of what is meant by a function and extending the calculus to this broader class of *generalized functions* (also called *distributions*), boundary value problems can be considered in the distributional sense. A distribution that satisfies the BVP is called a weak solution. It is "weak" because it may not be an ordinary function, or even if it is, it may not be a differentiable function, and so its derivatives have to be considered in a generalized way (see below).

Having proven existence and uniqueness of weak solutions to BVPs, one then proceeds, where possible, to prove that the distribution that is the weak solution is actually a *strong* (or classical) solution; i.e., it is an ordinary function possessing a certain degree of differentiability (or regularity). The theorems for this are called *regularity theorems* in the literature. This may seem like a somewhat roundabout manner to deal with boundary value problems, but it has proven to be very powerful and effective. There are many important circumstances, like turbulent fluid flow modeled by the Navier–Stokes equations, where weak solutions are known to exist theoretically, but the proof that these weak solutions are strong solutions is still lacking. Also, as discussed in the exercises in the chapter on Maxwell's equations, there are situations where the only way to interpret the solution is in the weak sense.

We will use the notation and concepts introduced in Exercise 3 from Exercise Set 9.2. You may wish to read the discussion there if you have not done so already). Thus $\mathcal{D}(\Omega)$ denotes the set of (infinitely differentiable) scalar fields $\psi$ on $\Omega$ that vanish outside a compact subset of $\Omega$, and $\mathcal{D}^*(\Omega)$ is the set of generalized scalar fields (or distributions) $\Gamma$ on $\Omega$. Recall that these are linear functionals $\Gamma : \mathcal{D}(\Omega) \to \mathbb{R}$, and that any ordinary scalar field $v$ (assuming that it is integrable over $\Omega$) can be considered as a generalized scalar field $\tilde{v} : \mathcal{D}(\Omega) \to \mathbb{R}$, via the definition

$$\tilde{v}(\psi) \equiv \int_\Omega v\psi \, dV,$$

for $\psi \in \mathcal{D}(\Omega)$. The extension of the partial differentiation operation to generalized scalar fields $\Gamma$ is done in such a way as to give the usual partial derivatives when $\Gamma$ is an ordinary (differentiable) scalar field. Thus the partial derivative $\partial \Gamma / \partial x$ is the generalized scalar field defined by

$$\frac{\partial \Gamma}{\partial x}(\psi) \equiv -\Gamma\left(\frac{\partial \psi}{\partial x}\right).$$

The reason for the minus sign (and motivation for the definition) comes from using integration by parts, actually Gauss's divergence theorem, to show that in the special case where $v$ is a differentiable scalar field, then

$$\frac{\partial \tilde{v}}{\partial x} = \widetilde{\frac{\partial v}{\partial x}}$$

(exercise). The definitions of the other partial derivatives $\partial \Gamma / \partial y$ and $\partial \Gamma / \partial z$ are similar. With these definitions, it is important to note that any generalized scalar field has partial derivatives to all orders (is infinitely differentiable). The same is true for ordinary scalar fields $v$, *but note:* The derivatives are in the generalized sense; $\partial \tilde{v} / \partial x$ always exists, even when $\partial v / \partial x$ may not.

We let $L^2(\Omega)$ denote the set of scalar fields $v$ on $\Omega$ that are square integrable:

$$L^2(\Omega) = \{ v \mid \int_\Omega v^2 \, dV < \infty \}.$$

One can show that if $u, v \in L^2(\Omega)$, then the product $uv$ is integrable over $\Omega$, and this gives rise to a bilinear function

$$\langle u, v \rangle = \int_\Omega uv \, dV$$

on $L^2(\Omega)$ that makes it an *inner product space* (cf. Exercise 5 in Exercises 10.5). The corresponding *norm* on $L^2(\Omega)$ is $\|v\| = \langle v, v \rangle^{1/2}$, and with respect to this norm, $L^2(\Omega)$ is *complete*. Thus $L^2(\Omega)$ is a *Hilbert space* (a complete inner product space).

The concept of a Hilbert space is central to the techniques used in proving existence and uniqueness of weak solutions (like the Lax–Milgram theorem below). In addition to $L^2(\Omega)$, the other Hilbert spaces important to the theory are $H^k(\Omega)$, for $k = 1, 2, \ldots$. These special Hilbert spaces are known as *Sobolev spaces*, after the mathematician who first introduced them. By definition, $H^k(\Omega)$ is the set of all generalized scalar fields $v$ whose partial derivatives up to order $k$ are in $L^2(\Omega)$. Specifically, this means that $v$ is an ordinary scalar field (but not necessarily differentiable) and is in $L^2(\Omega)$, that

$$\frac{\partial \tilde{v}}{\partial x}, \ \frac{\partial \tilde{v}}{\partial y}, \ \frac{\partial \tilde{v}}{\partial z},$$

are ordinary scalar fields (but not necessarily differentiable) and are in $L^2(\Omega)$; and similarly, that all the higher-order distributional partial derivatives up to the $k$th order are ordinary scalar fields and are in $L^2(\Omega)$.

Since we will be dealing only with the Laplace operator $\nabla^2$, which is a second order differential operator, we will need only the Sobolev space $H^1(\Omega)$ to formulate the weak problem. The bilinear form on this space that gives the inner product is

$$\langle u, v \rangle_1 \equiv \int_\Omega [\, uv + \nabla u \cdot \nabla v \,]\, dV,$$

and the corresponding norm is

$$\|v\|_1 = \langle v, v \rangle_1^{1/2} = \left( \int_\Omega [\, v^2 + |\nabla v|^2 \,]\, dV \right)^{1/2}.$$

One further Hilbert space will be important to us here. It is the space $H_0^1(\Omega)$ and is defined as follows. Observe that the spaces of functions we have introduced so far are related by the set inclusions

$$\mathcal{D}(\Omega) \subset H^1(\Omega) \subset L^2(\Omega).$$

The latter two are Hilbert spaces, while $\mathcal{D}(\Omega)$ is not. However, if we let

$$H_0^1(\Omega) = \overline{\mathcal{D}(\Omega)},$$

we get a Hilbert subspace of $H^1(\Omega)$. The overbar in this definition stands for the closure with respect to the $H^1(\Omega)$ norm. Namely, the elements of $H_0^1(\Omega)$ are those $v \in H^1(\Omega)$ that are limits of sequences $\{\psi_n\}$ of elements $\psi_n \in \mathcal{D}(\Omega)$.

To discuss the BVP (13.1)–(13.2) within the context of weak solutions, we divide into two cases. This is convenient for two reasons. First, the Dirichlet problem is the simplest pedagogically, and second, each case uses a different *Dirichlet form*. The notion of a Dirichlet form is central to formulating the weak version (also called the variational or distributional version) of the problem.

## 13.1.1.  Case ($a = 0$): The Dirichlet Problem

Here we consider the BVP

$$\nabla^2 u = f \qquad \text{in } \Omega, \tag{13.3}$$

$$u = 0 \qquad \text{on } \partial\Omega. \tag{13.4},$$

Notice that the boundary condition is homogeneous. However, there is no loss of generality, since we can always transform to homogeneous BCs (exercise).

Heuristically, the weak version of the above Dirichlet problem is the integral equation that arises by multiplying both sides of equation (13.3) by a suitable function $v$ and integrating both sides over $\Omega$:

$$\int_\Omega (\nabla^2 u) v \, dV = \int_\Omega f v \, dV. \tag{13.5}$$

Now we rewrite this integral equation by using integration by parts on the left side (see the proposition below), assuming $v = 0$ on $\partial\Omega$, and then express it in terms of a *Dirichlet form*. Here the Dirichlet form is the bilinear functional on $H^1(\Omega)$ defined by

$$D(u, v) \equiv -\int_\Omega \nabla u \cdot \nabla v \, dV.$$

With this we then get

**The Weak Version of the Dirichlet Problem:**    Determine $u \in H_0^1(\Omega)$ such that

$$D(u, v) = \langle f, v \rangle, \tag{13.6}$$

for all $v \in H_0^1(\Omega)$.

Here $f$ is a given function in $L^2(\Omega)$, and the bracket $\langle \cdot, \cdot \rangle$ is the inner product in this space. If we write out equation (13.6) explictly, it is

$$-\int_\Omega \nabla u \cdot \nabla v \, dV = \int_\Omega f v \, dV, \tag{13.7}$$

for all $v \in H_0^1(\Omega)$. Thus the weak version of the Dirichlet problem involves an integral equation (often called the integral version of the problem). It is important to note that the weak version requires a solution $u$ that is in $H_0^1(\Omega)$, and any function in this latter space is automatically zero on the boundary $\partial \Omega$. Thus the BC is built into the weak version and for this reason is called a *natural*, or *free*, boundary condition.

The solutions of the weak problem (13.6) are ordinary functions, but they need *not* be differentiable. However, the next proposition says that any weak solution that is suitably differentiable is a strong solution, i.e., satisfies Poisson's equation $\nabla^2 u = f$.

**Proposition 13.1**   *Suppose $u$ is a scalar field with continuous partial derivatives up to the second order and $v$ has continuous first-order partial derivatives. Then the* **integration by parts** *formula holds:*

$$\int_\Omega (\nabla^2 u) v \, dV = \int_{\partial \Omega} v \nabla u \cdot n \, dA - \int_\Omega \nabla u \cdot \nabla v \, dV. \tag{13.8}$$

*Thus in particular, for every $\psi \in \mathcal{D}(\Omega)$ we have*

$$\int_\Omega (\nabla^2 u) \psi \, dV = -\int_\Omega \nabla u \cdot \nabla \psi \, dV. \tag{13.9}$$

*Consequently, if $u \in H_0^1(\Omega)$ is a solution of the weak version (13.6) of the Dirichlet problem and if $u$ has continuous partial derivatives up to the second order, then $\nabla^2 u = f$ in $\Omega$ and $u = 0$ on $\partial \Omega$.*

**Proof**   The proofs of equations (13.8)–(13.9) are left as an exercise. Now suppose $u \in H_0^1(\Omega)$ has continuous derivatives up to the second order. Then by (13.9), we easily see that for every $\psi \in \mathcal{D}(\Omega)$,

$$\int_\Omega (\nabla^2 u) \psi \, dV = D(u, \psi).$$

Thus if $u$ is a solution of the weak version of the Dirichlet problem, $D(u, v) = \langle f, v \rangle$ for every $v \in H_0^1(\Omega)$, then in particular, $D(u, \psi) = \langle f, \psi \rangle$ for every

$\psi \in \mathcal{D}(\Omega)$. Otherwise stated,

$$\int_{\Omega} (\nabla^2 u)\psi \, dV = \int_{\Omega} f\psi \, dV,$$

for every $\psi \in \mathcal{D}(\Omega)$. Since $\psi \in \mathcal{D}(\Omega)$ is arbitrary, this forces $\nabla^2 u = f$ at each point of $\Omega$. We have already commented that any $u \in H_0^1(\Omega)$ is zero on the boundary of $\Omega$. Thus $u$ is a strong solution of the Dirichlet problem.

The existence of a unique solution to the weak Dirichlet problem is a direct result of the following more abstract result:

**Theorem 13.1 (Lax–Milgram Lemma)**    *Suppose $H$ is a Hilbert space and $D$ is a bilinear form on $H$ for which there are two constants $\alpha, \beta > 0$ such that*

$$|D(u, v)| \leq \alpha \|u\| \|v\| \qquad \text{(continuity)}, \qquad (13.10)$$

$$|D(v, v)| \geq \beta \|v\|^2 \quad \text{(coercivity)}, \qquad (13.11)$$

*for every $u, v \in H$. Then for each given $f \in H$, there exists a unique $u \in H$ that satisfies*

$$D(u, v) = \langle f, v \rangle \qquad \text{for all } v \in H. \qquad (13.12)$$

*Furthermore, if for a given $f$ the solution $u$ of (13.12) is denoted by*

$$u = Gf,$$

*then this association $G$ defines a continuous linear map $G : H \to H$.*

**Remarks:**    The Lax–Milgram result is the guiding principle in analyzing the existence and uniqueness question for boundary value problems. One first determines an appropriate Dirichlet form $D$ on a suitable Hilbert space $H$ for the BVP, and continuity of $D$ is then verified. The hard part is demonstrating coercivity for $D$. The constant $\beta$ in (13.11) is called a *coercion constant*. The linear map $G$ in the theorem is often given by an integral with integrand involving a fixed function, called a *Green's function*, multiplied by the given function $f$.

**Proof**    The proof is actually an elementary application of the techniques from a standard course on Hilbert spaces (the existence of a weak solution follows from Riesz's lemma, and the continuity of $G$ is easy to demonstrate). Since we do not assume this background here, we merely prove the uniqueness of the solution to the weak problem and the linearity of $G$.

To prove uniqueness, suppose that $u_1$ and $u_2$ are solutions to (13.12). Then

$$D(u_1, v) = \langle f, v \rangle = D(u_2, v),$$

for every $v \in H$, and consequently,

$$0 = D(u_1, v) - D(u_2, v) = D(u_1 - u_2, v),$$

for every $v \in H$. The last part here follows from the bilinearity of $D$. Since the last equation holds for any $v \in H$, it holds in particular for $v = u_1 - u_2$. Using this and the coercivity inequality gives

$$0 = D(u_1 - u_2, u_1 - u_2) \geq \beta \|u_1 - u_2\| \geq 0.$$

Hence $\|u_1 - u_2\| = 0$, and by one of the basic properties of any norm, this gives $u_1 - u_2 = 0$. Thus any two solutions of the weak problem coincide, and this proves uniqueness.

The uniqueness allows us to define the map $G$. Namely, for a given $f \in H$ let $Gf$ be the unique element of $H$ that satisfies (13.12), i.e.,

$$D(Gf, v) = \langle f, v \rangle,$$

for every $v \in H$. We prove the linearity of $G$ as follows. Suppose $f_1, f_2 \in H$, and $a, b$ are constants. Then by definition, $Gf_1$ and $Gf_2$ satisfy $D(Gf_1, v) = \langle f_1, v \rangle$ and $D(Gf_2, v) = \langle f_2, v \rangle$, for every $v \in H$. Consequently, since $D$ and $\langle \cdot, \cdot \rangle$ are bilinear functions, we get

$$D(aGf_1 + bGf_2, v) = aD(Gf_1, v) + bD(Gf_2, v)$$
$$= a\langle f_1, v \rangle + b\langle f_2, v \rangle$$
$$= \langle af_1 + bf_2, v \rangle$$

for every $v \in H$. Thus $aGf_1 + bGf_2$ is *the* solution of the equation $D(u, v) = \langle af_1 + bf_2, v \rangle$, for every $v \in H$. By definition of $G$, this gives $G(af_1 + bf_2) = aGf_1 + bGf_2$.

We now apply the Lax–Milgram lemma to the weak version of the Dirichlet problem to get existence and uniqueness of solutions. The Hilbert space is $H_0^1(\Omega)$. The first order of business is to show that the Dirichlet form

$$D(u, v) \equiv -\int_\Omega \nabla u \cdot \nabla v \, dV$$

satisfies the continuity and coercivity inequalities (13.10)–(13.11). To prove continuity we need the famous

**Schwarz's Inequality:**    In any inner product space $H$, the inner product and the norm satisfy

$$|\langle u, v \rangle| \leq \|u\| \, \|v\|,$$

for all $u, v \in H$.

**Examples:**

(a) For the space $L^2(\Omega)$, the Schwarz inequality looks like

$$\left| \int_\Omega uv \, dV \right| \leq \left[ \int_\Omega u^2 dV \right]^{1/2} \left[ \int_\Omega v^2 dV \right]^{1/2}.$$

Technically, this is the original version of the Schwarz inequality. From this it is easy to show (exercise) that the stronger inequality

$$\int_\Omega |uv| \, dV \leq \left[ \int_\Omega u^2 dV \right]^{1/2} \left[ \int_\Omega v^2 dV \right]^{1/2}$$

holds.

(b) For the space $\mathbb{R}^3$ with the dot product as the inner product, the Schwarz inequality is

$$|a \cdot b| \leq |a||b|,$$

or written out explicity in terms of the components of $a = (a_1, a_2, a_3)$ and $b = (b_1, b_2, b_3)$,

$$|a_1 b_1 + a_2 b_2 + a_3 b_3| \leq (a_1^2 + a_2^2 + a_3^2)^{1/2} (b_1^2 + b_2^2 + b_3^2)^{1/2}.$$

This inequality is often called *Cauchy's Inequality*.

(c) (*An application*) For two scalar fields $u, v$ on $\Omega$, note that the Schwarz inequality gives

$$|\nabla u \cdot \nabla v| \leq |\nabla u||\nabla v|,$$

at all points in $\Omega$.

The continuity of the Dirichlet form follows directly from the Schwarz inequality, applied in various ways. Suppose $u, v \in H_0^1(\Omega) \subset L^2(\Omega)$. Then

$$
\begin{aligned}
|D(u, v)| &= \left| \int_\Omega \nabla u \cdot \nabla v \, dV \right| \\
&\leq \int_\Omega |\nabla u \cdot \nabla v| \, dV \\
&\leq \int_\Omega |\nabla u| \, |\nabla v| \, dV \\
&\leq \left[ \int_\Omega |\nabla u|^2 dV \right]^{1/2} \left[ \int_\Omega |\nabla v|^2 dV \right]^{1/2} \\
&\leq \left[ \int_\Omega (u^2 + |\nabla u|^2) dV \right]^{1/2} \left[ \int_\Omega (v^2 + |\nabla v|^2) dV \right]^{1/2} \\
&= \|u\|_1 \, \|v\|_1 .
\end{aligned}
$$

Thus we can take $\alpha = 1$ to get the continuity of $D$ (cf. (13.10)).

Finally, we need to show that the Dirichlet form for the Dirichlet problem is coercive. This follows directly from the (first) Poincaré inequality

**Proposition 13.2 (Poincaré's Inequality)**    *Since $\Omega$ is bounded, there is a constant $c$ (depending only on the diameter of $\Omega$) such that*

$$
\int_\Omega v^2 dV \leq c \int_\Omega |\nabla v|^2 dV, \tag{13.13}
$$

*for all $v \in H_0^1(\Omega)$.*

Using this, we get coercivity of $D$ from

$$
\begin{aligned}
\|v\|_1^2 &= \int_\Omega (v^2 + |\nabla v|^2) dV \\
&\leq c \int_\Omega |\nabla v|^2 dV + \int_\Omega |\nabla v|^2 dV \\
&= (c + 1)|D(v, v)|.
\end{aligned}
$$

Thus we can take $\beta = (c + 1)^{-1}$ in (13.11) to get the coercivity.

This completes the verification that the Dirichlet form $D$ on $H_0^1(\Omega)$ is continuous and coercive. Thus by the Lax–Milgram lemma, there exists a unique solution $u$ to the weak version of the Dirichlet problem: $D(u, v) = \langle f, v \rangle$, for every $v \in H_0^1(\Omega)$. We will not discuss here results concerning the regularity (i.e., differentiablity) of $u$ and thus the fact that it is a solution in the classical sense.

### 13.1.2.  Case $(a \neq 0)$: The Neumann and Robin Problems

Dividing equation (13.1) by $a$, letting $\gamma = b/a$, and relabeling $g/a$ by $g$, we write the Neumann and Robin problems as

$$\nabla^2 u = f \qquad \text{in } \Omega, \tag{13.14}$$

$$\nabla u \cdot n + \gamma u = g \qquad \text{on } \partial\Omega. \tag{13.15},$$

The case $\gamma = 0$ is the Neumann problem, and $\gamma \neq 0$ is the Robin problem. We will consider both problems at once. The only distinction that needs to be made is that the Neumann problem (as we have seen before) requires a solvability condition on the given functions $f$, $g$:

**Solvability Condition for the Neumann Problem:**

$$\int_{\partial\Omega} g \, dA = \int_{\Omega} f \, dV. \tag{13.16}$$

If this condition is not satisfied, then there is no (classical) solution to the Neumann problem (exercise). More generally, after formulating the weak version of the problem, we will see that there is no weak solution if condition (13.16) is not satisfied.

The weak version of the Neumann and Robin problems has several features that are different than from the weak version of the Dirichlet problem. First, the Hilbert space to use is $H^1(\Omega)$, and second, the *Dirichlet form* has a boundary integral

$$D(u, v) \equiv -\int_{\Omega} \nabla u \cdot \nabla v \, dV - \int_{\partial\Omega} \gamma u v \, dA, \tag{13.17}$$

for $u, v \in H^1(\Omega)$. Additionally, there is a boundary integral in the weak version of the problem:

**The Weak Version of the Neumann and Robin Problems:**

Determine $u \in H^1(\Omega)$ such that

$$D(u, v) = \langle f, v \rangle - \int_{\partial\Omega} gv \, dA, \qquad (13.18)$$

for all $v \in H^1(\Omega)$.

Here $f$ is a given function in $L^2(\Omega)$, and the bracket $\langle \cdot, \cdot \rangle$ denotes the inner product in this space. Also, $g$ is a given function in $L^2(\partial\Omega)$. As before, the boundary conditions are *natural boundary conditions*, since they are built into the weak version. To see this and to verify as well that equation (13.18) is the appropriate generalization of the classical problem, we prove the following result.

**Proposition 13.3**   *If $u \in H^1(\Omega)$ is a solution of the weak version (13.18) of the Neumann/Robin problems and if $u$ has continuous partial derivatives up to the second order, then $\nabla^2 u = f$ in $\Omega$ and $\nabla u \cdot n + \gamma u = 0$ on $\partial\Omega$.*

**Proof**   Suppose $v \in H^1(\Omega)$ and has continous first-order partial derivatives (in the classical sense). Then applying the integration by parts formula to the definition of the Dirichlet form in (13.17), we easily get

$$D(u, v) = \int_{\Omega} (\nabla^2 u) v \, dV - \int_{\partial\Omega} [\nabla u \cdot n + \gamma u] v \, dA.$$

Thus since $u$ satisfies the weak equation (13.18), we get that

$$\int_{\Omega} (\nabla^2 u) v \, dV - \int_{\partial\Omega} [\nabla u \cdot n + \gamma u] v \, dA = \int_{\Omega} f v \, dV - \int_{\partial\Omega} gv \, dA, \quad (13.19)$$

for all $v \in H^1(\Omega)$ with continuous first-order partials. In particular, since each $v \in \mathcal{D}(\Omega) \subset H^1(\Omega)$ is zero on $\partial\Omega$, we get from equation (13.19) that

$$\int_{\Omega} (\nabla^2 u) v \, dV = \int_{\Omega} f v \, dV,$$

for all $v \in \mathcal{D}(\Omega)$. This leads to $\nabla^2 u = f$ on $\Omega$. We can now cancel the integrals over $\Omega$ in equation (13.19) to be left with

$$-\int_{\partial\Omega} [\nabla u \cdot n + \gamma u] v \, dA = -\int_{\partial\Omega} gv \, dA,$$

for all $v \in H^1(\Omega)$ with continuuous first-order partials. From this it follows that $\nabla u \cdot n + \gamma u = g$ on $\partial\Omega$.

The verification that the Dirichlet forms for the Neumann and Robin problems are continuous and coercive is left as an exercise. With this verified, the Lax–Milgram lemma gives us the existence of a unique weak solution of the weak version of these problems. As with the Dirichlet problem, we will not discuss here the differentiablity of these solutions.

## Exercises 13.1

1. Suppose $v : \Omega \to \mathbb{R}$ is a scalar field on $\Omega$ that is continuously differentiable. Use Gauss's divergence theorem to show that the distributional partial derivatives of $v$ are the "same" as the usual partial derivatives. That is, show that

$$\frac{\partial \tilde{v}}{\partial x} = \frac{\widetilde{\partial v}}{\partial x}, \quad \frac{\partial \tilde{v}}{\partial y} = \frac{\widetilde{\partial v}}{\partial y}, \quad \frac{\partial \tilde{v}}{\partial z} = \frac{\widetilde{\partial v}}{\partial z}.$$

2. As a simple illustration of the concept of distributional derivatives, consider the the 1-D case where $\Omega = (-1, 3)$. The triangular function

$$v(x) = \begin{cases} 0 & \text{if } x < 0, \\ x & \text{if } 0 \le x < 1, \\ 2 - x & \text{if } 1 \le x < 2, \\ 0 & \text{if } x \ge 2, \end{cases}$$

is *not* a differentiable function in the classical sense. It fails to have derivatives at $x = 0$, $x = 1$, and $x = 2$. Considered as a distrubution $\tilde{v}$, the derivative $\tilde{v}'$ always exists.

(a) Show that this distributional derivative is a function $\tilde{v}' = \tilde{w}$, where $w$ is the function defined by

$$w(x) = \begin{cases} 0 & \text{if } x < 0, \\ 1 & \text{if } 0 \le x < 1, \\ -1 & \text{if } 1 \le x < 2, \\ 0 & \text{if } x \ge 2. \end{cases}$$

(b) Compute the norm $\|v\|_1$ of $v$ in the space $H^1(\Omega)$.

(c) Compute the second derivative of $v$ (in the distributional sense). it should involve Dirac delta functions. The *Dirac delta function* at a point $c$ in $\Omega$, denoted by $\delta_c$, is the linear function defined by $\delta_c(\psi) \equiv \psi(c)$.

3. Discuss how to transform the BVP

$$\nabla^2 u = f \qquad \text{in } \Omega, \tag{13.20}$$

$$u = g \qquad \text{on } \partial\Omega, \tag{13.21}$$

into one with homogeneous boundary conditions. This is a very simple exercise if you assume (and you may so assume) that the function $g$ extends to a differentiable function on $\Omega$. That is, there is a function $S$ defined on $\Omega \cup \partial\Omega$ that is differentiable on $\Omega$, and $S = g$ on $\partial\Omega$. This last fact comes from the theory of *traces* (or trace theorems). Look this up in [Na 67] and add this to your discussion.

4. Prove the integration by parts formula 13.8 in Proposition 13.1. Also show that 13.9 holds.

5. (*Eigenvalues of the Laplacian*) Consider the homogeneous heat equation with homogeneous boundary conditions

$$u_t = \alpha^2 \nabla^2 u \qquad \text{in } \Omega, \text{ for } t > 0,$$

$$a\nabla u \cdot n + bu = 0. \qquad \text{on } \partial\Omega \text{ for } t > 0.$$

Using the separation of variables technique, we can separate out the time dependence by looking for solutions of the PDE + BCs of the form $u = vT$, where $v$ is a time-independent scalar field on $\Omega$, and $T$ is a function of $t$ only.

(a) Show, by the usual separation of variables argument, that this leads to the equations $\nabla^2 v = kv$ and $T' = k\alpha^2 T$, for some constant $k$. The separation constant $k$ is also known as an *eigenvalue of the Laplacian*. Here's why. Let $\mathcal{V}$ be the vector space of scalar functions $v$ on $\Omega$ that satisfy the boundary conditions $a\nabla v \cdot n + bv = 0$. Then viewing the Laplacian $\nabla^2$ as a linear operator, the PDE $\nabla^2 v = kv$ is precisely the eigenvalue/eigenvector equation from linear algebra. The eigenvalues of $\nabla^2$ are those numbers $k$ for which there is a nonzero solution $v \in \mathcal{V}$ of this equation.

(b) Show that for Dirichlet boundary conditions ($a = 0$) and for Robin boundary conditions ($a \neq 0$ and $b \neq 0$), the eigenvalues are negative, $k < 0$ (you may assume that $k$ is real). Show that for the Neumann boundary conditions ($a \neq 0$ and $b = 0$), the eigenvalues are nonpositive, $k \leq 0$, and that for the zero eigenvalue case ($k = 0$), any corresponding eigenfunction $v$ is a constant function. use the integration by parts formula (13.8).

6. Use the Schwarz inequality for $u, v \in L^2(\Omega)$ to prove the following stronger version of it:

$$\int_\Omega |uv|\, dV \leq \left[\int_\Omega u^2 dV\right]^{1/2} \left[\int_\Omega v^2 dV\right]^{1/2}.$$

7. Prove that the solvability condition for the Neumann problem

$$\int_{\partial\Omega} g\, dA = \int_\Omega f\, dV$$

is a necessary condition for the existence of a (classical) solution to the Neumann problem. Also prove that this is a necessary condition for existence of solutions to the weak version (13.18) of the Neumann problem. Discuss why no solvability condition is necessary in the Robin problem ($\gamma \neq 0$).

8. Use *Friedrich's inequality* (cf. [Ne 67], p. 20)

$$\int_\Omega |\nabla v|^2 dV + \int_{\partial\Omega} v^2 dA \geq k\|v\|_1^2,$$

for all $v \in H^1(\Omega)$, to prove that the Dirichlet form $D$ for the Neumann/Robin problem (cf. (13.17)) is coercive. The $k$ in Friedrich's inequality is a positive constant. Also prove the continuity of $D$ by using the fact that there exists a constant $m$ such that

$$\int_{\partial\Omega} v^2 dA \leq m\|v\|_1^2,$$

for all $v \in H^1(\Omega)$ (see [Ne 67], p. 84).

## 13.2. Solution of the Heat IBVP

The solution of the heat IBVP for general domains follows the pattern exhibited above for BVPs. First, the problem is recast in terms of a weak, or distributional, version. Then the existence of weak solutions is proven. Here the key tool is not the Lax–Milgram Lemma, but rather the Galerkin (or Ritz–Galerkin) method. This latter method not only proves the existence of weak solutions, but also exhibits a way of constructing approximations to the solutions numerically. The application of the Galerkin method to the heat problem here follows the same lines as those in Chapter 10 for the Navier–Stokes equations and in Chapter 11 for the Navier equation. While the Lax–Milgram lemma provides for both the existence and uniqueness of solutions, the Galerkin method proves only existence of solutions. The uniqueness must be proven by other means. Finally, the question of when a weak solution is actually differentiable in the classical sense (regularity) is addressed. We will not discuss this regularity question here, but rather just give a rough description of how the existence and uniqueness of weak solutions is proven.

### 13.2.1. Weak Solutions of the Heat Equation

We consider the non-homogeneous heat equation with (homogeneous) Dirichlet boundary conditions

$$u_t = \alpha^2 \nabla^2 u + f \qquad \text{in } \Omega, \text{ for } t > 0, \tag{13.22}$$

$$u = 0 \qquad \text{on } \partial\Omega, \text{ for } t > 0, \tag{13.23}$$

$$u = \phi \qquad \text{in } \Omega, \text{ for } t = 0. \tag{13.24}$$

Notice that we are considering only homogeneous BCs. The more general case can always be reduced to this (exercise). The discussion we give for this problem can be modified to apply to heat problems with Neumann or Robin boundary conditions.

As with the BVPs discussed above, the weak version of the heat IBVP arises (heuristically) by multiplying equation (13.22) by a function $w$ that is zero on $\partial\Omega$ and integrating:

$$\int_\Omega \frac{\partial u}{\partial t} w \, dV = \alpha^2 \int_\Omega (\nabla^2 u) w \, dV + \int_\Omega f w \, dV$$

$$= -\alpha^2 \int_\Omega \nabla u \cdot \nabla w \, dV + \int_\Omega f w \, dV.$$

Here we have used the integration by parts formula with the assumption that $w = 0$ on $\partial\Omega$. Thus the weak version of this heat IBVP should use the Hilbert space $H_0^1(\Omega)$ and the Dirichlet form $D(u, w) \equiv -\int_\Omega \nabla u \cdot \nabla w \, dV$. Specifically:

**The Weak Version of the Heat Problem**   Determine a function $u$ that satisfies

$$\frac{d}{dt} \langle u, w \rangle = \alpha^2 D(u, w) + \langle f, w \rangle, \tag{13.25}$$

$$u(0) = \phi, \tag{13.26}$$

for all $w \in H_0^1(\Omega)$ and for all $t \in (0, T]$.

To be more specific, $u$ is a function defined on the time interval $[0, T]$ with values $u(t)$ in the Hilbert space $H_0^1(\Omega)$ (i.e., a function $u : [0, T] \rightarrow H_0^1(\Omega)$), and initially $u(0) = \phi$, where $\phi$ is a given function in $H_0^1(\Omega)$. Also, $f : [0, T] \rightarrow L^2(\Omega)$ is a given function on $[0, T]$ with values in $L^2(\Omega)$, and the bracket $\langle \cdot, \cdot \rangle$ is the inner product in this space. As is customary, in equation (13.25) we have suppressed the fact that $u$ and $f$ depend on $t$. To be more precise, we should write

$$\frac{d}{dt} \langle u(t), w \rangle = \alpha^2 D(u(t), w) + \langle f(t), w \rangle,$$

for all $w$ in $H_0^1(\Omega)$. Additionally, the derivative with respect to $t$ on the left side of equation (13.25) should be interpreted in the distributional sense. However, we will not deal with this in any precise way here.

## 13.2.2.   The Galerkin Method

To prove the existence of a solution $u$ of the weak version (13.25)–(13.26) of the heat problem, we first show the existence of approximate solutions to this problem.

Thus suppose $\{w_n\}_{n=1}^N$ is a linearly independent subset of functions in $H_0^1(\Omega)$, and let $\mathcal{W}_N \subseteq H_0^1(\Omega)$ denote the subspace spanned by these vector fields. We consider the problem of finding a solution $u = u^N$ of equation (13.25) that takes values in $\mathcal{W}_N$; i.e., $u^N(t) \in \mathcal{W}_N$ for $t$ in $[0, T]$. This amounts to finding scalar functions $b_n = b_n(t), n = 1, \ldots, N$, such that the function defined by

$$u^N(t) \equiv \sum_{n=1}^N b_n(t) w_n \tag{13.27}$$

satisfies equation (13.25) for every $w \in \mathcal{W}_N$. Because equation (13.25) is linear in $w$, it will be satisfied for every $w$ in $\mathcal{W}_N$ if it is satisfied for each $w_k, k = 1, \ldots, N$. Substituting the expression (13.27) for $u$ and substituting $w_k$ for $w$ in equation (13.25), we get the following equations:

$$\sum_{n=1}^N b_n'(t)\langle w_n, w_k \rangle = \alpha^2 \sum_{n=1}^N b_n(t) D(w_n, w_k) + \langle f(t), w_k \rangle, \tag{13.28}$$

for $k = 1, \ldots, N$. These constitute a *linear* system of ordinary differential equations involving the unknown functions $b_1(t), \ldots, b_N(t)$ (all the other quantities are known constants, computed from the given functions $w_n$ and $f$ according to the above definitions). We get one of these DE's for each $k = 1, \ldots, N$, and the whole collection constitutes a system of $N$ differential equations for $N$ unknown functions. Because of the linear independence of $\{w_n\}_{n=1}^N$, we can rewrite this system in normal form, $b_n'(t) = H_n(b(t), t)$ (exercise) and then apply the general existence and uniqueness theorem (see the Introduction), which says that for given initial values $b_n(0) = \beta_n, n = 1, \ldots, N$, there is one and only one set $b_1(t), \ldots, b_N(t)$ of solutions that satisfy the system of DE's and the given initial values.

In the present circumstance, since we are seeking to approximate $u(t)$ with $u^N(t)$, and the initial temperature $u(0) = \phi$ is assumed given, the initial values $\beta_n, n = 1, \ldots, N$, will be chosen such that $u^N(0)$ is the *projection* of $u(0)$ onto the subspace $\mathcal{W}_N$. (See Exercise 5 in Exercises 10.5 for how to do this.)

This shows how to find an *approximate* (weak) solution to the heat problem on $\mathcal{W}_N = \text{span}\{w_1, \ldots, w_N\}$, given linearly independent functions $w_n \in H_0^1(\Omega), n = 1, \ldots, N$. To use this to find a weak solution $u(t)$ (at least theoretically), one would construct, for each $N = 1, 2, \ldots$, an approximate solution $u^N(t)$ as above and then obtain $u(t)$ as a limit:

$$u(t) = \lim_{N \to \infty} u^N(t).$$

To discuss the details of this in a rigorous and precise way requires an under-standing of certain topics in functional analysis, which is not assumed as prior knowledge for this book. However, the above discussion should give the reader an intuitive idea of some aspects of the theory for existence of solutions to the heat equation on general domains $\Omega$.

When the heat source function $f$ does not depend on time, the above discussion also contains, as a corollary, the construction of approximate solutions to the Dirichlet problem $\nabla^2 u = -f/\alpha^2$ in $\Omega$ and $u = 0$ on $\partial\Omega$ by the Galerkin method. It also provides an alternative proof of existence of solutions, but the Lax–Milgram method is usually preferred for this. The solutions of this Dirichlet problem are just the *steady-states*, i.e., time-independent solutions, of the heat equation and boundary condition. The weak version is the same as in the next-to-last section: Determine a function $u$ in $H_0^1(\Omega)$ that satisfies

$$D(u, w) = -\alpha^{-2}\langle f, w\rangle,$$

for all $w$ in $H_0^1(\Omega)$. The Galerkin method for solving this by approximations,

$$u^N \equiv \sum_{n=1}^{N} b_n w_n,$$

leads to the system of linear *algebraic* equations

$$\sum_{n=1}^{N} b_n D(w_n, w_k) = -\alpha^{-2}\langle f, w_k\rangle,$$

for $k = 1, \ldots, N$, for the unknown constants $b_1, \ldots, b_N$. The coefficient matrix $G_{nk} \equiv D(w_n, w_k)$ can be shown to be nonsingular (exercise), and thus the system has a unique solution. This, then, gives us the approximate solution $u^N$ to the Dirichlet problem. The question of how good these approximations are for various values of $N$ and the sense in which $\lim_{N \to \infty} u^N$ gives an exact solution will not be discussed here.

## 13.2.3.    Uniqueness of Solutions

To prove the uniqueness of the solution to the weak version of the heat problem (13.25)–(13.26), suppose that $u_1$ and $u_2$ are two solutions of it. We let $u = u_1 - u_2$

and proceed to show that $u = 0$. First, note that since $u_1, u_2$ satisfy equations (13.25)–(13.26), we have that $u$ satisfies

$$\frac{d}{dt}\langle u, w \rangle = \alpha^2 D(u, w), \tag{13.29}$$

$$u(0) = 0, \tag{13.30}$$

for all $w \in H_0^1(\Omega)$. This comes from the linearity of the equations as follows:

$$\frac{d}{dt}\langle u, w \rangle - \alpha^2 D(u, w)$$

$$= \frac{d}{dt}\langle u_1, w \rangle - \frac{d}{dt}\langle u_2, w \rangle - \left[\alpha^2 D(u_1, w) - \alpha^2 D(u_2, w)\right]$$

$$= \left[\frac{d}{dt}\langle u_1, w \rangle - \alpha^2 D(u_1, w)\right] - \left[\frac{d}{dt}\langle u_2, w \rangle - \alpha^2 D(u_2, w)\right]$$

$$= \langle f, w \rangle - \langle f, w \rangle$$

$$= 0.$$

Next, since equation (13.29) holds for all $w$ in $H_0^1(\Omega)$, it holds in particular for $u(t)$, which is in $H_0^1(\Omega)$ for all $t \in [0, T]$. Now, in equation (13.29) we can replace $w$ on the right side by $u(t)$, but we cannot do so directly on the left side because of the time derivative $d/dt$. This is a rather subtle point, but it turns out that with a little work we get the following:

$$\frac{1}{2}\frac{d}{dt}\langle u, u \rangle = \alpha^2 D(u, u),$$

for all $t \in [0, T]$ (exercise). Now integrate both sides of this from $0$ to $T$ to get

$$\frac{1}{2}\langle u(T), u(T) \rangle = \alpha^2 \int_0^T D(u, u)\,dt$$

$$= -\alpha^2 \int_0^T \int_\Omega \nabla u \cdot \nabla u \, dV \, dt$$

$$\leq \frac{-\alpha^2}{c} \int_0^T \int_\Omega u^2 dV \, dt$$

$$\leq 0.$$

Here, the next-to-last line comes from the Poincaré inequality (13.13). Now, since $\frac{1}{2}\langle u(T), u(T) \rangle \geq 0$, each of the terms in the above set of equations and inequalties must be zero. Thus, in particular,

$$\frac{-\alpha^2}{c} \int_0^T \int_\Omega u^2 dV \, dt = 0.$$

From this we get that $u = 0$ at almost all points and times. This proves the uniqueness.

### Exercises 13.2

This exercise set is, with the exception of Exercises 1 and 2, devoted to introducing some additional details on the Galerkin method. This method is also called the *Ritz method* and the *Ritz–Galerkin method*. Technically, Ritz first developed the method for Dirichlet forms $D$ that are symmetric, $D(v, w) = D(w, v)$, and bilinear (which is the case in this chapter), and later Galerkin studied the more general case where $D$ is not symmetric or not bilinear. In addition, when the Ritz–Galerkin method is implemented with certain particular types of basis elements $\{w_n\}_{n=1}^N$, the method becomes what is known as the *finite element method*.

1. In the discussion on the Galerkin method for the Dirichlet problem $\nabla^2 u = f$, it is crucial to know that the matrix $G = \{G_{nk}\}$ is invertible, where $G_{nk} = D(w_n, w_k)$, and $D(u, v) = -\int_\Omega \nabla u \cdot \nabla v \, dV$ is the Dirichlet form for the problem. Show that this is indeed the case. (Use the coerciveness of $D$ and see the ideas in Exercise 5 in Section 10.5.) Use the same ideas to show that the matrix $B = \{B_{n,k}\}$ is invertible, where $B_{nk} = \langle w_n, w_k \rangle$, and thus the system of ODEs (13.28) can be written in normal form $b'_n(t) = H_n(b(t), t)$, for $n = 1, \ldots, N$.

2. Consider the heat IBVP

$$u_t = \alpha^2 \nabla^2 u \qquad \text{in } \Omega, \text{ for } t > 0,$$

$$u = 0 \qquad \text{on } \partial\Omega, \text{ for } t > 0,$$

$$u = 0 \qquad \text{in } \Omega, \text{ for } t = 0.$$

Clearly, the identically zero function $u(x, y, z, t) = 0$, for all $(x, y, z) \in \Omega$ and all $t$, is a solution of the IBVP. Show that any other function $u : \Omega \times [0, T] \to \mathbb{R}$ that is a classical solution to the IBVP (i.e., $u$ is twice continuously differentiable in the spatial variables and continuously differentiable in $t$) is identically zero. Do this by showing that

$$\frac{1}{2} \frac{d}{dt} \langle u, u \rangle = \langle u_t, u \rangle,$$

where as usual, $\langle u, u \rangle = \int_\Omega u^2 dV$ is the inner product in $L^2(\Omega)$. Then use arguments similar to those in the text for weak solutions. The above identity is more tricky to prove when $u$ is a weak solution and the derivatives are in the distributional sense (see [Te 72], p. 261). However, with it, the proof of uniqueness of weak solutions in the text is now complete.

3. (*Continuous, piecewise linear functions*) This exercise and the next serve to introduce the Ritz–Galerkin method in 1 dimension, where, say, $\Omega = [0, 1]$ is the unit interval. First, we look at some useful results concerning continuous piecewise linear functions.

Figure 13.1 shows an example of a typical piecewise linear function $u$ on $[0, 1]$ that is also continuous. For such a function there is a partition $P = \{0 < a_1 < a_2 \ldots < a_N < 1\}$ of $[0, 1]$ into subintervals $I_n \equiv [a_{n-1}, a_n)$ for $n = 1, \ldots, N$ and $I_{N+1} = [a_N, 1]$, such that $u$ is linear when restricted to each subinterval. For convenience, we have taken $a_0 = 0$ and $a_{N+1} = 1$. The use of $N + 1$ intervals instead of $N$ intervals is so that the notation will match that in the above discussion of the Galerkin method. A line in the plane is uniquely determined by two distinct points $(a, b)$, $(a', b')$ on it via the formula

$$y = b + \frac{b' - b}{a' - a}(x - a),$$

and thus for a fixed partition $P$, a continuous piecewise linear function $u$ is completely determined by its values $b_n = u(a_n), n = 0, \ldots, N + 1$, at the division points $a_n$. Indeed:

(a) If $b_0, b_1, \ldots, b_{N+1}$ are any given numbers, show that a formula for the continuous piecewise linear function $u = u(\cdot, b)$ with these values at the division points of $P$ is

$$u(x) = u(x, b) \equiv \sum_{n=1}^{N+1} \left[ b_{n-1} + \frac{b_n - b_{n-1}}{a_n - a_{n-1}}(x - a_{n-1}) \right] 1_{I_n}(x). \tag{13.31}$$

Here $1_E$ denotes the *characteristic function* for the set $E \subseteq [0, 1]$. It is defined by

$$1_E(x) = \begin{cases} 1 & \text{if } x \in E, \\ 0 & \text{if } x \notin E. \end{cases}$$

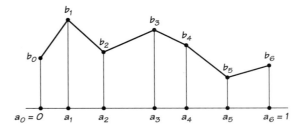

Figure 13.1.  A continuous piecewise linear function on $[0, 1]$

(b) We are denoting the function in equation (13.31) by $u(x, b)$ in order to indicate explicity its dependence on the vector of parameters $b = (b_0, \ldots, b_{N+1}) \in \mathbb{R}^{N+2}$. Show that $u(x, b)$ is a linear function of $b$; i.e.,

$$u(x, \lambda b + \tilde{\lambda}\tilde{b}) = \lambda u(x, b) + \tilde{\lambda}u(x, \tilde{b}),$$

for all $b, \tilde{b} \in \mathbb{R}^{N+2}$ and scalars $\lambda, \tilde{\lambda}$. Indeed, by manipulating the sum in Eq.(13.31) show that it can be written as:

$$u(b, x) = \sum_{n=0}^{N+1} b_n w_n(x). \tag{13.32}$$

(c) Write out explicitly the formulas for the functions $\{w_n\}_{n=0}^{N+1}$ in equation (13.32) in terms of characteristic functions. If we let $\mathcal{W}(P)$ denote the collection of continuous piecewise linear functions on $[0, 1]$ relative to the partition $P$, then the above work shows that the functions $\{w_n\}_{n=0}^{N+1}$ constitute a basis for $\mathcal{W}(P)$, and thus this space is a vector space of dimension $N + 2$. The graphs of the basis functions are shown in Figure 13.2. Note that $u(b, x)$ is a general element of $\mathcal{W}(P)$ whose coordinates with respect to the basis $\{w_n\}_{n=0}^{N+1}$ are $b_0, \ldots, b_{N+1}$. Also,

$$w_0(x) = u(x, 1, 0, 0, \ldots, 0),$$

$$w_1(x) = u(x, 0, 1, 0, \ldots, 0), \tag{13.33}$$

$$\vdots \quad \vdots$$

$$w_{N+1}(x) = u(x, 0, 0, 0, \ldots, 1).$$

(d) The standard approximation of a continuous function $f$ on $[0, 1]$ by a function from $\mathcal{W}(P)$ is the one whose coordinates are the values of $f$ at the division points of $P$, i.e., the function $u(\cdot, b)$ with $b_n = f(a_n), n = 0, \ldots, N+1$. These approximations become better as the partitions $P$ become finer. For simplicty, suppose the partitions $P = P_N$ have division points that are equally spaced: $a_n = n/(N+1)$. For each of the following functions $f$, use a computer to graph $f$

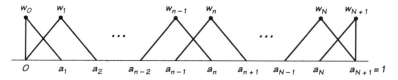

Figure 13.2.   Graphs of the basis functions: $\{w_n\}_{n=0}^{N+1}$

and its standard continuous piecewise linear approximations for $N = 2, 4, 6, 8$.

$$f(x) = e^{-x} \sin 4\pi x, \quad f(x) = x^{-1} \sin 4\pi x, \quad f(x) = x^{\sin 4\pi x}.$$

For example, in *Maple* the following code defines the general element of $\mathcal{W}(P_N)$ as given by equation (13.31).

```
with(linalg):
alias(h=Heaviside);
a:=vector(100); b:=vector(100);
for n from 0 to N+1 do a[n]:=n/(N+1) od;
u:=proc(x,b,N) if x=1 then s:= b[N+1] else
     s:=0; for n to N+1 do
     s:=s+(b[n-1]+(b[n]-b[n-1])/(a[n]-a[n-1])
     *(x-a[n-1]))*(h(x-a[n-1])-h(x-a[n]));
od; fi; end;
```

There are several things to note here. First, *Maple* denotes Heaviside's function

$$\text{Heaviside}(x) = 1_{[0,\infty)}(x) = \begin{cases} 0 & \text{if } x < 0, \\ 1 & \text{if } x \geq 0, \end{cases}$$

by his actual name, and we want to use something shorter, like $h$, for it. This is the reason for the `alias` statement in the second line of code. The Heaviside function is defined on all of $\mathbb{R}$. You should verify that the characteristic function of the interval $[a, a')$ can be written as

$$1_{[a,a')}(x) = h(x - a) - h(x - a').$$

also, note that the `with(linalg)` command includes the linear algebra package in the *Maple* session. We do not need this here, but in the next exercise, we will augment the above code with some code for doing Gaussian elimination, and then we will need the linear algebra package. To use the above code to plot the $N = 4$ approximation to $f(x) = \cos 2\pi x$, we include

```
f:=x->cos(2*Pi*x);
for n from 0 to N+1 do b[n]:=f(a[n]) od;
plot({u(x,b,N),f(x)},x=0..1,color=black);
```

Figure 13.3 shows the resulting plots.

4. (*The Ritz–Galerkin method*) Consider the analogue of the Dirichlet problem in 1 dimension:

$$u'' = f \qquad \text{in } [0, 1] \tag{13.35}$$

$$u(0) = 0, \qquad u(1) = 0. \tag{13.36}$$

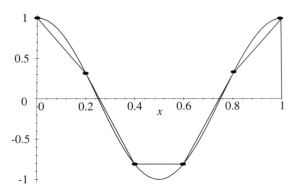

Figure 13.3.   A standard piecewise linear approximation to $f(x) = \cos 2\pi x$.

While this problem is easy to solve directly without the Galerkin method, it serves to illustrate all the main features of this method in a simplified setting. The exercises below deal with the method in 2 dimensions where the notation becomes more complicated.

(a) Assuming that $f$ is continuous, show that the solution to the boundary value problem (13.35)-(13.36) is

$$u(x) = \int_0^1 K(x, p)f(p)dp,$$

where

$$K(x, p) = (x - p)1_{[0,x)}(p) - (1 - p)x.$$

Exhibit all the steps used in deducing this result by integrating equation (13.35), including two constants of integration, and then choosing the constants such that the BCs hold. The function $K$ is known as the *Green's function* or *kernel function* for the problem.

(b) Multiplying the DE (13.35) by a function $w$ that is zero on the boundary ($w(0) = 0 = w(1)$) and integrating by parts gives

$$-\int_0^1 u'w'dx = \int_0^1 fw\,dx,$$

where the derivatives are in the distributional sense. We write this equation

$$D(u, w) = \langle f, w \rangle,$$

and as in the 3-D case, we seek a solution $u \in H_0^1([0, 1])$ such that the last equations holds for all $w \in H_0^1([0, 1])$. With the notation from the last exercise, we implement the Galerkin method with the approximating subspace $\mathcal{W}_0(P_N)$ of piecewise linear continuous functions that are *zero on the boundary:* $w(0) = 0 = w(1)$. As we saw in the last exercise, $\mathcal{W}(P_N)$ is $(N+2)$ dimensional. Here we see that its subspace $\mathcal{W}_0(P_N)$ is $N$-dimensional, since the boundary conditions dictate that the general element $u(x, b)$ (given by equation (13.32)) is in this subspace if and only if $b_0 = 0 = b_{N+1}$. Further, it is easy to see that $\{w_1, \ldots, w_N\}$ constitute a basis for $\mathcal{W}_0(P_N)$ (see equation (13.33) and Figure 13.2). The Galerkin method in this setting is to approximate the solution to the BVP (13.35)-(13.36) by

$$u(x, b) = \sum_{n=1}^{N} b_n w_n(x),$$

where $b$ is the solution of the system of linear equations

$$Gb = c.$$

Here $G = \{G_{nm}\}$ is the $N \times N$ matrix with entries

$$G_{nk} = D(w_n, w_k) = -\int_0^1 w'_n(x) w'_k(x) dx,$$

and $c = (c_1, \ldots, c_N)$ is the vector with components

$$c_k = \langle f, w_k \rangle = \int_0^1 f(x) w_k(x) dx.$$

Show that the matrix $G$ is

$$G = \begin{bmatrix} -2N & N & 0 & \cdots & 0 & 0 \\ N & -2N & N & \cdots & 0 & 0 \\ 0 & N & -2N & \cdots & 0 & 0 \\ \vdots & \vdots & \vdots & \ddots & \vdots & \vdots \\ 0 & 0 & 0 & \cdots & -2N & N \\ 0 & 0 & 0 & \cdots & N & -2N \end{bmatrix}. \qquad (13.37)$$

The matrix $G$ is *tridiagonal*, i.e., has nonzero entries on its diagonal, subdiagonal, and supradiagonal, and zero entries elsewhere. This makes solving the system $Gb = c$ considerably quicker when $N$ is large.

For each of the functions

$$f(x) = 12x^2 - 9x + 11/8, \qquad f(x) = 100e^{-x} \sin 10x,$$

$$f(x) = 10x^{-1} \sin 10x, \qquad (x) = 100e^{-x^2/2} \sin 10x,$$

(i) Use a computer to compute the Galerkin approximations $u(x, b)$ to the solution of the BVP (13.35)–(13.36). Do this for $N = 5, 10, 20, 40$.

(ii) Where possible, use calculus (integrate twice and determine the constants of integration) to find the exact solution of the BVP (13.35)-(13.36). Note that this is not possible for the fourth function.

(iii) Use a computer to plot the approximate solutions and the exact solution (where possible).

The following is an example of how to do this in *Maple* for the function $f(x) = \sin x$. The exact solution of $u'' = \sin(x)$ with $u(0) = 0 = u(1)$ is easily calculated to be $u(x) = -\sin(x) + \sin(1)x$. To compute the approximate solutions, we first augment the above *Maple* code with the following code, which computes the basis functions $w_1, \ldots, w_N$ (see equation(13.33)):

```
w:=array(1..100);
for n to N do
    for j to N do if j=n then b[j]:=1; else b[j]:=0 fi; od;
    w[n]:=unapply(u(x,b,N),x) od;
```

We used the command unapply in the above code to take the symbolic expression $u(x, b, N)$, involving the variable $x$, and make it into a function of $x$. Next, we compute the components of $c$ and set up the matrix $G$:

```
c:=vector(N);G:=matrix(N,N);
f:=x->sin(x);
for n to N do c[n]:=evalf(int(f(x)*w[n](x),x=0..1)) od;
for n to N do
 for k from n to N do
     if k=n then G[n,k]:=-2*N
        elseif k=n+1 then G[n,k]:=N
           else G[n,k]:=0 fi od;
 G[k,n]:=G[n,k] od;
```

Now we use *Maple*'s  rref  command to bring the augmented matrix $AM = [G : c]$ to reduced row echelon form. Then the solution $b$ will be the last column of the resulting matrix:

```
AM:=augment(G,c);
RR:=rref(AM);
for n to N do b[n]:=RR[n,N+1] od;
```

Using all of this code with $N = 4$ (so that there are 5 division points) gives the approximate solution $u(b, x, 4)$ shown in Figure 13.4.

The above is an implementation of the Galerkin method with $\mathcal{W}_0(P_N)$ being the space of continuous piecewise linear functions on $P_N$ that vanish on the boundary. This can be generalized by using spaces of continuous piecewise polynomial functions (of a fixed degree) that vanish on the boundary. These implementations of the Galerkin

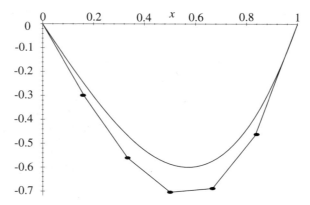

Figure 13.4.   Plot of the Galerkin approximation $u(x, b, 5)$ and the exact solution $u(x) = -10 \sin(x) + 10 \sin(1)x$ to the BVP $u'' = 10 \sin(x)$, $u(0) = 0 = u(1)$

method, where the basis elements $w_n$, $n = 1, \ldots, N$, have a particularly simple form, constitute what is known as the *finite element method* for solving PDEs. See [AB 84].

5. (*Piecewise linear functions on triangular complexes*) The real need for the Ritz/Galerkin/finite element method is for BVPs on 2-D and 3-D domains $\Omega$. The geometry and notation are more complicated than what we encountered in the 1-D case, but some of the main features can be briefly described in this and the next exercise.

We will deal only with the case where the planar domain $\Omega$ is approximated by a triangular complex $\Omega_r$. Figure 13.5 shows a simple example of what we mean by this. Thus $\Omega_r$ is composed of a finite number of triangles $T_1, \ldots, T_r$ with the intersection of any two of these triangles being either an edge, a vertex, or empty. Suppose $a_1, \ldots, a_K$ is a labeling, in some order, of all the vertices in the triangular complex. Each vertex $a_n$ is a point in the plane, so let its coordinate expression be $a_n = (a_{n1}, a_{n2})$.

A piecewise linear function $u$ on $\Omega_r$ is one that is a linear function on each piece $T_j$ of $\Omega$. Pictorially, the graph of $u$ is a part of a plane over each piece. When $u$ is also

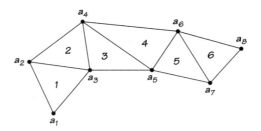

Figure 13.5.   A planar triangular complex $\Omega_r$ that approximates the domain $\Omega$.

continuous, it is completely determined by its values $b_n = u(a_n)$, $n = 1, \ldots, K$, at all the vertices in the complex. See Figure 13.6. This is the analogue of the 1-D case in Exercise 3. Here the graph of $u$ consists of triangular facets, and each of these is determined by its 3 vertices as follows:

(a) For 3 points $p = (a_1, a_2, b)$, $p' = (a'_1, a'_2, b')$, and $p'' = (a''_1, a''_2, b'')$ in $\mathbb{R}^3$, the vector methods from calculus give that the plane through these points has equation $n \cdot [(x, y, z) - p] = 0$, where $n = (p' - p) \times (p'' - p)$ is a normal to the plane. Show that the same equation can be written as the following determinant:

$$\begin{vmatrix} x & y & z & 1 \\ a_1 & a_2 & b & 1 \\ a'_1 & a'_2 & b' & 1 \\ a''_1 & a''_2 & b'' & 1 \end{vmatrix} = 0. \tag{13.38}$$

(b) Use the Laplace expansion for determinants to show that equation (13.38) can be written in the form

$$z = \frac{1}{D(a, a', a'')} \begin{vmatrix} 1 & x & y & 0 \\ 1 & a_1 & a_2 & b \\ 1 & a'_1 & a'_2 & b' \\ 1 & a''_1 & a''_2 & b'' \end{vmatrix}, \tag{13.39}$$

where

$$D(a, a', a'') = \begin{vmatrix} a_1 & a_2 & 1 \\ a'_1 & a'_2 & 1 \\ a''_1 & a''_2 & 1 \end{vmatrix}$$

is $\pm$ the area of the triangle with vertices $a, a', a''$ (verify this latter assertion too).

(c) Show that the expression (13.39) is invariant (does not change) under permutations of the points $p = (a_1, a_2, b)$, $p' = (a'_1, a'_2, b')$, and $p'' = (a''_1, a''_2, b'')$.

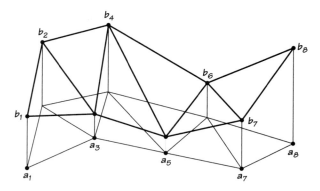

Figure 13.6.   A graph of a continuous piecewise linear function $u$ over the triangular complex $\Omega_r$.

This will be important later when we have a complex composed of a number of triangular facets each of which is the graph of part of the planar surface with an equation like (13.39). We can label or order the vertices in the total complex however we wish and still get the correct formula.

(d) Show that a formula for the continuous piecewise linear function $u$ on the triangular complex $\Omega_r$ with values $u(a_n) = b_n, n = 1, \ldots, K$, at the vertices $a_n$ can be written as follows. We assume that the triangles $T_j$, as sets, are disjoint, $T_j \cap T_k = \emptyset$, and that the union of all of them is $\Omega_r$. There are numerous ways to achieve this by including the edges and vertices of the complex in some triangles and not others. No matter how we do this, the $u$ we define below will be the same. For ease of reference, let $m_{1j}, m_{2j}, m_{3j}$ denote the indices for the three vertices in the $j$th triangle $T_j$. Thus the vertices of $T_j$ are $a_{m_{1j}}, a_{m_{2j}}, a_{m_{3j}}$, and the values that $u$ is to have at these vertices are $b_{m_{1j}}, b_{m_{2j}}, b_{m_{3j}}$. This gives us a $3 \times r$ matrix $m := \{m_{ij}\}_{i=1\ldots3, j=1\ldots r}$, which enables us to refer to data for each triangle. For example, a matrix $m$ for the complex shown in Figure 13.5 is

$$m = \begin{bmatrix} 1 & 3 & 3 & 4 & 5 & 6 \\ 2 & 2 & 4 & 5 & 6 & 7 \\ 3 & 4 & 5 & 6 & 7 & 8 \end{bmatrix}.$$

With this notation, a formula for $u$ is

**The continuous piecewise linear function with values $\{b_n\}_{n=1}^{K}$ at the vertices $\{a_n\}_{n=1}^{K}$**

$$u(x, y, b) = \sum_{j=1}^{r} \begin{vmatrix} 1 & x & y & 0 \\ 1 & a_{m_{1j}1} & a_{m_{1j}2} & b_{m_{1j}} \\ 1 & a_{m_{2j}1} & a_{m_{2j}2} & b_{m_{2j}} \\ 1 & a_{m_{3j}1} & a_{m_{3j}2} & b_{m_{3j}} \end{vmatrix} \frac{1_{T_j}(x, y)}{D(a_{m_{1j}}, a_{m_{2j}}, a_{m_{3j}})}. \tag{13.39}$$

Because of the result in the last exercise, the above formula does not depend on how the vertices $a_n$ in the complex are labeled.

(e) Let $\{\varepsilon_n\}_{n=1}^{N}$ be the standard unit vector basis for $\mathbb{R}^K$; i.e., $\varepsilon_n = (\varepsilon_{n1}, \ldots, \varepsilon_{nK})$, where $\varepsilon_{nn} = 1$ and $\varepsilon_{nk} = 0$ for $k \neq n$. Since $u(x, y, b)$ is a linear function of $b = (b_1, \ldots, b_K) \in \mathbb{R}^K$ (verify this!), we can write it as

$$u(x, y, b) = \sum_{n=1}^{K} b_n w_n(x, y),$$

where

$$w_n(x, y) \equiv u(x, y, \varepsilon_n).$$

(Verify this!). Find a formula for the functions $w_n$. These functions constitute a basis $\{w_n\}_{n=1}^{K}$ for the space $\mathcal{W}(\Omega_r)$ of continuous piecewise linear functions

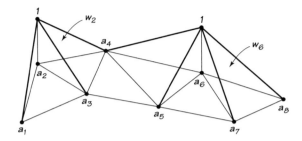

Figure 13.7.   Two of the basis elements: $w_2$, $w_6$ from the basis for the space of continuous piecewise linear functions on the complex $\Omega_r$

on $\Omega_r$ By definition, $w_n$ is the continuous, piecewise linear function with values $w_n(a_n) = 1$ and $w_n(a_k) = 0$, for $k \neq n$, at the vertices of the complex. Thus $w_n$ is zero everywhere except for points in one of the triangles having $a_n$ as a vertex. For example, Figure 13.7 shows two such basis elements for the complex from Figure 13.5.

6. (*The finite element method for 2-D domains*) Consider the Dirichlet problem $\nabla^2 u = f$ on $\Omega$, with $u = 0$ on $\partial\Omega$, where $\Omega$ is a planar domain. To use the Ritz–Galerkin method to obtain approximate solutions to this, we first use approximations to $\Omega$ by triangular complexes $\Omega_r$. As $r$ (the number of triangles) increases, the triangles should become smaller, and the approximation to the boundary of $\Omega$ should become better. The techniques involved in choosing the approximations $\Omega_r$ are known as *grid generation* and are too complicated to present here briefly. This exercise just indicates some aspects of the Galerkin method applied to a given complex $\Omega_r$, as for example shown in Figure 13.8. For convenience, we label the vertices such that $a_1, \ldots, a_N$ are the interior vertices, while $a_{N+1}, \ldots, a_K$ are the vertices on the boundary. Then

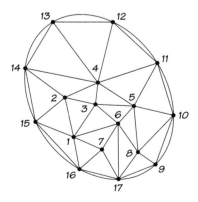

Figure 13.8.   A triangular complex approximating the planar region $\Omega$

one can show that the space $\mathcal{W}_0(\Omega_r)$ of continuous piecewise linear functions on $\Omega_r$ that are *zero on the boundary* is in fact an $N$-dimensional vector space with functions $\{w_n\}_{n=1}^{N}$, defined in the last exercise, as a basis (do this!). This is the approximating space for the Galerkin method. In the last exercise you should have found that the $w_n$'s have the form

$$w_n(x, y) = \sum_{j \in J_n}(A_j^n x + B_j^n y + C_j^n)1_{T_j}(x, y),$$

where $J_n$ is the set of indices $j$ such that $a_n \in T_j$. One can show (extra credit) that the gradient of $w_n$, in the distributional sense, is the (discontinuous) vector field

$$\nabla w_n(x, y) = \sum_{j \in J_n}(A_j^n, B_j^n)1_{T_j}(x, y).$$

Given this, do the following:

(a) Show that

$$\nabla w_n \cdot \nabla w_k = \sum_{j \in J_n \cap J_k}(A_j^n A_j^k + B_j^n B_j^k)1_{T_j}$$

(b) For $k \neq n$, show that $J_n \cap J_k$ consists of two indicies if the vertices $a_n$, $a_k$ belong to the same triangle, while $J_n \cap J_k = \emptyset$ if $a_n$, $a_k$ belong to different triangles.

(c) From the above, show that

$$G_{nk} \equiv D(w_n, w_k) = \sum_{j \in J_n \cap J_k}(A_j^n A_j^k + B_j^n B_j^k)\text{Area}(T_j),$$

and thus the entries in the $n$th row of the matrix $G$ are zero except for those $k$ that are *adjacent* to $n$ (i.e., $a_k$ and $a_n$ are connected by an edge in the complex). This means, generally, that for $N$ large, $G$ is a *sparse* matrix. Compute the matrix $G$ for the complex shown in Figure 13.8.

# A Vector Analysis

We discuss here a few more topics from vector analysis, line, surface, and volume integrals, and several important theorems, Stokes's theorem, Gauss's divergence theorem, and the change of variables formula. To keep the discussion brief and elementary, we take the point of view that curves and surfaces are *maps* (or more generally collections of maps) rather than the usual intrinsically geometric objects: 1- and 2-dimensional manifolds. The manifold point of view requires a more extensive development than we wish to present here ([AM 78], [AMR 82], [Boo 75]).

## A.1.  Curves and Line Integrals

A *smooth curve* $C$ in $\mathbb{R}^3$ is by definition a set $C = \{\alpha\}$ consisting of a single element $\alpha$, where $\alpha : I \to \mathbb{R}^3$ is a differentiable map, defined on an interval $I = [a, b]$. Saying that $\alpha$ is differentiable at the endpoints $a, b$ means that $\alpha$ extends to a differentiable function on a larger open interval containing $[a, b]$. If $\alpha$ also satisfies $\alpha'(s) \neq 0$ for all $s \in I$, then $C$ is called a *regular* smooth curve. This last requirement is just the statement that the tangent vector field

$$e_1 \equiv \alpha'$$

along the curve is nonzero at each point. Then we can normalize to get a unit tangent vector field

$$\tau \equiv e_1/|e_1|$$

along the curve.

Often we think of a smooth curve $C$ geometrically in terms of its *image* $\alpha(I) = \{\alpha(t) | t \in I\}$, and in this view $\alpha$ is called the parametrization of the curve (the parameter is $t$ in this case). Bear in mind, however, that different curves $C = \{\alpha\}$, $\widetilde{C} = \{\widetilde{\alpha}\}$, with $\alpha \neq \widetilde{\alpha}$, can have the same image: $\alpha(I) = \widetilde{\alpha}(\widetilde{I})$.

**Example A.1**    The map $\alpha(\theta) = (R \cos \theta, R \sin \theta, 0)$, for $\theta \in I = [0, 2\pi]$, defines a regular smooth curve $C = \{\alpha\}$, since

$$\alpha'(\theta) = (-R \sin \theta, R \cos \theta, 0) \neq 0,$$

for any $\theta$. This map parametrizes a circle $\alpha(I)$ of radius $R$ in the $x$-$y$ plane. This circle is also parametrized by $\beta(\theta) = (R \cos \theta, R \sin \theta, 0)$, $\theta \in J = [0, 4\pi]$, and by $\gamma(\theta) = (R \sin \theta, R \cos \theta, 0)$, $\theta \in K = [0, 2\pi]$.

Suppose $E$ is a vector field on an open set containing the image of a smooth curve $C = \{\alpha\}$. Then we define the *line integral* of $E$ along $C$ by

$$\int_C E \cdot \tau \, dL \equiv \int_a^b E(\alpha(s)) \cdot \alpha'(s) \, ds. \tag{A.1}$$

The notation on the left side of (A.1) is just that, *notation*, and we will use this notation even when $C$ is not regular (so that the unit tangential field $\tau$ is not necessarily defined). For example, if $\alpha$ is a constant map, $\alpha(s) = k$, for every $s \in \mathbb{R}$, then $\int_C E \cdot \tau \, dL = 0$, for any vector field $E$. The symbology in the notation can be explained heuristically as follows: defining the *length* of a smooth curve $C$ by

$$L(C) \equiv \int_a^b |\alpha'(s)| \, ds. \tag{A.2}$$

Then we can symbolically write the differential element of length as $dL = |\alpha'| \, ds$, and thus also,

$$E \cdot \tau \, dL = E \cdot \frac{\alpha'}{|\alpha'|} |\alpha'| \, ds = E \cdot \alpha' \, ds.$$

This formal manipulation, while having no rigor to it, serves to motivate the choice of notation for the line integral (A.1).

Besides the line integral introduced above, we will also need the notion of the integral of a scalar function $g : I \to \mathbb{R}$ along $C$. The notation for and definition of this type of integral is

$$\int_C g \, dA = \int_a^b g(s) |\alpha'(s)| \, ds. \tag{A.3}$$

If a scalar function $\tilde{g}$ is defined on the image of $\alpha$, then its line integral along $C$ is the one above with $g = \tilde{g} \circ \alpha$.

The integral on the right-hand sides of equations (A.1)–(A.3) is the usual Lebesgue integral (and with suitable assumptions on the integrands, the integral exists). For an interval $I = [a, b]$ and an integrable function $f$ on $I$, we will use the various notations

$$\int_a^b f(x)dx = \int_{[a,b]} f(x)dx = \int_I f(x)dx$$

to denote the value of the Lebesgue integral of $f$ over $I$. We caution that this integral does *not* have the property that for $a < b$,

$$\int_b^a f(x)dx = -\int_a^b f(x)dx,$$

mainly because we have assigned no meaning to the symbol on the left side of this equation.

The above notion of a curve, i.e., of a smooth curve, is too restrictive to cover all that we want to do here. As it stands, we cannot evaluate a line integral around the boundary of a square, since this boundary is not a smooth curve. However, it does consist of four pieces that are smooth curves. This motivates the following definition.

A *piecewise smooth curve* $C$ in $\mathbb{R}^3$ is a finite collection of maps

$$C = \{\alpha_1, \ldots, \alpha_m\},$$

where each $\alpha_j : I_j \to \mathbb{R}^3$ is a differentiable map defined on an interval $I_j = [a_j, b_j]$ in $\mathbb{R}$. Technically, $C$ is a *multiset* rather than a set, since we wish to allow for the case where some of the maps $\alpha_j$ may be repeated in the listing $\alpha_1, \ldots, \alpha_m$. Thus $C$ is the (multiset) union

$$C = \bigcup_{i=1}^m C_i$$

of finitely many smooth curves $C_1, \ldots, C_m$, where $C_j = \{\alpha_j\}$. The line integral of a vector field $E$ over $C$ is naturally defined by

$$\int_C E \cdot \tau \, dL \equiv \sum_{i=1}^m \int_{C_i} E \cdot \tau \, dL. \tag{A.4}$$

**Convention:**   To simplify the terminology, we will refer to a piecewise smooth curve $C$ as being simply a *curve*.

## A.2.   Surfaces and Surface Integrals

For the most part, the discussion of surfaces here parallels that for curves, and you could probably guess what many of the corresponding definitions would be. However, since the geometry is more complicated for surfaces, we go through all the details for the sake of definiteness.

A *smooth surface* $S$ in $\mathbb{R}^3$ is by definition a singleton set $S = \{\alpha\}$, where $\alpha$ is a differentiable map $\alpha : U \to \mathbb{R}^3$ defined on a rectangle $U = [a, b] \times [c, d]$ in $\mathbb{R}^2$. Here $\alpha$ is thought of as the map that parametrizes $S$, and $U$ is called the parameter domain. The points in $U$ are denoted by $s = (s_1, s_2)$, with $s_1$ and $s_2$ being the two parameters that determine the points $p = \alpha(s)$ on the (image) of the surface. In terms of components, $\alpha$ has the form

$$\alpha(s) = (\alpha^1(s), \alpha^2(s), \alpha^3(s)).$$

The *image* of $S$ is the image $\alpha(U)$ of the parameter map $\alpha$, and it is this image that is usually thought of when one thinks of $S$ geometrically. If $\alpha$ also has the property that its Jacobian matrix

$$\alpha'(s) = \begin{bmatrix} \alpha^1_{s_1}(s) & \alpha^1_{s_2}(s) \\ \alpha^2_{s_1}(s) & \alpha^2_{s_2}(s) \\ \alpha^3_{s_1}(s) & \alpha^3_{s_2}(s) \end{bmatrix} \tag{A.5}$$

has rank 2 at each point $s$ in $U$, then $S$ is called a *regular* smooth surface. An alternative way of saying this is that the two columns of the Jacobian matrix (A.5), when viewed as vectors in $\mathbb{R}^3$, are linearly independent at each point $s$. We use the following notation for these two vector fields:

$$e_1 = \frac{\partial \alpha}{\partial s_1} = (\alpha^1_{s_1}, \alpha^2_{s_1}, \alpha^3_{s_1}), \tag{A.6}$$

$$e_2 = \frac{\partial \alpha}{\partial s_2} = (\alpha^1_{s_2}, \alpha^2_{s_2}, \alpha^3_{s_2}). \tag{A.7}$$

Note that for a point $s = (s_1, s_2)$ in $U$, the map $t \mapsto \alpha(t, s_2)$ is a curve lying on $S$, and $e_1(s)$ is a tangent vector to this curve at the point $\alpha(s)$. Similarly, the map $t \mapsto \alpha(s_1, t)$ is a curve lying on $S$, and $e_2(s)$ is a tangent vector to this curve at the point $\alpha(s)$. In the case where $S$ is a regular surface, these two curves on $S$ are called the *coordinate curves*, and we see that at the point $\alpha(s)$ on $S$, the tangent plane to $S$ is spanned by the vectors $e_1(s)$, $e_2(s)$. Consequently, $e_1(s) \times e_2(s)$ is normal (perpendicular) to $S$ at the point $\alpha(s)$. This observation gives a unit *normal* vector field

$$n \equiv \frac{e_1 \times e_2}{|e_1 \times e_2|},$$

defined at all points along the surface.

**Example A.2 (Sphere of radius $R$)**    Consider the map

$$\alpha(\theta, \phi) = (R \sin \phi \cos \theta, \, R \sin \phi \cos \theta, \, R \cos \phi),$$

with $(\theta, \phi) \in U \equiv [0, 2\pi] \times [0, \pi]$. The surface $S = \{\alpha\}$ is the sphere of radius $R$ with center at the origin. The two parameters $\theta$ and $\phi$ are shown geometrically in Figure A.1, and they are called the latitudinal and longitudinal angles respectively. This so because the coordinate curves obtained by varying $\theta$ while holding $\phi$ fixed are latitude lines on the sphere, and similarly, the longitudinal lines arise by varying $\phi$ while holding $\theta$ fixed. It is easy to calculate the tangential vector fields $e_1$, $e_2$ and unit normal $n$ for this example (see the exercises) and see that $n$ is inwardly directed at each point along the surface. Other choices of $\alpha$ lead to an outwardly pointing unit normal (exercise).

If $E$ is a vector field defined on an open set containing $S$, we define the *surface integral* of $E$ along $S$ by

$$\int_S E \cdot n \, dA \equiv \int_U E(\alpha(s)) \cdot [e_1(s) \times e_2(s)] \, ds_1 ds_2. \tag{A.8}$$

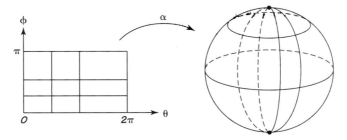

Figure A.1.    A sphere of radius $R$. Note that horizontal lines in the parameter domain $U$ are mapped onto latitudinal circles on the sphere, while vertical lines in $U$ are mapped onto longitudinal semicircles. These are the coordinate curves for this surface.

**Caution:** The notation on the left side of this definition is just notation, and the definition applies even when the surface is not regular (in which case the unit normal vector field $n$ does not exist). As with line integrals, the notation can be motivated as follows. Define the *surface area* of $S$ by

$$A(S) \equiv \int_U |e_1(s) \times e_2(s)| \, ds_1 ds_2. \tag{A.9}$$

Then this suggests that the element of surface area is heuristically given by $dA = |e_1 \times e_2| \, ds_1 ds_2$, and thus $E \cdot n \, dA$ gives the integrand on the right side of equation (A.8).

We also have need of the notion of the integral of a scalar function $g : U \to \mathbb{R}$ over the surface $S$. This is defined by

$$\int_S g \, dA = \int_U g(s)|e_1(s) \times e_2(s)| \, ds_1 ds_2. \tag{A.10}$$

If a scalar function $\tilde{g}$ is defined on the image of $\alpha$, then its surface integral over $S$ is the one above with $g = \tilde{g} \circ \alpha$.

Again, in all these definitions (A.8)–(A.10), we are using the Lebesgue integral (double integral over $U$).

As in the discussion for curves, we need a broader notion of a surface for later purposes, since the boundary of a cube is not a smooth surface, but rather is the union of six smooth surfaces. The generalization is as follows. A *piecewise smooth surface* $S$ is a multiset

$$S = \{\alpha_1, \ldots, \alpha_m\}$$

of differentiable maps $\alpha_j : U_j \to \mathbb{R}^3$, for $j = 1, \ldots, m$. Thus $S$ is the union of finitely many smooth surfaces,

$$S = \bigcup_{j=1}^m S_j,$$

where $S_j = \{\alpha_j\}$. For simplicity we will refer to a piecewise smooth surface simply as a *surface*. The surface integral of a vector field $E$ along such a surface $S$ is naturally defined as

$$\int_S E \cdot n \, dA = \sum_{j=1}^m \int_{S_j} E \cdot n \, dA.$$

## A.3.    Regions and Volume Integrals

In analogy with the above development of curves and surfaces as maps, or more generally as multisets of maps, it will be convenient to define a *smooth region* $\Omega$ in $\mathbb{R}^3$ as singleton set $\Omega = \{\alpha\}$, where $\alpha : U \to \mathbb{R}^3$ is a differentiable map defined on a rectangular box $U \subset \mathbb{R}^3$. The map

$$\alpha(s) = (\alpha^1(s), \alpha^2(s), \alpha^3(s)),$$

with $s = (s_1, s_2, s_3) \in U$, is thought of as parametrizing the image $\alpha(U)$ of the region $\Omega$. Often, $\Omega$ and its image are thought of as one and the same. However, this is proper only when $\Omega$ is a *regular* smooth region. This means that the Jacobian matrix

$$\alpha'(s) = \begin{bmatrix} \alpha^1_{s_1}(s) & \alpha^1_{s_2}(s) & \alpha^1_{s_3}(s) \\ \alpha^2_{s_1}(s) & \alpha^2_{s_2}(s) & \alpha^2_{s_3}(s) \\ \alpha^3_{s_1}(s) & \alpha^3_{s_2}(s) & \alpha^3_{s_3}(s) \end{bmatrix} \tag{A.11}$$

has rank 3 at each point $s = (s_1, s_2, s_3)$ in $U$. An alternative way of saying this is that $\det(\alpha'(s)) \neq 0$, at each point $s$ in $U$. Thus the columns of the Jacobian matrix give us three vector fields,

$$e_1 = \frac{\partial \alpha}{\partial s_1} = (\alpha^1_{s_1}, \alpha^2_{s_1}, \alpha^3_{s_1}), \tag{A.12}$$

$$e_2 = \frac{\partial \alpha}{\partial s_2} = (\alpha^1_{s_2}, \alpha^2_{s_2}, \alpha^3_{s_2}), \tag{A.13}$$

$$e_3 = \frac{\partial \alpha}{\partial s_3} = (\alpha^1_{s_3}, \alpha^2_{s_3}, \alpha^3_{s_3}), \tag{A.14}$$

which are linearly independent at each point $\alpha(s)$. Note that for a point $s = (s_1, s_2, s_3)$ in $U$, the map $t \mapsto \alpha(t, s_2, s_3)$ is a curve in $\Omega$, and $e_1(s)$ is a tangent vector to this curve at the point $\alpha(s)$. Similarly, the maps $t \mapsto \alpha(s_1, t, s_3)$ and $t \mapsto \alpha(s_1, s_2, t)$ are curves in $\Omega$, and $e_2(s)$, $e_3(s)$ are tangent vectors to these respective curves at the point $\alpha(s)$. These three curves in $\Omega$ are called the *coordinate curves*.

If $f$ is a scalar field defined on an open set containing $\Omega$, we define the *volume integral* of $f$ over $\Omega$ by

$$\int_\Omega f \, dV \equiv \int_U f(\alpha(s)) \det(\alpha'(s)) \, ds_1 ds_2 ds_3. \tag{A.15}$$

The integral on the left-hand side of this definition is the Lebesgue integral (triple integral over $U$). Note that by definition, the *volume* of $\Omega$ is

$$V(\Omega) = \int_U |\det(\alpha'(s))| \, ds_1 ds_2 ds_3.$$

The notion of a *piecewise* smooth region $\Omega$ as a multiset of smooth maps and the definition of the volume integral of scalar fields over such regions is analogous to that for curves and surfaces.

The volume intergal is in a certain sense related to the change of variables formula from multivariable calculus, which may be stated as follows:

**Theorem A.1 (Change of Variables Formula)**   *Suppose D and E are two bounded, open sets in $\mathbb{R}^n$ and that $\phi : D \to E$ is a 1-to-1, differentiable function that maps D onto E. If $f : E \to \mathbb{R}$ is a continuous function, then*

$$\int_{\phi(D)} f(y) \, dy = \int_D f(\phi(x))| \det(\phi'(x))| \, dx. \tag{A.16}$$

*The integrals here are Lebesgue integrals, and $dx = dx_1 dx_2 \cdots dx_n$, $dy = dy_1 dy_2 \cdots dy_n$.*

**Exercises A.1**

1. A helix (with parameters $a$, $b$) is a curve defined by

$$\alpha(t) = (a \cos t, a \sin t, bt), \tag{A.17}$$

with $t \in \mathbb{R}$. Actually this is a right-handed helix (as opposed to a left-handed helix), so called because of the way it winds around the cylinder

$$x^2 + y^2 = a^2.$$

It makes one turn about the cylinder for $t$ in $I = [0, 2\pi]$. Considering just this portion of the helix $C = \{\alpha\}$, find:

(a) The length of $C$.

(b) The value of the line integral $\int_C E \cdot \tau \, dL$, where $E$ is the vector field

$$E(x, y, z) = (x^2, xy, z^2).$$

2. Suppose $C = \{\alpha\}$ is a smooth curve, with $I$ the domain of $\alpha$, and $\phi : J \to I$ is a differentiable map from an interval $J$ onto $I$. Assume that either (a) $\phi'(s) > 0$ for

all $s \in J$, or (b) $\phi'(s) < 0$ for all $s \in J$. Let $\widetilde{C} = \{\alpha \circ \phi\}$. This curve is called a *reparametrization* of $C$ by the map $\phi$. Show that for any vector field $E$,

$$\int_{\widetilde{C}} E \cdot \tau \, dL = \pm \int_C E \cdot \tau \, dL,$$

where the plus sign is used in case (a) and the minus sign in case (b).

3. We discussed the sphere of radius $R$ in a previous example. Some of the other standard surfaces from geometry are:

   (a) **A cylinder:** defined by the map

   $$\alpha(\theta, z) = (R \cos \theta, R \sin \theta, z),$$

   for $(\theta, z) \in [0, 2\pi] \times [0, h]$, where $h > 0$.

   (b) **A cone:** defined by the map

   $$\alpha(\theta, t) = (t \cos \theta, t \sin \theta, mt),$$

   for $(\theta, t) \in [0, 2\pi] \times [0, k]$, where $m, k$ are positive.

   (c) **A torus:** defined by the map

   $$\alpha(\theta, \phi) = ((a + b \cos \phi) \cos \theta, (a + b \cos \phi) \sin \theta, b \sin \phi),$$

   for $(\theta, \phi) \in [0, 2\pi] \times [0, 2\pi]$. Here $a > b > 0$.

   (d) **A Möbius strip:** defined by the map

   $$\alpha(\theta, t) = ((a + t \cos(\theta/2)) \cos \theta, (a + t \cos(\theta/2)) \sin \theta, t \sin(\theta/2)),$$

   for $(\theta, t) \in [0, 2\pi] \times [-k, k]$, where $a > k > 0$.

   For each of these surfaces (including the sphere) do the following:

   (i) Draw a few of the coordinate curves.

   (ii) Calculate the tangent vector fields $e_1, e_2$ to the surface, and the normal vector field $e_1 \times e_2$.

   (iii) Find the area of the surface.

   (iv) Calculate the surface integral $\int_S E \cdot n \, dA$, where $E$ is the vector field

   $$E(x, y, z) = (x, y, z).$$

4. Parametrize the sphere of radius $R$ so that $n$ points outward.

## A.4.   Boundaries

From a geometrical viewpoint, the boundary of a curve should be the set of its endpoints, the boundary of a surface should be the set of curves that bound it, and the boundary of a region should be the set of surfaces that bound it. Since we are viewing things in terms of maps, the definition of the boundary will have to be phrased in terms of multisets of maps. We begin with the definition of the boundary of a region and work our way down to boundaries of curves.

**Regions:**   If $\Omega = \{\alpha\}$ is a smooth region in $\mathbb{R}^3$, its *boundary* is the piecewise smooth surface

$$\partial\Omega = \{\gamma_1, \gamma_2, \gamma_3, \gamma_4, \gamma_5, \gamma_6\},$$

where the maps $\gamma_j$, $j = 1, \ldots, 6$, are defined as follows. For simplicity, we assume that the domain of $\alpha$ is $U = [0, a] \times [0, b] \times [0, c]$. Then the surfaces comprising the boundary are defined by

$$\gamma_1(s_1, s_2) = \alpha(s_2, s_1, 0), \qquad (s_1, s_2) \in [0, a] \times [0, b],$$

$$\gamma_2(s_1, s_2) = \alpha(s_1, s_2, c), \qquad (s_1, s_2) \in [0, a] \times [0, b],$$

$$\gamma_3(s_1, s_2) = \alpha(0, s_2.s_1), \qquad (s_1, s_2) \in [0, b] \times [0, c],$$

$$\gamma_4(s_1, s_2) = \alpha(a, s_1, s_2), \qquad (s_1, s_2) \in [0, b] \times [0, c],$$

$$\gamma_5(s_1, s_2) = \alpha(s_2, 0, s_1), \qquad (s_1, s_2) \in [0, a] \times [0, c],$$

$$\gamma_6(s_1, s_2) = \alpha(s_1, b, s_2), \qquad (s_1, s_2) \in [0, a] \times [0, c].$$

Geometrically each of the $\gamma_j$'s can be viewed as the composition of $\alpha$ with a map from a rectangle onto one of the six faces of the parameter domain $U$ of $\alpha$. For example, $\gamma_2 = \alpha \circ \beta_2$, where $\beta_2 : U_1 \to U$, is given by $\beta_2(s_1, s_2) = (s_1, s_2, c)$. This is exhibited in Figure A.2.

*Note*: It is important to observe that the definition of the $\gamma_j$'s is chosen so that the version of Gauss's divergence theorem given below holds. In particular, note that each face $F_j = \{\beta_j\}$, $j = 1, \ldots, 6$, has its unit normal $n$ pointing toward the exterior of $U$.

The boundary of a piecewise smooth region

$$\Omega = \bigcup_{j=1}^{m} \Omega_j$$

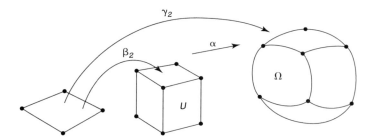

Figure A.2.  The boundary of a smooth region $\Omega$ is the piecewise smooth surface $\partial\Omega = \{\gamma_1, \ldots, \gamma_6\}$. The figure illustrates that $\gamma_2$ is the composition of the map $\alpha : U \to \mathbb{R}^3$ defining the region and a map $\beta_2 : [0, a] \times [0, b] \to U$ from a rectangle onto the top face of $U$.

is naturally defined as the union of the boundaries of each of its pieces:

$$\partial\Omega = \bigcup_{j=1}^{m} \partial\Omega_j.$$

**Surfaces:**   If $S = \{\alpha\}$ is a smooth curve, the *boundary* of $S$ is the piecewise smooth curve

$$\partial S = \{\gamma_1, \gamma_2, \gamma_3, \gamma_4\},$$

where the maps $\gamma_j$, $j = 1, 2, 3, 4$ are defined as follows. For simplicity we assume that the domain of $\alpha$ is $U = [0, a] \times [0, b]$. Then $\gamma_1$, $\gamma_3$, with domain $[0, a]$, and $\gamma_2$, $\gamma_4$, with domain $[0, b]$, are defined by

$$\gamma_1(s) = \alpha(s, 0), \qquad s \in [0, a],$$

$$\gamma_2(s) = \alpha(a, s), \qquad s \in [0, b],$$

$$\gamma_3(s) = \alpha(a - s, b), \qquad s \in [0, a],$$

$$\gamma_4(s) = \alpha(0, b - s), \qquad s \in [0, b],$$

Geometrically, each of the $\gamma_j$'s can be viewed as the composition of $\alpha$ with a map from a rectangle onto one of the four edges of the parameter domain $U$ of $\alpha$. For example, $\gamma_3 = \alpha \circ \beta_3$, where $\beta_3 : [0, a] \to U$, is given by $\beta_3(s) = (a - s, b)$. This is exhibited in Figure A.3.

   *Note*: It is important to observe that the definition of the $\gamma_j$'s is chosen so that the version of Stokes's theorem given below holds. In particular note that each edge $E_j = \{\beta_j\}$, $j = 1, 2, 3, 4$, has its unit tangent $\tau$ pointing in the direction of

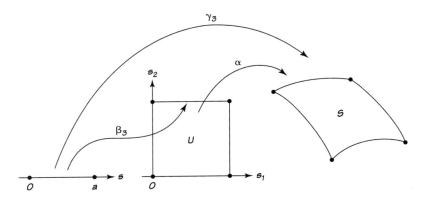

Figure A.3.    The boundary of a surface $S$ is the piecewise smooth curve $\partial S = \{\gamma_1, \gamma_2, \gamma_3, \gamma_4\}$. The figure illustrates that $\gamma_3$ is the composition of the map $\alpha : U \to \mathbb{R}^3$ defining the surface and a map $\beta_3 : [0, a] \to U$ from an interval onto the top edge of $U$.

counterclockwise flow around the boundary of $U$. The definition of the boundary extends naturally to piecewise smooth surfaces. Thus if

$$S = \bigcup_{j=1}^{m} S_j,$$

where each $S_j = \{\alpha_j\}$ is smooth, then the boundary is

$$\partial S = \bigcup_{j=1}^{m} \partial S_j.$$

**Curves:**    Finally, we come to the definition of the boundary of a curve. This would actually be the simplest to define (the boundary is the set of end points) except for two things: (1) In keeping with our overall viewpoint, the boundary should be a multiset of maps, and (2) for the sake of convenience, the "point integral" of a scalar function over a multiset of "point maps" needs to be introduced. Thus, before defining the boundary of a curve, we make the following definitions. A *point map* is a map $\alpha : \{s\} \to \mathbb{R}^3$ where $s = \pm 1$. Thus point maps come in two flavors (or signs) depending on whether the domain of the map is $\{1\}$ or $\{-1\}$. A *signed point* $P$ is a singleton set consisting of a point map $P = \{\alpha\}$, and geometrially, it is thought of as the point $\alpha(s)$ in $\mathbb{R}^3$ with a plus or minus sign attached. If $f$ is a scalar field on $\mathbb{R}^3$, the *point integral* of $f$ over $P$ is defined by

$$\int_P f = sf(\alpha(s)).$$

This integral is just the $\pm 1$ times $f$ evaluated at $\alpha(s)$. The extension of this to the case where $P$ is a multiset of point maps,

$$P = \bigcup_{j=1}^{m} P_j,$$

with $P_j = \{\alpha_j\}$, is naturally defined as

$$\int_P f = \sum_{j=1}^{m} \int_{P_j} f.$$

With these preliminaries out of the way, we define the boundary of a curve $C$ as follows. First, suppose $C = \{\alpha\}$ is smooth. Then its boundary is the multiset of point maps

$$\partial C = \{\gamma_1, \gamma_2\},$$

where $\gamma_1 : \{-1\} \to \mathbb{R}^3$ and $\gamma_2 : \{1\} \to \mathbb{R}^3$ are defined by

$$\gamma_1(-1) = \alpha(a),$$

$$\gamma_2(1) = \alpha(b).$$

Here the domain of $\alpha$ is the interval $I = [a, b]$. Geometrically, each of the $\gamma_j$'s can be viewed as the composition of $\alpha$ with a point map $\beta_j$ from $\{\pm 1\}$ onto one of the two vertices of the line seqment $[a, b]$. This is exhibited in Figure A.4. Note that the definition gives the desired result that

$$\int_{\partial C} f = f(\alpha(b)) - f(\alpha(a)).$$

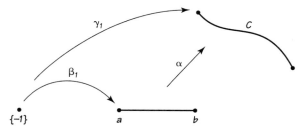

Figure A.4.    The boundary of a curve $C$ is the multiset of point maps: $\partial C = \{\gamma_1, \gamma_2\}$. The figure illustrates that $\gamma_1$ is the composition of the map $\alpha : [a, b] \to \mathbb{R}^3$, definning the curve, and the point map $\beta_1 : \{-1\} \to [a, b]$, onto left vertex of $[a, b]$.

The extension of the definition of the boundary to piecewise smooth curves is accomplished naturally as follows. If

$$C = \bigcup_{j=1}^{m} C_j,$$

with $C_j = \{\alpha_j\}$, is a piecewise smooth curve, then its boundary is

$$\partial C = \bigcup_{j=1}^{m} \partial C_j.$$

## A.5.   Closed Curves and Surfaces

The common geometrical conception of a closed curve or a closed surface is that the boundary of the curve or surface is empty. Since our viewpoint here is that curves and surfaces are multisets of maps, we define the notion of being closed as follows: A curve $C$ is *closed* if

$$\int_{\partial C} f = 0, \qquad \forall \text{ scalar field } f.$$

Similarly, a surface $S$ is called *closed* if

$$\int_{\partial S} E \cdot \tau \, dL = 0, \qquad \forall \text{ vector field } E.$$

While these definitions may seem somewhat indirect, they are quite general and cover a variety of cases.

**Example A.3**   Consider the sphere $S = \{\alpha\}$ discussed in the example above. It is straightforward to compute the boundary $\partial S = \{\gamma_1, \gamma_2, \gamma_3, \gamma_4\}$ and get

$$\gamma_1(s) = (0, 0, R), \qquad\qquad\qquad\qquad s \in [0, 2\pi],$$

$$\gamma_2(s) = (R \sin(s), 0, R \cos(s)), \qquad\qquad s \in [0, \pi],$$

$$\gamma_3(s) = (0, 0, -R), \qquad\qquad\qquad\qquad s \in [0, 2\pi],$$

$$\gamma_4(s) = (R \sin(\pi - s), 0, R \cos(\pi - s)), \qquad s \in [0, \pi].$$

Let $C_j = \{\gamma_j\}$, $j = 1, \ldots, 4$, be the four boundary curves. As shown in Figure A.5, the maps $\gamma_1, \gamma_3$ are constant maps with images the north and south poles

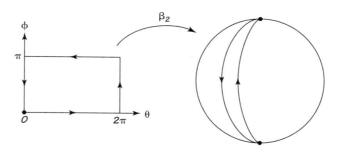

Figure A.5.   The boundary of the sphere consists of two constant curves (mapping onto the north and south poles respectively) and two oppositely directed longitudinal semicircles.

respectively. Since they are constant maps, $\gamma_1'(s) = 0$ and $\gamma_3'(s) = 0$ for every $s$, and consequently, $\int_{C_1} E \cdot \tau \, dL = 0$ and $\int_{C_3} E \cdot \tau \, dL = 0$, for any vector field $E$. On the other hand, the images of $\gamma_2, \gamma_4$ are semicircles on the sphere running from the north to south pole and south to north pole respectively, and in fact, $\gamma_4(s) = \gamma_2(\pi - s)$. Consequently a direct calculation using the change of variables formula gives

$$\int_{C_4} E \cdot \tau \, dL = -\int_{C_2} E \cdot \tau \, dL,$$

for every vector field $E$. This also follows from a general result, since $C_4$ is just a reparametrization of $C_2$ (see Exercise 2 in Exercises A.1). Thus we see that

$$\int_{\partial S} E \cdot \tau \, dL = \sum_{j=1}^{4} \int_{C_j} E \cdot \tau \, dL = 0,$$

for every vector field $E$, and thus the sphere $S$ is a closed surface.

## A.6.   The Theorems of Gauss, Green, and Stokes

Two of the most important theorems in vector analysis are Gauss's divergence theorem and Stokes's theorem. They have many uses and can be interpreted in various ways. A basic interpretation of them is that they are just generalizations of the second part of the fundamental theorem of calculus,

$$\int_a^b f'(x)dx = f(b) - f(a),$$

where $f$ is a continuously differentiable function on $[a, b]$. Green's theorem is actually a special case of Stokes's theorem.

**Theorem A.2 (Gauss, Stokes, and Gradient Theorems)**   *Suppose E is a vector field and f is a scalar field, each of which is defined (and continuously differentiable) on an open set $\mathcal{O}$ in $\mathbb{R}^3$. Then for a region $\Omega$, a surface S, and a curve C (each of which is piecewise differentiable and has its image contained in $\mathcal{O}$), we have the following:*

**(1) Gauss's Divergence Theorem:**

$$\int_\Omega \mathrm{div}(E)\, dV = \int_{\partial\Omega} E \cdot n \, dA.$$

**(2) Stokes's Theorem:**

$$\int_S \mathrm{curl}(E) \cdot n \, dA = \int_{\partial S} E \cdot \tau \, dL.$$

**(3) Gradient Theorem:**

$$\int_C \nabla f \cdot \tau \, dL = \int_{\partial C} f.$$

**Proof**   We just prove Gauss's divergence theorem and leave the rest for exercises. The proof is actualy rather elementary in that it uses just the fundamental theorem of calculus and a few tricky manipulations. We assume that the region is smooth, $\Omega = \{\alpha\}$. The piecewise smooth case will follow from this by linearity of the integrals. We also assume that $E$ has the form $E = (E^1, 0, 0)$. The general case will be the same but will occupy a great deal more space to write out. Also, for convenience, we assume that the domain $U$ of $\alpha$ is the unit cube $U = [0, 1] \times [0, 1] \times [0, 1]$. Based on these assumptions, we have to show that the volume integral

$$\int_\Omega \mathrm{div}(E)\, dV \equiv \int_U E_x^1(\alpha(s)) \det(\alpha'(s))\, ds_1 ds_2 ds_3 \qquad (A.18)$$

can be rewritten as a surface integral.

To see this, we first make a few general observations. In terms of components, we write the map $\alpha$ as $\alpha(s) = (\alpha^1(s), \alpha^2(s), \alpha^3(s))$, for $s \in U$. Using the Laplace expansion of the determinant with respect to the first row, we can write

$$\det(\alpha'(s)) = \alpha_{s_1}^1(s) A(s) - \alpha_{s_2}^1(s) B(s) + \alpha_{s_3}^1(s) C(s),$$

where $A(s)$, $B(s)$, $C(s)$ denote the determinants of the $2 \times 2$ submatrices of $\alpha'(s)$ obtained by deleting the first row and columns 1, 2, and 3 respectively. Also, observe that the three rows of $\alpha'(s)$ are $\nabla\alpha^1(s)$, $\nabla\alpha^2(s)$, and $\nabla\alpha^3(s)$. From this we get two identities. First: for any vector $w = (w^1, w^2, w^3) \in \mathbb{R}^3$,

$$w^1 A(s) - w^2 B(s) + w^3 C(s) = \det(w, \nabla\alpha^2(s), \nabla\alpha^3(s)), \qquad (A.19)$$

which again is just the Laplace expansion of the determinant about its first row. Second:

$$\nabla\alpha^2(s) \times \nabla\alpha^3(s) = (A(s), -B(s), C(s)).$$

As particular cases of the first identity we take $w = \nabla\alpha^2(s)$ or $w = \nabla\alpha^3(s)$ and use the fact that a determinant with two equal rows is zero to get

$$\alpha^2_{s_1}(s)A(s) - \alpha^2_{s_2}(s)B(s) + \alpha^2_{s_3}(s)C(s) = 0,$$

$$\alpha^3_{s_3}(s)A(s) - \alpha^3_{s_2}(s)B(s) + \alpha^3_{s_3}(s)C(s) = 0.$$

Finally, note that $\mathrm{div}(\nabla f \times \nabla g) = 0$ is a general identity that holds for any two scalar fields $f$, $g$, and so by the second identity,

$$A_{s_1} - B_{s_2} + C_{s_3} = \mathrm{div}(\nabla\alpha^2 \times \nabla\alpha^3) = 0.$$

We now use all of these identities in deriving the following identity, which is at the core of the theorem. We start with the integrand from the volume intergal in equation (A.18) and rewrite it so that we can use the fundamental theorem of calculus. Thus

$$E^1_x(\alpha(s)) \det(\alpha'(s))$$

$$= E^1_x(\alpha(s))[\alpha^1_{s_1}(s)A(s) - \alpha^1_{s_2}(s)B(s) + \alpha^1_{s_3}(s)C(s)]$$

$$= E^1_x(\alpha(s))[\alpha^1_{s_1}(s)A(s) - \alpha^1_{s_2}(s)B(s) + \alpha^1_{s_3}(s)C(s)]$$

$$\quad + E^1_y(\alpha(s))[\alpha^2_{s_1}(s)A(s) - \alpha^2_{s_2}(s)B(s) + \alpha^2_{s_3}(s)C(s)]$$

$$\quad + E^1_z(\alpha(s))[\alpha^3_{s_1}(s)A(s) - \alpha^3_{s_2}(s)B(s) + \alpha^3_{s_3}(s)C(s)]$$

$$= \frac{\partial}{\partial s_1}\left(E^1(\alpha(s))\right)A(s) - \frac{\partial}{\partial s_2}\left(E^1(\alpha(s))\right)B(s) + \frac{\partial}{\partial s_3}\left(E^1(\alpha(s))\right)C(s)$$

$$= \frac{\partial}{\partial s_1}\left(E^1(\alpha(s)A(s))\right) - \frac{\partial}{\partial s_2}\left(E^1(\alpha(s)B(s))\right) + \frac{\partial}{\partial s_3}\left(E^1(\alpha(s)C(s))\right).$$

Now let $R = [0, 1] \times [0, 1]$ be the unit square. Then from the last identity and the fundamental theorem of calculus we easily get

$$
\int_\Omega \mathrm{div}(E)\, dV
$$

$$
= \int_R \Big[ E^1(\alpha(1, s_2, s_3)) A(1, s_2, s_3) - E^1(\alpha(0, s_2, s_3)) A(0, s_2, s_3) \Big] ds_2 ds_3
$$

$$
+ \int_R \Big[ E^1(\alpha(s_1, 0, s_3)) B(s_1, 0, s_3) - E^1(\alpha(s_1, 1, s_3)) B(s_1, 1, s_3) \Big] ds_1 ds_3
$$

$$
+ \int_R \Big[ E^1(\alpha(s_1, s_2, 1)) C(s_1, s_2, 1) - E^1(\alpha(s_1, s_2, 0)) C(s_1, s_2, 0) \Big] ds_1 ds_2
$$

$$
= \sum_{j=1}^{6} \int_{S_j} E \cdot n\, dA = \int_{\partial\Omega} E \cdot n\, dA.
$$

This completes the proof of Gauss's divergence theorem.

**Corollary A.1 (Green's Theorem)**  *Suppose the surface $S = \{\alpha\}$ is a planar region and $E$ is a planar vector field; i.e., $\alpha(s) = (\alpha_1(s), \alpha_2(s), 0)$, for $s \in U$, and $E(x, y, z) = (E^1(x, y), E^2(x, y), 0)$, for $(x, y, z) \in \mathcal{O}$ (an open set containing $S$). Then*

$$
\int_S \left( \frac{\partial E^2}{\partial x} - \frac{\partial E^1}{\partial y} \right) dx\, dy = \int_{\partial S} E \cdot \tau\, dL. \tag{A.20}
$$

*The integral on the left side of this equation is the Lebesgue (double) integral over the planar region $S = \alpha(U)$.*

**Proof**   Because of the form of $\alpha$, we find that

$$
e_1 = \left( \frac{\partial \alpha_1}{\partial s_1}, \frac{\partial \alpha_2}{\partial s_1}, 0 \right),
$$

$$
e_2 = \left( \frac{\partial \alpha_1}{\partial s_2}, \frac{\partial \alpha_2}{\partial s_2}, 0 \right),
$$

and so $e_1 \times e_2 = (0, 0, \det(\alpha'))$. (Here, in computing the Jacobian matrix $\alpha'$, we consider $\alpha$ as a map from $U$ into $\mathbb{R}^2$.) Hence $n = (0, 0, 1)$. Also, because of the form of $E$, it is easy to see that

$$\text{curl}(E) = \left(0, 0, \frac{\partial E^2}{\partial x} - \frac{\partial E^1}{\partial y}\right)$$

Now, applying Stokes's theorem, we get

$$\int_{\partial S} E \cdot \tau \, dL = \int_S \left(\frac{\partial E^2}{\partial x} - \frac{\partial E^1}{\partial y}\right) dA.$$

Finally, using the change of variables on the surface integral above allows us to write it in the form stated in Green's formula.

**Exercises A.6**

1. Show that the torus is a closed surface (Problem 3(c) in Exercise Set A.1)
2. Show that the cylinder is not a closed surface (Problem 3(a) in Exercise Set A.1). Show how to add a top and bottom to the cylinder so that the resulting surface is closed.
3. Prove Stokes's theorem and the gradient theorem.

# A.7.   Curvatures and Fundamental Forms

The curvatures and fundamental forms for curves and surfaces play a key role in the geometry of these objects and also are an essential part of the equations of continuum mechanics for 1- and 2-dimensional continua. We describe the geometry here and discuss the application of this to continuum mechanics in Appendix B. For simplicity we consider only smooth curves and surfaces. The piecewise smooth case is an easy extension of this. For the sake of brevity, little attempt is made to motivate the concepts of curvature and fundamental forms. For pedagogial reasons we begin the discussion with surfaces.

## A.7.1.   Curvatures for Surfaces

Consider a smooth surface $S = \{\alpha\}$, with $\alpha : U \to \mathbb{R}^3$, and assume that $S$ is regular; i.e., the tangential vector fields $e_1(s)$, $e_2(s)$ are linearly independent for each $s \in U$. Consequently, the matrix

$$G = \{G_{ij}\}_{i,j=1,2},$$

where

$$G_{ij} = e_i \cdot e_j,$$

is invertible (for each $s$) (exercise). Then entries $G_{ij}$ of this matrix are called the *components of the metric (first fundamental form)* for the surface. One can easily show (exercise) that

$$\det G^{1/2} = |e_1 \times e_2|,$$

and so the surface area can be expressed in terms of the metric components $G_{ij}$. The entries of the inverse matrix $G^{-1}$ will be denoted by superscripts,

$$G^{-1} = \{G^{ij}\}_{i,j=1,2},$$

and in terms of these entries we can define the *reciprocal tangential vector fields* $e^1, e^2$ by

$$e^i \equiv G^{ij} e_j.$$

**Note: (summation convention):**   In the above expression (and elsewhere in the text) there is implied summation on the repeated index, which in this case is $j$. Thus $G^{ij}e_j$ is a convenient abbreviation for $G^{i1}e_1 + G^{i2}e_2$. It is an easy exercise to show that $e^1, e^2$ are also linearly independent at each point, and

$$e^i \cdot e_j = 0, \quad e^i \cdot e_i = 1,$$

for every $i \neq j$. One calls $\{e_1, e_2, n\}$ a *moving frame* along the surface, since $\{e_1(s), e_2(s)\}$, at each point $s$, span the tangent plane to the surface at the point $\alpha(s)$, and $n(s)$ spans the normal line to the surface at the point. The reciprocal moving frame is $\{e^1, e^2, n\}$.

Next we introduce the matrix

$$H = \{H_{ij}\}_{i,j=1,2},$$

where the entries are defined by

$$H_{ij} = \frac{\partial^2 \alpha}{\partial s_i \partial s_j} \cdot n.$$

The entries $H_{ij}$ of the matrix $H$ are called the *components of the second fundamental form* for the surface. The second fundamental form, together with the metric, determines the various curvatures for the surface. The two that we need elsewhere in the text are

**Gaussian Curvature:**

$$\kappa = \det(G^{-1}H) = \frac{\det(H)}{\det(G)}.$$ (A.21)

**Mean Curvature:**

$$\mu = \operatorname{tr}(G^{-1}H).$$ (A.22)

*Note*: In the mean curvature, tr denotes the trace operation on matrices. Also, historically, the mean curvature was introduced as an average of two other curvatures, and so there usually is a $1/2$ in the definition of $\mu$. However, we have chosen to omit this in order to simplify the notation.

## A.7.2. Frenet Frames, Curvature, and Torsion for Curves

Consider a smooth curve $C = \{\alpha\}$, with $\alpha : I \rightarrow \mathbb{R}^3$, and assume that $C$ is regular. In analogy with the development above, we define the *component of the metric* for the curve to be the function ($1 \times 1$ matrix)

$$G = G_{11} = e_1 \cdot e_1,$$

where $e_1 = \alpha'$. Note that $G$ is a positive function, and

$$\det G^{1/2} = G^{1/2} = |\alpha'|,$$

which exhibits how the metric for the curve is connected with the measurement of its length. In analogy to what we did for surfaces, we denote the inverse of $G$ by $G^{-1} = G^{11} \equiv 1/G_{11}$, and then the reciprocal tangential vector field is $e^1 = G^{11}e_1$. More simply stated,

$$e^1 = \frac{\alpha'}{|\alpha'|^2}.$$

Thus clearly, $e^1 \cdot e_1 = 1$ at each point along the curve.

If we assume in addition to regularity of the curve, that also it is such that the velocity $\alpha'(s)$ and acceleration $\alpha''(s)$ are linearly independent for each $s \in I$, then the *Frenet frame* can be defined along the curve. This is the moving frame

$\{T, N, B\}$ defined by

$$T = \frac{\alpha'}{|\alpha'|},$$

$$N = \frac{\alpha'}{|\alpha'|} \times \frac{\alpha' \times \alpha''}{|\alpha' \times \alpha''|},$$

$$B = \frac{\alpha' \times \alpha''}{|\alpha' \times \alpha''|}.$$

These vector fields along the curve are called the unit *tangent, normal,* and *binormal,* respectively. They are of unit length and orthogonal to each other. Note that the binormal is perpendicular to the plane $P$ determined by the velocity and acceleration vectors (indeed $B$ is the normalization of their cross product), and the normal is $N = T \times B$ and thus lies in the plane $P$. The plane $P$ is known as the *osculating plane* for the curve.

There are two important curvature functions for the curve $C$, called the *curvature* and *torsion* respectively. They are defined by

**Curvature:**

$$\kappa = \frac{|\alpha' \times \alpha''|}{|\alpha'|^3}. \tag{A.23}$$

**Torsion:**

$$\tau = \frac{\det(\alpha', \alpha'', \alpha''')}{|\alpha' \times \alpha''|^2}. \tag{A.24}$$

## A.7.3.   Metrics and Integration

As we have seen for smooth curves and surfaces in $\mathbb{R}^3$, the components $G_{ij}$ of their metrics can be used to express the elements of length and area:

(1)  For curves, $\det G^{1/2} = |\alpha'|$.

(2)  For surfaces, $\det G^{1/2} = |\frac{\partial \alpha}{\partial s_1} \times \frac{\partial \alpha}{\partial s_2}|$.

The components of the metric for a smooth region $\Omega = \{\alpha\}$ in $\mathbb{R}^3$ are, by definition, the entries in the matrix

$$G = \alpha'^* \alpha',$$

and it is easy to see that

(3) For regions $\det G^{1/2} = |\det \alpha'|$.

Thus in all cases, the integral of a scalar field $g$, defined along a curve, surface, or region, can be expressed by

$$\int_U g(s) \det G^{1/2}(s)\, ds,$$

where $U$ is the parameter domain.

## Exercises A.7

1. Show that the matrix $G$ of metric components for a regular surface is invertible and that $\det G^{1/2} = |e_1 \times e_2|$.

2. Show that the reciprocal tangential vector fields $e^1, e^2$ are linearly independent at each point $s \in U$ and that $e^i \cdot e_j = 0$, for $i \neq j$, while $e^i \cdot e_i = 1$.

3. Calculate the Gauss curvature and mean curvature of the sphere, cylinder, cone, torus, and Möbius strip.

4. Calculate the Frenet frame $\{T, N, B\}$, curvature, and torsion for the helix.

# B Continuum Mechanics

Continuum mechanics, of which fluid mechanics is a special branch, is an old area of physics, dating back at least to Newton's day and receiving in some respects an almost definitive formulation by Euler in the 1700s. It provides us with a rich source of nonlinear systems of PDEs and continues to be an area of intense modern research, both theoretically and computationally.

Here we derive the central equations in continuum mechanics in a somewhat general setting. These equations are the equations of motion, or system of PDEs, that a continuum — a solid, liquid, or gas — must obey as it moves and deforms in space subject to various external and internal forces. It is most natural to begin first with the laws of motion formulated in terms of certain integrals and then from these derive the PDE version of these laws. See [Fu 65], [MH 83], [Tru 77], [Gur 81], [HM 76].

The various types of continua that we wish to study can be quite diverse in makeup, varying from gases in containers, to fluids in vessels (or tanks), to elastic solids, to flexible rods (ideally modeled as 1-dimensional continua), to vibrating membranes like drumheads (modeled as 2-dimensional continua). These latter two types, 1- and 2-dimensional continua, are somewhat special, and we defer treatment of them until the end. Thus we begin our discussion with 3-dimensional continua.

## B.1. The Eulerian Description

There are two alternative methods of describing the motion of a (3-dimensional) continuum that in a certain sense are equivalent. These are called the *Eulerian description* and the *Lagrangian description*, named after their respective founders. We adopt the Eulerian description at the outset.

Thus suppose the continuum, or substance, is in motion and occupies a region $\Omega$ of $\mathbb{R}^3$. If you are studying a fluid, you can think of $\Omega$ as the vessel in which the fluid circulates. In general, the substance of study need not occupy the entire region $\Omega$. At each point $(x, y, z)$ in $\Omega$, we let

$$v(x, y, z, t) = \left\{ \begin{array}{c} \text{the velocity of the substance flowing} \\ \text{through } (x, y, z) \text{ at time } t \end{array} \right\}. \qquad \text{(B.2)}$$

This gives a time-dependent vector field $v$ on $\Omega$, called the *velocity vector field* of the continuum. In terms of components,

$$v(x, y, z, t) = \left( v^1(x, y, z, t), v^2(x, y, z, t), v^3(x, y, z, t) \right).$$

Note that if the continuum does not fill up the region $\Omega$, the velocity vector field is discontinuous, dropping to 0 at points where there is no substance. To keep things simple, we will often overlook this technicality.

We let $\phi : \mathcal{D} \subseteq \Omega \times \mathbb{R} \to \Omega$ be the flow generated by the vector field $v$. Thus $\phi$ satisfies

$$\frac{\partial \phi}{\partial t}(x, y, z, t) = v(\phi(x, y, z, t), t), \qquad \text{(B.3)}$$

$$\phi(x, y, z, 0) = (x, y, z), \qquad \text{(B.3)}$$

for each $(x, y, z) \in \Omega$ and all $t \in I_{(x,y,z)}$ (the maximum interval of existence for the integral curve starting at $(x, y, z)$ at $t = 0$, see the Introduction). The flow $\phi$ is is the main conceptual tool for analyzing the motion of the continuum. Thus suppose $B \subseteq \Omega$ is a portion of the continuum that we wish to follow as it moves and deforms, and suppose $I$ is an interval of times (containing 0) such that $B \times I \subseteq \mathcal{D}$. Then we get a one-parameter family $\{\phi_t\}_{t \in I}$ of transformations $\phi_t : B \to \Omega$ that models the motion of $B$. Assume that $B$ is a smooth, regular region $B = \{\alpha_0\}$, with $\alpha_0 : U \to \mathbb{R}^3$ (we identify $B$ and the image of $\alpha_0$, i.e.: $B = \alpha_0(U)$). For each $t \in I$ let

$$\alpha_t = \phi_t \circ \alpha_0,$$

which gives a regular, smooth region $B_t = \{\alpha_t\}$, which represents the region into which $B = B_0$ deforms over the interval of time from 0 to $t$. Indeed, we consider $\phi_t$ as transforming $B$ into $B_t$ (and write $B_t = \phi_t(B)$). See Figure B.1. For technical reasons it is convenient to assume that the flow does not "collapse" $B$ during the motion. Thus we make the

**Assumption:**    $\det(\phi'_t(x, y, z)) > 0$, for all $(x, y, z) \in B, t \in I$.

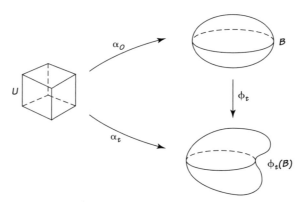

Figure B.1. Motion of the portion $B$ of the continuum under the flow $\phi$ generated by the velocity vector field $v$

We also assume that $B$ is positively oriented, that is, $\det(\alpha_0'(s)) > 0$, for every $s \in B$.

Next, we consider a time-dependent scalar field $\rho$ on $\Omega$, known as the *mass density function* for the continuum:

$$\rho(x, y, z, t) = \left\{ \begin{array}{c} \text{the mass per unit volume of the substance} \\ \text{flowing through } (x, y, z) \text{ at time } t \end{array} \right\}. \tag{B.4}$$

Thus $\rho(x, y, z, t)dV$ represents the infinitesimal amount of mass at the point $(x, y, z, t)$ at time $t$. More properly, for each portion $B$ of the continuum, the volume integral

$$m(B_t, t) = \int_{B_t} \rho\, dV \tag{B.5}$$

gives the total mass of $B_t$ at time $t$.

In the Eulerian description of the motion, the velocity component functions $v_1, v_2, v_3$, and the density function $\rho$ are the fundamental unknowns, and the basic laws of physics are used to derive a system of PDEs that these four functions must satisfy. The derivation goes as follows.

For each portion $B$ of the continuum, the motion and deformation of $B$ under the flow must be such that the following three postulates hold:

(1) The mass of $B_t$ should not change over time.

(2) The time rate of change of the linear momentum of $B_t$ should equal the forces acting on $B_t$.

(3) The time rate of change of the angular momentum of $B_t$ should equal the torques applied to $B_t$.

To phrase these postulates in analytic terms and also to interpret them geometrically, note that the linear and angular momenta are represented as volume integrals (like the expression (B.5) above for the mass of $B_t$ at time $t$) The linear momentum of $B_t$ at time $t$ is

$$\mathcal{L}(B_t, t) = \int_{B_t} \rho v \, dV,$$

which can be viewed as the sum over all the linear momenta $\rho v \, dV$ of the infinitesimal parts comprising $B_t$. In this sense, one can think of $\int_{B_t} \rho v \, dV$ as the total, or net, linear momentum of $B_t$. It is a vector, and can be plotted with its initial point at the center of mass of $B_t$. See Figure B.2. The angular momentum of $B_t$ at time $t$ is

$$\mathcal{A}(B_t, t) = \int_{B_t} [\vec{x} \times \rho v] dV,$$

where as before, we use the notation

$$\vec{x} \equiv (x, y, z),$$

to denote the position vector from the origin to the point $(x, y, z)$, and $\times$ denotes the cross product of vectors. Again, this can be viewed as the sum of all the angular momenta $\vec{x} \times \rho v dV$ of the infinitesimal parts comprising $B_t$. We should say a few things about why this cross product accurately expresses the angular momentum. Recall that the vector $\vec{x} \times \rho v$ is perpendicular to the the plane of the

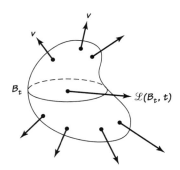

Figure B.2.   Total linear momentum $\mathcal{L}(B_t, t)$ of the portion $B_t$ of the continuum at time $t$. This vector $\mathcal{L}(B_t, t)$, is plotted with its initial point at the center of mass of $B_t$.

triangle formed by the vectors $\vec{x}$ and $\rho v$ and has length equal to twice the area of this triangle. See Figure B.3. Thus at time $t$, the angular momentum of the infinitesimal part at $(x, y, z) \in B_t$ consists of a spin (or rotation) about the line through $(0, 0, 0)$ in the direction of the vector $\vec{x} \times \rho v$. The magnitude of this vector is the magnitude of the spin, and the spin is in the direction indicated by the right hand rule. Integrating (summing) over all the infinitesimal parts gives the (total, or net) angular momentum $\mathcal{A}(B_t, t) = \int_{B_t} [\vec{x} \times \rho v] dV$, which measures the total spin (or rotation) that $B_t$ exerts about the line through the origin in the direction of the vector $\mathcal{A}(B_t, t)$.

With these interpretations, the above postulates are analytically phrased in the following three equations:

**(1)  Conservation of Mass:**

$$\frac{d}{dt} \int_{B_t} \rho \, dV = 0. \tag{B.6}$$

**(2)  Linear Momentum Equation:**

$$\frac{d}{dt} \int_{B_t} \rho v \, dV = \mathcal{F}. \tag{B.7}$$

**(3)  Angular Momentum Equation:**

$$\frac{d}{dt} \int_{B_t} [\vec{x} \times \rho v] \, dV = \mathcal{T}. \tag{B.8}$$

In equation (B.7), the $\mathcal{F}$ stands for the total force applied to $B_t$ at time $t$, and in equation (B.8), the $\mathcal{T}$ stands for the total torque applied to $B_t$ at time $t$.

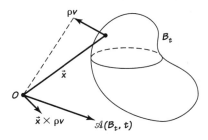

Figure B.3.   Total angular momentum $\mathcal{A}(B_t, t)$ of the portion $B_t$ of the continuum at time $t$. This vector $\mathcal{A}(B_t, t)$ is plotted at the origin and gives the axis, direction, and magnitude of the spin that $B_t$ exerts on the contiuum.

The exact specification of $\mathcal{F}$ and $\mathcal{T}$ depends on the material nature of the continuum, as well the external applied forces. Thus, to make equations (B.6)–(B.8) more specific, we need to discuss the finer details of our mathematical model of the continuum.

## B.1.1.   The Stress Tensor T

The forces acting on the continuum are divided into external and internal forces:

$$\mathcal{F} = \mathcal{F}_{\text{ext}} + \mathcal{F}_{\text{int}}.$$

The external force $\mathcal{F}_{\text{ext}}$ does not depend on the particular internal structure of the continuum (the way it responds to deformations) and consists of, or combines, all the various forces like gravity and electromagnetism. We assume that it is given by a density

$$\rho F = \rho F(x, y, z, t) = \rho\left( F^1(x, y, z, t), \ F^2(x, y, z, t), \ F^3(x, y, z, t) \right),$$

so that the external force acting on $B_t$ is given by

$$\mathcal{F}_{\text{ext}}(B_t, t) = \int_{B_t} \rho F \, dV. \tag{B.9}$$

The internal force $\mathcal{F}_{\text{int}}$ is more complicated to describe, since it arises from the very deformation (or motion) of the continuum itself and depends on the particular constitution of the continuum, that is, the material of which it is made. When a particular material is stressed due to some deformation, its response depends, at the microscopic level, on the specific atomic and molecular structure of the material. Thus the responses due to compression of a volume of gas, a volume of liquid, a piece of wood, a piece of clay, or a piece of foam rubber, are all quite different. Likewise, attempts to deform these materials by an expansive or stretching motion will produce different internal reactions, or stresses, with the various parts of each material interacting based on the molecular bonding. The internal force $\mathcal{F}_{\text{int}}$ arises from these stresses, and we wish to formulate a reasonable model for it at the macroscopic level, without going into all the details present down at the microscopic level. This will involve the stress tensor, which we now describe.

We consider the portion $B_t$ of the continuum at time $t$ during the motion and ask for a description of the force $\mathcal{F}_{\text{int}}$ exerted on it by the rest of the continuum surrounding it. See Figure B.4

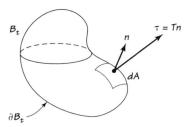

Figure B.4.   The internal force exerted on $B_t$ by its surroundings is due to stresses (force/unit area) acting across its boundary.

We postulate that this force is comprised of stresses $\tau$ (force per unit area) acting at each point along the boundary surface $\partial B_t$:

$$\mathcal{F}_{\text{int}}(B_t, t) = \int_{\partial B_t} \tau \, dA. \tag{B.10}$$

There are some general principles (due to Cauchy) that lead to a specific form for $\tau$. For the sake of brevity we will not go into this here, but merely make the assumption that the stress tensor $\tau$ has the form

$$\tau = Tn \tag{B.11}$$

$$= \left( T^1 \cdot n, \ T^2 \cdot n, \ T^3 \cdot n \right), \tag{B.12}$$

where $n$ is the unit normal to the surface $\partial B_t$ and $T$ is a $3 \times 3$ matrix with rows $T^1, T^2$, and $T^3$. The matrix $T$ is also called the *stress tensor*. It depends on $(x, y, z, t)$ and the nature of the material for the continuum. In the text, we will be more specific on how to further specify $T$ to model a particular material, but for now we work with a general $T$.

By using assumption (B.11) on the form of the stress tensor, we arrive at the expression for the internal force exerted on $B_t$ by the surrounding material:

$$\mathcal{F}_{\text{int}}(B_t, t) = \int_{\partial B_t} Tn \, dA$$

$$= \left( \int_{\partial B_t} T^1 \cdot n \, dA, \ \int_{\partial B_t} T^2 \cdot n \, dA, \ \int_{\partial B_t} T^3 \cdot n \, dA \right)$$

$$= \left( \int_{B_t} \text{div}(T^1) \, dV, \ \int_{B_t} \text{div}(T^2) \, dV, \ \int_{B_t} \text{div}(T^3) \, dV \right)$$

$$= \int_{B_t} \mathrm{div}(T)\, dV.$$

In the above, we have used Gauss's divergence theorem to rewrite the boundary integrals as volume integrals. Also, the last line involves the natural extension of the divergence operator to tensor fields.

Next we discuss the torque $\mathcal{T}$ and, just as we did for $\mathcal{F}$, we split this into two parts:

$$\mathcal{T} = \mathcal{T}_{\mathrm{ext}} + \mathcal{T}_{\mathrm{int}}. \tag{B.13}$$

The part $\mathcal{T}_{\mathrm{ext}}$ is the torque arising from the external force $\mathcal{F}_{\mathrm{ext}}$, and one can easily reason that it is given by

$$\mathcal{T}_{\mathrm{ext}}(B_t, t) = \int_{B_t} [\vec{x} \times \rho F]\, dV, \tag{B.14}$$

where again we are using the notation $\vec{x} \equiv (x, y, z)$. In terms of infinitesimals, the infinitesimal force $\rho F\, dV$ applied at $(x, y, z)$ creates a turning moment about the axis through the origin in the direction of $\vec{x} \times \rho F\, dV$, and this moment has magnitude $|\vec{x} \times \rho F\, dV|$. Then equation (B.14) just expresses the fact that $\mathcal{T}_{\mathrm{ext}}(B_t, t)$ is the sum (actually integral) of all these infinitesimal torques.

In a similar vein, the torque $\mathcal{T}_{\mathrm{int}}$ is reasoned to be given by the surface integral

$$\mathcal{T}_{\mathrm{int}}(B_t, t) = \int_{\partial B_t} [\vec{x} \times (Tn)]\, dA. \tag{B.15}$$

As before, it will be convenient to convert this into a volume integral. For this, note that

$$\vec{x} \times (Tn) = \left( yT^3 \cdot n - zT^2 \cdot n,\ zT^1 \cdot n - xT^3 \cdot n,\ xT^2 \cdot n - yT^1 \cdot n \right)$$

$$= Sn,$$

where $S$ is the matrix with rows

$$S^1 = yT^3 - zT^2,$$

$$S^2 = zT^1 - xT^3, \tag{B.16}$$

$$S^3 = xT^2 - yT^1,$$

(exercise). An easy calculation of div($S$), gives

$$\text{div}(S) = u + \vec{x} \times \text{div}(T), \tag{B.17}$$

where $u$ is the vector field

$$u \equiv (T^{32} - T^{23}, T^{13} - T^{31}, T^{21} - T^{12}) \tag{B.18}$$

(exercise). Finally, a use of Gauss's divergence theorem allows us to express (B.15) in terms of a volume integral:

$$\begin{aligned}
\mathcal{T}_{\text{int}}(B_t, t) &= \int_{\partial B_t} [\vec{x} \times (Tn)] \, dA \\
&= \int_{\partial B_t} Sn \, dA \\
&= \int_{B_t} \text{div}(S) \, dV \\
&= \int_{B_t} \left[ u + \vec{x} \times \text{div}(T) \right] dV.
\end{aligned} \tag{B.19}$$

## B.1.2.  The Eulerian Equations of Motion

We can now fully express the three basic equations — conservation of mass, linear momentum equation, and angular momentum equation — based on the above model for the forces and torques. Doing so yields three integral equations. and from these we would like to derive corresponding PDEs. The key tool for this derivation is the following theorem.

**Theorem B.1 (Transport Theorem)**    *Suppose $f = f(x, y, z, t)$ is a time dependent scalar field on $\Omega$. Then*

$$\frac{d}{dt} \int_{B_t} f \, dV = \int_{B_t} \left[ \frac{\partial f}{\partial t} + \nabla_v f + \text{div}(v) f \right] dV. \tag{B.20}$$

**Proof**    To simplify the notation in the proof, we will use the notation

$$\vec{x} \equiv (x, y, z).$$

The proof is a straightforward application of the definition for volume integrals, differentiation under the integral sign, the chain rule, and a rule for differentiating a determinant. Thus as a first step we get

$$\frac{d}{dt}\int_{B_t} f\, dV = \frac{d}{dt}\int_B f\big(\phi(\vec{x},t),t\big)\det\big(\phi'(\vec{x},t)\big)dV$$

$$= \int_B \frac{d}{dt}\big[f\big(\phi(\vec{x},t),\ t\big)\det\big(\phi'(\vec{x},t)\big)\big]dV. \qquad (B.21)$$

Next, we use the product rule to differentiate the product in equation (B.21). The derivative of the first function is (by the chain rule):

$$\frac{d}{dt}f\big(\phi(\vec{x},t),\ t\big)$$

$$= \frac{\partial f}{\partial t}\big(\phi(\vec{x},t),\ t\big) + \nabla f\big(\phi(\vec{x},t),t\big)\cdot\frac{\partial\phi}{\partial t}(\vec{x},t) \qquad (B.22)$$

$$= \frac{\partial f}{\partial t}\big(\phi(\vec{x},t),\ t\big) + \nabla f\big(\phi(\vec{x},t),\ t\big)\cdot v\big(\phi(\vec{x},t),\ t\big) \qquad (B.23)$$

$$= \big[\frac{\partial f}{\partial t} + \nabla_v f\big]\big(\phi(\vec{x},t),\ t\big). \qquad (B.24)$$

In the above, the step from (B.22) to (B.23) used the basic definition of the flow, equation (B.3). The derivative of the determinant term in (B.21) is calculated to be (exercise)

$$\frac{d}{dt}\det\big(\phi'(\vec{x},t)\big) = \mathrm{div}(v)\big(\phi(\vec{x},t),t\big)\det\big(\phi'(\vec{x},t)\big). \qquad (B.25)$$

Using these results now in Equation (B.21), we see that the last expression there is

$$\int_B \Big[\frac{\partial f}{\partial t} + \nabla_v f + \mathrm{div}(v)f\Big]\big(\phi(\vec{x},t),t\big)\det(\phi'(\vec{x},t))dV$$

$$= \int_{B_t}\Big[\frac{\partial f}{\partial t} + \nabla_v f + \mathrm{div}(v)f\Big]dV. \qquad (B.26)$$

This completes the proof of the theorem.

The transport theorem was stated for a scalar field $f$, but it is also true for vector fields. Specifically,

**Corollary B.1**   *If $w = w(x, y, z, t)$ is a time-dependent vector field on $\Omega$, then*

$$\frac{d}{dt} \int_{B_t} w \, dV = \int_{B_t} \left[ \frac{\partial w}{\partial t} + \nabla_v w + w \, \text{div}(v) \right] dV. \qquad \text{(B.27)}$$

We now apply these transport theorems to the three laws for the motion of the continuum. First, the conservation of mass can be written

$$0 = \frac{d}{dt} \int_{B_t} \rho \, dV = \int_{B_t} \left[ \frac{\partial \rho}{\partial t} + \nabla_v \rho + \rho \, \text{div}(v) \right] dV.$$

Since this holds for each portion $B$ of the continuum, one can argue that the integrand must necessarily be zero, i.e.,

$$\frac{\partial \rho}{\partial t} + \nabla_v \rho + \rho \, \text{div}(v) = 0,$$

on $\Omega$. This is the PDE version of the conservation of mass law. Using this, the above corollary, and some differentiation identities for $\nabla_v$ and div from previous exercises (see Chapter 1), we see that the rate of change of linear momentum can be written as

$$\frac{d}{dt} \int_{B_t} \rho v \, dV = \int_{B_t} \left\{ \left[ \frac{\partial \rho}{\partial t} + \nabla_v \rho + \rho \, \text{div}(v) \right] v + \rho \left[ \frac{\partial v}{\partial t} + \nabla_v v \right] \right\} dV$$

$$= \int_{B_t} \rho \left[ \frac{\partial v}{\partial t} + \nabla_v v \right] dV.$$

In a similar way, we get the following for the rate of change of the angular momentum:

$$\frac{d}{dt} \int_{B_t} \vec{x} \times \rho v \, dV = \int_{B_t} \vec{x} \times \left\{ \left[ \frac{\partial \rho}{\partial t} + \nabla_v \rho + \rho \, \text{div}(v) \right] v + \rho \left[ \frac{\partial v}{\partial t} + \nabla_v v \right] \right\} dV$$

$$= \int_{B_t} \vec{x} \times \rho \left[ \frac{\partial v}{\partial t} + \nabla_v v \right] dV.$$

From these two calculations, we see that the linear and angular momentum equations (in integral form) can be written as:

$$\int_{B_t} \rho \left[ \frac{\partial v}{\partial t} + \nabla_v v \right] dV = \int_{B_t} \left[ \rho F + \text{div}(T) \right] dV, \qquad \text{(B.28)}$$

$$\int_{B_t} \vec{x} \times \rho \left[ \frac{\partial v}{\partial t} + \nabla_v v \right] dV = \int_{B_t} \left[ \vec{x} \times [\rho F + \mathrm{div}(T)] + u \right] dV. \quad \text{(B.29)}$$

In these equations, $B$ is an arbitrary reference piece of the continuum, and $t$ is any time. Because of this arbitrariness, one can argue that the above equations hold if and only if the integrands on each side are equal; i.e.,

$$\rho \left[ \frac{\partial v}{\partial t} + \nabla_v v \right] = \rho F + \mathrm{div}(T), \quad \text{(B.30)}$$

$$\vec{x} \times \rho \left[ \frac{\partial v}{\partial t} + \nabla_v v \right] = \vec{x} \times [\rho F + \mathrm{div}(T)] + u. \quad \text{(B.31)}$$

We see immediately from (B.30) that equation (B.31) reduces to

$$u = 0,$$

which from the definition of $u$ in (B.18) merely expresses the requirement that the stress tensor be symmetric:

$$T^{ij} = T^{ji} \qquad \text{for every } i, j. \quad \text{(B.32)}$$

In the text, we will look at some of the details of how the stress tensor $T$ is chosen to model certain materials. We will specify $T$ in all cases so as to satisfy the symmetry requirement (B.32). We will also see that $T$ depends on the velocity $v$ and other functions, like the pressure $p$, as well. Consequently, the main PDEs governing the motion of the continuum are following the two equations.

**The (Eulerian) Continuum Equations:**

$$\frac{\partial \rho}{\partial t} + \nabla_v \rho + \rho \, \mathrm{div}(v) = 0, \quad \text{(B.33)}$$

$$\rho \left[ \frac{\partial v}{\partial t} + \nabla_v v \right] = \rho F + \mathrm{div}(T). \quad \text{(B.34)}$$

This is a nonlinear system of four PDEs involving four unknown functions $\rho, v^1, v^2$, and $v^3$ explicitly. As we just mentioned, $T$ depends implicitly on $v^1, v^2, v^3$, and on other functions as well.

**Exercises B.1**

1. Prove the identities (B.16)–(B.18).

2. Prove the identity (B.25) concerning the time derivative of the determinant of $\phi'(\vec{x}, t)$.

3. In this exercise we will derive the heat equation *with convection* from the first law of thermodynamics. There is an alternative way of doing this that is conceptually simpler, and this will be discussed as well.

   The general setup is for a continuum (like a gas or fluid)) circulating in the region $\Omega$, with velocity vector field $v$ and corresponding flow $\phi$. As in the derivation of the heat equation without convection, $u$ is the temperature function, giving the temperature at the various points in $\Omega$ at various times. The mass density is denoted by $\rho$, and the density for the internal sources/sinks is $f$. As in the discussion for the continuum equations, we let $F$ denote the density for the external forces and let $T$ be the stress tensor used in describing the internal forces.

   We concentrate attention on a subregion $B \subseteq \Omega$ as it moves and deforms under the flow over time into the subregion $B_t = \phi_t(B)$ at time $t$. Viewing $B$ as the reference region that we follow over time, the quantities of interest that pertain to it at time $t$ are as follows:

**Kinetic Energy:**

$$K = \int_{B_t} \rho \frac{|v|^2}{2} \, dV. \tag{B.35}$$

Letting $e$ denote the density for the internal energy, the total internal energy of $B$ at time $t$ is

**Internal Energy:**

$$E = \int_{B_t} \rho e \, dV. \tag{B.36}$$

The other quantities we need in the derivation are

**Mechanical Power (or Work Done):**

$$W = \int_{B_t} \rho F \cdot v \, dV + \int_{\partial B_t} (Tn) \cdot v \, dA. \tag{B.37}$$

**Nonmechanical Power (Heat Supplied):**

$$W = \int_{B_t} \rho f \, dV + \int_{\partial B_t} \kappa \nabla u \cdot n \, dA. \tag{B.38}$$

The first law of thermodynamics postulates that the time rate of change of the total energy inside $B_t$ is equal to the work done plus the heat supplied:

**First Law of Thermodynamics:**

$$\frac{d}{dt}(E + K) = W + Q. \tag{B.39}$$

Based upon this, your exercise is to derive the PDE version of this law as follows:

(a) Show that $(Tn) \cdot v = (v^*T) \cdot n$, where $*$ denotes matrix transpose.

(b) Use the transport theorem together with the continuum equations (B.33)–(B.34) to show that the first law of thermodynamics implies that

$$\rho\left(\frac{\partial e}{\partial t} + \nabla_v e\right) = \kappa \nabla^2 u + \rho f + \text{tr}(v'T). \tag{B.40}$$

Here for simplicity we have assumed that $\kappa$ is a constant.

(c) Assume that the internal energy density $e$ is proportional to the temperature: $e = cu$, where $c$ is a constant. Suppose also that $T = -pI$ and $\text{div}(v) = 0$. Show that equation (B.40) reduces to

**Heat Equation with Convection:**

$$\frac{\partial u}{\partial t} + \nabla_v u = \alpha^2 \nabla^2 u + c^{-1} f, \tag{B.41}$$

where $\alpha^2 = \kappa/(\rho c)$.

An alternative, perhaps naive, way of deriving the heat equation with convection is the following. Our derivation of the heat equation *without convection* in Chapter 2 was based upon the relation

$$\frac{d}{dt} \int_B c\rho u \, dV = \int_{\partial B} \kappa \nabla u \cdot n \, dA + \int_B f \, dV, \tag{B.42}$$

for each portion $B \subseteq \Omega$. This equates the rate of change of heat inside $B$ with that which flows in/out across its boundary plus that which is added/subtracted by internal sources/sinks. This relation does not take into account that the motion of the continuum inside $\Omega$ will naturally carry (convect) the heat/cold from one part of $D$ to another. Thus if the initial reference subregion is $B$, then its position at time $t$ is $B_t$, and over the next instant it moves to position $B_{t+dt}$. Thus it seem natural to replace the above relation with

$$\frac{d}{dt} \int_{B_t} c\rho u \, dV = \int_{\partial B_t} \kappa \nabla u \cdot n \, dA + \int_{B_t} f \, dV \tag{B.43}$$

Using this and the transport theorem,

(d) Derive the heat equation with convection.

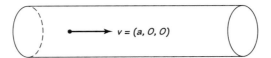

Figure B.5.    A heated fluid in a cylinder.

As a special case of the heat equation with convection, consider the situation where $\Omega$ is the cylindrical region shown in Figure B.5, and the continuum is an incompressible fluid flowing in $\Omega$ with velocity vector field

$$v = (a, 0, 0),$$

where $a$ is a nonnegative constant. Assume that there are no internal sources/sinks of heat ($f = 0$), the lateral surface is insulated, the left and right ends are held at 0 degrees, and that at time $t = 0$ the fluid in $\Omega$ is heated up with temperature distribution: $\phi(x) = x - x^2$.

(e) Show that for temperature distributions $u = u(x, t)$ that depend only on $x$ and $t$, the heat equation with convection (B.41) reduces to

$$u_t + au_x = \alpha^2 u_{xx}.$$

(f) Use separation of variables to find the series solution. What is the eventual temperature distribution?

4. (*Transport theorem for curves*) Suppose $C \subseteq B$ is a curve in $B$ parametrized by $\alpha : I \to B$, $C = \alpha(I)$, with component expression $\alpha(s) = (\alpha^1(s), \alpha^2(s), \alpha^3(s))$. Then $C$ is transported by the flow $\phi_t$ generated by the velocity vector field $v$. At time $t$ the curve $C_t \equiv \phi_t(C)$ is naturally parametrized by

$$\alpha_t(s) \equiv \phi_t(\alpha(s)) = \phi(\alpha(s), t).$$

Assuming the appropriate differentiablity of the curve $C$ and the flow map, derive the following versions of the transport theorem for curves:

(a) For a time dependent vector field $w$ on $B$,

$$\frac{d}{dt}\int_{C_t} w \cdot T \, dL = \int_{C_t}\left(\frac{\partial w}{\partial t} + \nabla_v w + (v')^* w\right) \cdot T \, dL. \qquad \text{(B.44)}$$

Here $*$ denotes matrix transpose.

(b) Prove the identity $(w')^* w = \nabla(\frac{|w|^2}{2})$.

(c) Show that if $C$ is a closed curve, then

$$\frac{d}{dt}\int_{C_t} v \cdot T \, dL = \int_{C_t}\left(\frac{\partial v}{\partial t} + \nabla_v v\right)\cdot T \, dL. \tag{B.45}$$

## B.2.   The Lagrangian Description

While the Eulerian description is often most convenient for the study of certain continua, like fluids, the Lagrangian description is in some cases, like for elastic solids, the preferred viewpoint. The two descriptions are (for 3 dimensional continua) equivalent ways of describing the motion.

To compare and clarify the relations between the two, we will now let $\tilde{\rho}$, $\tilde{F}$, $\tilde{v}$, and $\tilde{T}$ denote respectively the *Eulerian* mass and force densities, velocity vector field, and the stress tensor. The corresponding *Lagrangian* quantities will be denoted by $\rho$, $F$, $v$, and $T$. The former are fields defined at the points in the region $\Omega$ in which the continuum is moving, while the latter will be fields defined on some reference region, like the initial position region or (for us) the parameter domain. Specifically, let $B = B_0 = \alpha_0(U)$ be some smooth, regular region in $\Omega$ that we wish to observe over time. The flow $\phi_t : B \to \Omega$ generated by the Eulerian vector field $\tilde{v}$ moves and deforms $B_0$ over time $t$ into $B_t = \phi_t(B_0) = \phi_t \circ \alpha_0(U)$. We let (as before)

$$\alpha_t = \phi_t \circ \alpha_0,$$

and also use the notation $\alpha(s, t) = \alpha_t(s)$, with $s = (s_1, s_2, s_3)$ being a point in the parameter domain $U$. Often we will suppress the $t$ from the notation and just write $\alpha$ for $\alpha_t$. Then $U$ is viewed as a reference set of points for the motion of the body, which at time $t$ has position and shape given by $B_t = \alpha_t(U)$. All the quantities that in the Eulerian description are defined on the body in its present position $B_t$ will be referred or transferred back to corresponding quantities defined on $U$. Thus while $\tilde{F}$ is vector filed on $B_t$, its Lagrangian counterpart $F$ is a vector field on $U$ (in differential geometry $F$ is known as a vector field *along $B_t$* or *along the map $\alpha_t$*). Once this is done, the Lagrangian version of the continuum equations will be a set of PDEs involving $\alpha$ as the principal unknown.

For the Lagrangian formulation we will use the notation developed in the vector analysis discussion, except that now all the geometric quantities associated with the moving region depend on $t$. Often we will suppress the time dependence from the notation. Thus $\alpha' = \alpha_t'$ denotes the derivative (Jacobian matrix) of the map $\alpha_t : U \to \mathbb{R}^3$. We let $G = \{G_{ij}\}$ denote the matrix with entries

$$G_{ij} = e_i \cdot e_j,$$

where $e_1 = \partial \alpha / \partial s_1$, $e_2 = \partial \alpha / \partial s_2$, and $e_3 = \partial \alpha / \partial s_3$. We assume that the flow map satisfies $\det \phi'_t(x, y, z) > 0$ for all $t$ and all $(x, y, z)$ in the region $B_0$. From this and the other definitions, it follows that

$$\det G^{1/2} = |\det \alpha'| = \det \alpha'$$

(exercise).

**Definition B.1 (The Lagrangian Quantities)**   We define $\rho$, $F$, $v$, $T$ in terms of their Eulerian counterparts as follows:

$$\rho(s, t) = \tilde{\rho}(\alpha_t(s), t),$$

$$F(x, y, z, t) = \widetilde{F}(\alpha_t(s), t),$$

$$v(s, t) = \tilde{v}(\alpha_t(s), t),$$

$$T(s, t) = \widetilde{T}(\alpha_t(s), t).$$

The above prescription applied to $\tilde{v}$ gives the expected result for the Lagrangian velocity $v$:

$$v(s, t) = \tilde{v}(\alpha_t(s), t) = \tilde{v}(\phi_t(\alpha_0(s)), t)$$

$$= \frac{\partial \phi}{\partial t}(\alpha_0(s), t)) = \frac{\partial \alpha}{\partial t}(s, t).$$

We will use $\partial \alpha / \partial t$ as the preferred form for the Lagrangian velocity.

Based on these definitions, we now have the task before us of translating the Eulerian version of the continuum equations into the Lagrangian version. This is easiest to do if we work with the conservation of mass and linear momentum equations in integral form. Thus from the above definitions we get

$$\int_{B_t} \tilde{\rho} \, dV = \int_U \rho \det G^{1/2} \, dV,$$

$$\int_{B_t} \tilde{\rho} \tilde{v} \, dV = \int_U \rho \frac{\partial \alpha}{\partial t} \det G^{1/2} \, dV.$$

These give the Lagrangian expressions for the total mass and linear momentum in terms of integrals over the reference body $U$. (Note that since we will view $U = \{id\}$ as parametrized by the identity map $id : U \rightarrow U$, this latter integral is just the Lebesgue (triple) integral over $U$, as a set.)

Thus the equations for the conservation of mass and linear momentum (in Lagrangian integral form) are

$$\frac{d}{dt}\int_U \rho \det G^{1/2}\, dV = 0,$$

$$\frac{d}{dt}\int_U \rho \frac{\partial \alpha}{\partial t}\det G^{1/2}\, dV = \int_U \rho F \det G^{1/2}\, dV + \int_U \operatorname{div}(\widetilde{T}) \circ \alpha \det G^{1/2} dV.$$

Since each equation must also hold for any subdomain of $U$, we see that the conservation of mass equation leads to

$$\frac{\partial}{\partial t}[\rho \det G^{1/2}] = 0,$$

on $U$ for all $t$. Using this in the linear momentum equation leads to:

$$\rho \frac{\partial^2 \alpha}{\partial t^2} = \rho F + \operatorname{div}(\widetilde{T}) \circ \alpha,$$

on $U$ for all $t$. Note that this equation still has the Eulerian stress tensor $\widetilde{T}$ in it, and our last task is to rewrite it in terms of $T$. While it is true that $\widetilde{T} \circ \alpha = T$, a similar thing does *not* hold for $\operatorname{div}(\widetilde{T}) \circ \alpha$. We leave it as an exercise to prove that $\operatorname{div}(\widetilde{T}) \circ \alpha = \operatorname{div}_B(T)$, where $\operatorname{div}_B(T)$ is defined as follows:

**Definition B.2**   The divergence of $T$ *along* $B = \{\alpha\}$ is the vector field

$$\operatorname{div}_B(T) = \det G^{-1/2}\frac{\partial}{\partial s_i}(\det G^{1/2}T e^i) \tag{B.46}$$

along $B$. Here there is implied summation on the repeated index $i$, and $e^i = G^{ij}e_j$ is the $i$th reciprocal basis vector (again with implied summation on $j$ this time).

Those familiar with differential geometry will recognize that the expression for $\operatorname{div}_B(T)$ is just the divergence of $\widetilde{T}$ in the coordinate chart $(B, \alpha^{-1})$.

Based on the above development, we arrive at

**The (Lagrangian) Continuum Equations:**

$$\rho \det G^{1/2} = \rho_0 \det G_0^{1/2}, \tag{B.47}$$

$$\rho \frac{\partial^2 \alpha}{\partial t^2} = \rho F + \operatorname{div}_B(T). \tag{B.48}$$

Here, for convenience, we have introduced $\rho_0(s) \equiv \rho(s, 0)$ and $G_0(s) \equiv G(s, 0)$. The linear momentum equation is a second-order PDE for $\alpha$ that also involves the unknown $\rho$. Since the initial density $\rho_0$ and initial placement $\alpha_0$ are considered as given, we can use the conservation of mass equation to eliminate $\rho$ from the linear momentum equation. Also bear in mind that the stress tensor $T$ involves $\alpha$ in some way that expresses the constitution of the particular continuum being modeled. This is further discussed in the text for elastic materials.

## B.3.   Two-Dimensional Materials

Of the several models for treating the continuum mechanics of two-dimensional materials, we employ here the one that views such a continuum as a surface (two dimensional manifold) moving and deforming in $\mathbb{R}^3$ over time. The Lagrangian viewpoint is the most convenient for this model, because to formulate the theory from the Eulerian viewpoint requires the theory of distributions (generalized functions).

We let $S_t \subseteq \mathbb{R}^3$ denote the surface representing the continuum at time $t$ and assume that $S_0$ is the initial postion of the continuum. To keep the formulation simple, we assume that for each $t$ in an interval $I = [0, a]$, the surface $S_t$ is smooth and regular: $S_t = \{\alpha_t\}$, with parameter map

$$\alpha_t : U \to \mathbb{R}^3,$$

where $U \subseteq \mathbb{R}^2$ is the parameter domain. We identify $S_t$ with the image of the parameter map $S_t = \alpha_t(U)$. See Figure B.6. The points $s$ in the parameter domain $U$ are denoted by $s = (s_1, s_2)$. We will also use the notation $\alpha(s, t) = \alpha_t(s)$, and we assume that $\alpha$ depends smoothly on $t$ as well.

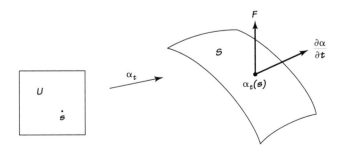

Figure B.6.   The motion of the two-dimensional continuum is modeled by surface parameter maps $\alpha_t : U \to \mathbb{R}^3$. Also shown are the external force density $F(s, t)$ and velocity $\partial\alpha(s, t)/\partial t$ at the point $\alpha(s, t)$ on the surface.

The basic differential geometry of surfaces from Appendix A will play a part in the equations of motion here. Bear in mind that the notation in Appendix A is for a single surface, whereas here we have a whole family of surfaces $\{S_t\}_{t\in I}$, one for each time $t$. We will use the same notation here, with the understanding that everything depends implicitly on the time.

The basic (Lagrangian) quantities associated with the motion of $S$ are

$$\rho(s, t) = \left\{ \begin{array}{c} \text{mass density per unit area of the surface} \\ S_t \text{ at the point } \alpha(s, t) \text{ at time } t \end{array} \right\},$$

$$\frac{\partial \alpha}{\partial t}(s, t) = \{ \text{velocity of the point } \alpha(s, t), \text{ on the surface } S_t \text{ at time } t \},$$

$$F(s, t) = \left\{ \begin{array}{c} \text{external force per unit area acting} \\ \text{on } S_t \text{ at the point } \alpha(s, t) \text{ at time } t \end{array} \right\}.$$

The external force and velocity are illustrated in Figure B.6.

The specification of the stress tensor for $S_t$ requires a little explanation. We first need a definition of the outward directed normal to the boundary of a given surface:

**Definition B.3**    For a smooth regular surface $S = \{\alpha\}$, the *outward-directed normal* to the boundary $\partial S = \{\gamma_1, \gamma_2, \gamma_3, \gamma_4\}$ is defined as follows. Recall (Appendix A) that we can write $\gamma_i = \alpha \circ \beta_i$, with $\beta_i : I_i \to U$, for $i = 1, 2, 3, 4$, parametrizing the boundary of $U$. Then $\nu : \cup_{i=1}^4 I_i \to \mathbb{R}^3$ is defined by

$$\nu(p) = \frac{\gamma_i'(p)}{|\gamma_i'(p)|} \times n(\beta_i(p)),$$

for $p \in I_i, i = 1, 2, 3, 4$, (see Figure B.7). This concept extends naturally to piecewise smooth regular surfaces.

We also need a definition of the stress tensor for the surface (membrane):

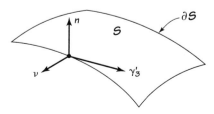

Figure B.7.    The outward-directed unit normal $\nu$ to the boundary of the surface $S$

**Definition B.4 (Stress Tensor for a 2-D Continuum)**   The *stress tensor* for $S$ is a $3 \times 3$ matrix-valued function $T$ of the points $s$ in the parameter domain

$$T(s) = \begin{bmatrix} T^{11}(s) & T^{12}(s) & T^{13}(s) \\ T^{21}(s) & T^{22}(s) & T^{23}(s) \\ T^{31}(s) & T^{32}(s) & T^{33}(s) \end{bmatrix}.$$

The stress tensor $T$ also depends on the deformation $\alpha$ as well, but we suppress this from the notation.

Now consider a (perhaps small) portion $S_t'$ of the total surface $S_t$, with, of course, $S_t' = \alpha_t(U')$, where $U' \subseteq U$. As shown in Figure B.8, there are forces acting along the boundary $C_t' = \partial S_t'$ of this portion $S_t'$ due to the material *outside* of $S_t'$ acting on it. The density for these forces of stress are modeled naturally by

$$T(\beta_i(p))\nu(p) = \left\{ \begin{array}{c} \text{stress force per unit length acting on the} \\ \text{boundary curve } C_t' \text{ to } S_t' \text{ at the point } \alpha(\beta_i(p)) \end{array} \right\},$$

where $p \in I_i$. Integrating this density around the boundary gives the total force due to stresses acting across the boundary:

$$\int_{\partial S_t'} T\nu \, dL = \left\{ \begin{array}{c} \text{total force due to stress acting on} \\ S_t' \text{ across its boundary } \partial S_t' \end{array} \right\}.$$

We are now able to formulate the laws of motion for the membrane from the Lagrangian point of view. These laws are naturally stated in integral form first, and then the corresponding PDE versions will result.

We follow the motion of a portion $S_t'$ of the moving membrane $S_t$ and require that its mass remain constant and the changes in its linear and angular momenta equal the forces and torques acting on it. This is formulated by the following equations:

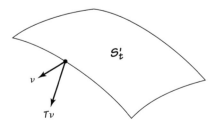

Figure B.8.   The stressing forces $T\nu$ acting on the portion $S_t'$ at points along its boundary.

**Conservation of Mass:**

$$\frac{d}{dt} \int_{S_t'} \rho \, dA = 0.$$  (B.49)

**Linear Momentum Equation:**

$$\frac{d}{dt} \int_{S_t'} \rho \frac{\partial \alpha}{\partial t} dA = \int_{S_t'} \rho F \, dA + \int_{\partial S_t'} Tv \, dL.$$  (B.50)

**Angular Momentum Equation:**

$$\frac{d}{dt} \int_{S_t'} [\alpha \times \rho \frac{\partial \alpha}{\partial t}] dA = \int_{S_t'} [\alpha \times \rho F] dA + \int_{\partial S_t'} [\alpha \times Tv] dL.$$  (B.51)

These are the appropriate laws describing how all the particles comprising the piece $S_t'$ of the continuum move under the applied forces and torques. Since these laws apply to any portion $S_t'$ of the total continuum, we can get a nonintegral version of these laws of motion just as we did for the 3-d case. However, to do so, we need the analogue of Gauss's divergence theorem:

**Lemma B.1 (Divergence Theorem for a Surface)**    *For a vector field* $w$ :
$U \to \mathbb{R}^3$,

$$\int_{\partial S} w \cdot v \, dL = \int_S \text{div}_S(w) \, dA,$$

*where*

$$\text{div}_S(w) = \det G^{-1/2} \left[ \frac{\partial}{\partial s_1} (w \cdot (e_2 \times n)) - \frac{\partial}{\partial s_2} (w \cdot (e_1 \times n)) \right]$$

$$= \det G^{-1/2} \frac{\partial}{\partial s_j} \left( \det G^{1/2} \, w \cdot e^j \right)$$

$$= \frac{\partial w}{\partial s_j} \cdot e^j + \mu \det G^{1/2} \, w \cdot n.$$

*In the second and third expressions for* $\text{div}_S(w)$ *there is implied summation on the index* $j$. *In the third expression,* $\mu$ *denotes the mean curvature for the surface.*

*Note*: The operator $\text{div}_S$ is called the *surface divergence* operator, and the first expression above for $\text{div}_S(w)$ is the definition, while the other two expressions will be shown to be equivalent forms for it. The second expression for $\mathrm{div}_S(w)$ is the customary one, while the others are often useful in calculations.

**Proof**   The proof uses the elementary facts that for three vectors $a, b, c \in \mathbb{R}^3$,

$$a \cdot (b \times c) = \det(a, b, c),$$

and the cyclical permutation of $a, b, c$ does not change the value of the determinant. In addition, we need the fact that if $B$ is a $3 \times 2$ matrix and $v \in \mathbb{R}^2$, then

$$a \cdot Bv = B^* a \cdot v,$$

with $*$ denoting the matrix transpose. Using these observations, we get

$$
\begin{aligned}
\int_{\partial S} w \cdot v \, dL &= \sum_{i=1}^{4} \int_{I_i} w(\beta_i(p)) \cdot \frac{\gamma_i'(p) \times n(\beta_i(p))}{|\gamma_i'(p))|} \, |\gamma_i'(p)| \, dp \\
&= \sum_{i=1}^{4} \int_{I_i} [n(\beta_i(p)) \times w(\beta_i(p))] \cdot \gamma_i'(p) \, dp \\
&= \sum_{i=1}^{4} \int_{I_i} [n(\beta_i(p)) \times w(\beta_i(p))] \cdot [\alpha'(\beta_i(p))\beta_i'(p)] \, dp \\
&= \sum_{i=1}^{4} \int_{I_i} \left( \alpha'(\beta_i(p))^* [n(\beta_i(p)) \times w(\beta_i(p))] \right) \cdot \beta_i'(p) \, dp \\
&= \sum_{i=1}^{4} \int_{I_i} E(\beta_i(p)) \cdot \beta_i'(p) \, dp \\
&= \int_{U} \left( \frac{\partial E^2}{\partial s_1}(s) - \frac{\partial E^1}{\partial s_2}(s) \right) ds_1 ds_2.
\end{aligned}
$$

In the last line we have used Green's theorem (applied to the parameter domain $U$ and its boundary $\partial U$), and for the sake of notation we have introduced the vector field $E$ on $U$ defined by

$$E(s) = (E^1(s), E^2(s))$$

$$\equiv \alpha'(s)^*[n(s) \times w(s)]$$

$$= \Big( e_1(s) \cdot [n(s) \times w(s)], \ e_2(s) \cdot [n(s) \times w(s)] \Big)$$

$$= \Big( w(s) \cdot [e_1(s) \times n(s)], \ w(s) \cdot [e_2(s) \times n(s)] \Big).$$

Thus we see from the definition of $\text{div}_S(w)$ that

$$= \int_U \left( \frac{\partial E^2}{\partial s_1}(s) - \frac{\partial E^1}{\partial s_2}(s) \right) ds_1 ds_2$$

$$= \int_U \det G^{-1/2}(s) \left( \frac{\partial E^2}{\partial s_1}(s) - \frac{\partial E^1}{\partial s_2}(s) \right) \det G^{1/2}(s) \, ds_1 ds_2$$

$$= \int_S \text{div}_S(w) \, dA.$$

This completes the first part of the theorem. The fact that $\text{div}_S(w)$ can alternatively be expressed in two other ways comes from the following observations. First, it is easy to show that

$$e_1 \times n = -\det G^{1/2} \, e^2,$$

$$e_2 \times n = \det G^{1/2} \, e^1$$

(exercise). This gives the first of the two alternative expressions for $\text{div}_S(w)$. The last expression for $\text{div}_S(w)$ follows easily from the first expression and the identity

$$\frac{\partial}{\partial s_1}(e_2 \times n) - \frac{\partial}{\partial s_1}(e_2 \times n) = \mu \det G^{1/2} . \tag{B.52}$$

Verification of this is left as an exercise.

We now can easily derive the PDE version of the equations of motion. As before, for convenience, we introduce $\rho_0(s) \equiv \rho(s, 0)$ and $G_0(s) \equiv G(s, 0)$.

**Theorem B.2 (Laws of Motion for a 2-D Continuum)**    *The above integral version of the laws of motion are equivalent to the following:*

**Conservation of Mass:**

$$\rho \det G^{1/2} = \rho_0 \det G_0^{1/2}.$$

**Linear Momentum Equation:**

$$\rho \frac{\partial^2 \alpha}{\partial t^2} = \rho F + \mathrm{div}_S(T).$$

**Angular Momentum Equation:**

$$T e^1 \cdot e^2 = T e^2 \cdot e^1,$$

$$T e^i \cdot n = 0, \qquad \text{for } i = 1, 2.$$

*These equations hold on U for all $t \in I$. In the angular momentum equation, the second condition means that T maps vectors in the tangent plane into vectors in the tangent plane.*

**Proof**    The conservation of mass equation for an arbitrary portion $S_t' = \alpha_t(U')$ can be written as

$$0 = \frac{d}{dt} \int_{S_t'} \rho \, dA$$

$$= \frac{d}{dt} \int_{U'} \rho(s, t) \det G^{1/2}(s, t) ds_1 ds_2$$

$$= \int_{U'} \frac{d}{dt} [\rho(s, t) \det G^{1/2}(s, t)] ds_1 ds_2.$$

Since $U'$ is arbitrary (can be shrunk to zero), the integrand in the last integral must be identically zero. This says that $\rho \det G^{1/2}$ does not depend on the time, and so we get the conservation of mass law as stated.

Using the conservation of mass law (i.e., that $\rho \det G^{1/2}$ does not depend on the time), it is easy to see that

$$\frac{d}{dt} \int_{S_t'} \rho \frac{\partial \alpha}{\partial t} \, dA = \int_{S_t'} \rho \frac{\partial^2 \alpha}{\partial t^2} \, dA.$$

Also, we can convert the line integral for the stress force into a surface integral

$$\int_{\partial S'_t} T\nu \, dL = \int_{S'_t} \operatorname{div}_S(T) \, dA,$$

using the divergence theorem for surfaces. Thus the nonintegral version of the linear momentum equation is apparent. The derivation of the identities for the angular momentum equation is left as an exercise.

## B.4.   One-Dimensional Materials

All the development in the last section for two-dimensional materials carries over in a natural way to one-dimensional materials. Thus we will be more brief with the explanations here.

Let $C_t \subseteq \mathbb{R}^3$ denote the curve representing the 1-D continuum at time $t$, and assume that $C_0$ is the initial position of the continuum. We assume that for each $t$ in an interval $I$, the curve $C_t$ is smooth and regular: $C_t = \{\alpha_t\}$, with parameter map

$$\alpha_t : J \to \mathbb{R}^3,$$

where $J \subseteq \mathbb{R}^3$ is an interval. The mass density, velocity, and external force density for the 1-D continuum are $\rho$, $\partial\alpha/\partial t$, and $F$ respectively. While these are functions on $J \times I$, their values at $(s, t)$ are referred to the curve $C_t$ in $\mathbb{R}^3$. See Figure B.9. The stress tensor $T$ is a $3 \times 3$ matrix-valued function defined on $J \times I$, and the model for the stressing forces is as follows. Consider a portion $C'_t$ of the total curve $C_t$, with $C'_t = \alpha_t(J')$, where $J' = [a, b] \subseteq J$. Then the force

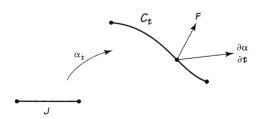

Figure B.9.   The external force density $F(s, t)$ and velocity $\partial\alpha(s, t)/\partial t$ at the point $\alpha(s, t)$ on the curve $C_t$

on $C_t'$ due to stresses acting across its boundary is by definition

$$T(b,t)\frac{\alpha_t'(b)}{|\alpha_t'(b)|} - T(a,t)\frac{\alpha_t'(a)}{|\alpha_t'(a)|} = T(s,t)\frac{\alpha_t'(s)}{|\alpha_t'(s)|}\bigg|_a^b$$

$$= Tv\bigg|_a^b$$

$$= \int_{\partial C_t'} Tv.$$

Here $v$ denotes the *outward-directed normal* to the boundary of $C_t'$. This vector and the above force are shown in Figure B.10. As you might expect, we can convert this expression for the stressing force on the boundary of piece $C_t'$ into a line integral over this curve. To do so requires a divergence theorem for curves, which we now present. In keeping with the notation involved in the geometry of regions and surfaces, we denote the single component of the metric for the curve $C_t$ by:

$$\det G^{1/2}(s) = |\alpha_t'(s)|,$$

and the reciprocal tangent vector field is $e^1 = G^{11}e_1$.

**Lemma B.2 (Divergence Theorem for a Curve)**    *Suppose $C = \{\alpha\}$ is a smooth regular curve, and $w : J \to \mathbb{R}^3$ is a vector field along $C$. Then*

$$\int_{\partial C} w \cdot v = \int_C \mathrm{div}_C(w)\, dL, \qquad (\text{B.53})$$

*where $v$ is the outward-directed normal to the boundary and $\mathrm{div}_C$ is the curve divergence operator defined by*

$$\mathrm{div}_C(w) = \det G^{-1/2}\frac{d}{ds}(\det G^{1/2}\, w \cdot e^1). \qquad (\text{B.54})$$

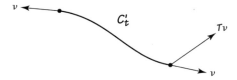

Figure B.10.    The outward directed normal $v$ to the boundary of the curve $C_t'$ and the two stressing forces $Tv$ at the boundary

*Note*: As you can perhaps now see, there is a general divergence theorem in differential geometry which gives the special cases $n = 1, 2, 3$ presented in this book.

**Proof**   This is very elementary and involves just applying the definitions and the fundamental theorem of calculus. Thus

$$\int_{\partial C} w \cdot v = w(s) \cdot \frac{\alpha'(s)}{|\alpha'(s)|}\bigg|_a^b$$

$$= \int_a^b \frac{d}{ds}\left(w(s) \cdot \frac{\alpha'(s)}{|\alpha'(s)|}\right) ds$$

$$= \int_a^b \det G^{-1/2} \frac{d}{ds}\left(\det G^{1/2} w \cdot e^1\right) \det G^{1/2} ds$$

$$= \int_C \operatorname{div}_C(w)\, dL.$$

If we now write down the *integral* equations for the conservation of mass linear momentum, and angular momentum for the portion $C_t'$ of the 1-D continuum, they will be identical to equations (B.49)–(B.51) for the portion $S_t'$ of the 2-D continuum (except that now there are line integrals instead of surface integrals and other obvious notational changes. Thus it is easy to derive the following PDE version of the equations of motion:

**Theorem B.3 (Laws of Motion for a 1-D Continuum)**   *The above integral version of the laws of motion equivalent to the following:*

**Conservation of Mass:**

$$\rho \det G^{1/2} = \rho_0 \det G_0^{1/2}.$$

**Linear Momentum Equation:**

$$\rho \frac{\partial^2 \alpha}{\partial t^2} = \rho F + \operatorname{div}_C(T).$$

**Angular Momentum Equation:**

$$T e^1 \cdot N = 0,$$

$$T e^1 \cdot B = 0.$$

*These equations hold on $J$ for all $t \in I$. In the angular momentum equation, $N$ and $B$ are the unit normal and unit binormal to the curve, and thus the two conditions say merely that $T e^1$ must be tangent to the curve at each point.*

### Exercises B.4

1. In the Lagrangian formulation in three dimensions, prove that:

    (a) $\det G^{1/2} = |\det \alpha'| = \det \alpha'$.

    (b) $\text{div}(\widetilde{T}) \circ \alpha = \text{div}_B(T)$.

2. In the Lagrangian formulation in two dimensions, prove that:

    (a) $e_1 \times n = -\det G^{1/2} e^2$ and $e_2 \times n = \det G^{1/2} e^1$.

    (b) $\frac{\partial}{\partial s_1}(e_2 \times n) - \frac{\partial}{\partial s_1}(e_2 \times n) = \mu \det G^{1/2}$.

    (c) The integral version of the angular momentum equation leads to the conditions

$$T e^1 \cdot e^2 = T e^2 \cdot e^1,$$

$$T e^i \cdot n = 0, \qquad \text{for } i = 1, 2,$$

on the stress tensor $T$.

3. Prove Theorem B.3.

# C Maple Reference Guide

For those not familar with *Maple*, this appendix serves to outline and briefly describe the various aspects of *Maple* used in the text. The text was designed in a way to allow the reader to learn some elementary *Maple* techniques easily while proceeding with the main task: learning something about PDEs. The intent was to present only a few (from the vast number) of features that *Maple* has to offer and to do so at a leisurely pace throughout the book.

In my experience, *Maple* is best assimilated in this fashion. Most people do not want to be overwhelmed at the outset by all that *Maple* can do, but rather wish to see some of the things it can do and then, at their own pace, add more syntax, programming, and graphical features to their base knowledge of *Maple*.

Even with this pedagogy, many of us, and especially the students I have taught, have a need for some sort of manual to refer to. There are numerous manuals and tutorials on *Maple* (cf. [BM 94], [HHR 96], [Mon 96], [Red 96]), but as an aid to mastering *Maple*, or at least what is needed to work the exercises in this text, this appendix provides an abbreviated reference manual. In addition, there are many extra features and tips about using *Maple* presented on the CD-ROM accompanying this text. Anyone with extra time to devote to learning more about *Maple* will benefit by working through the material there.

The interface with *Maple*, i.e., the way in which one enters commands to and receives output from *Maple*, has been in constant flux (and improvement) over the various releases of the software. This will be the case in the future as well, and so no attempt will be made to to describe the interface here. The various features of *Maple* presented below are common to *Release 3* and *Release 4*, with some minor differences between releases.

## C.1.   Mathematical Expressions and Functions

The symbols for the basic arithmetical operation are shown in Table C.1, and algebraic expressions are built from these in a way that is common to many programming languages. Thus for example, the expression

$$\frac{a[(x^2 + y^2)^3 - 5y]}{x^{-1} + y^{1/3}}$$

can be written in *Maple* code as

```
a*((x^2+y^2)^3-5*y)/(x^(-1)+y^(1/3))
```

This is known as a *Maple expression*, and the symbols a,x,y have no values or data types associated with them. The whole expression is built from simpler expressions using the arithmetical operators. *Note*: Grouping is accomplished in *Maple* by using parentheses: ),(. The use of brackets ],[ and braces },{ is not permitted in arithmetical expressions.

The operators in Table C.2 allow you to manipulate expressions and build more complicated expressions. The assignment operator allows you to assign a name to an expression, as, for example,

```
> r := a*((x^2+y^2)^3-5*y)/(x^(-1)+y^(1/3));
```

You can now use r as a shorthand way of referring to the expression on the right side of the above assignment statement. The arrow operator -> has the same meaning and use as $\mapsto$ (or sometimes $\to$) does in mathematics. Namely, in mathematics we usually indicate (or define) a function of $x$ by something like $x \mapsto 1/(1 + x^2)$, and we often name this function something like $f$. The definition of such a function in *Maple* is accomplished by the command

```
> f := x -> 1/(1+x^2);
```

*Maple* also uses the commonly accepted functional notation; i.e., f(1) denotes the value of the function f at $x = 1$. Note that by contrast, the above assignment for r does *not* make r a function of x and y. Thus the evaluation r(x,y) makes

Table C.1.   Symbols for the arithmetical operators

| | | |
|---|---|---|
| + | addition | example: a+b |
| − | subtraction | example: a−b |
| * | multiplcation | example: a*b |
| / | division | example: a/b |
| ^ | exponentiation | example: a^b |

Table C.2.    The assignment, arrow, and comparison operators

|  |  |  |
|---|---|---|
| := | assignment operator | example: a:=1 |
| -> | arrow operator | example: x ->5*x+x^2 |
| = | equal | example: x = y |
| < | less than | example: x < y |
| <= | less than or equal | example: x <= y |
| > | greater than | example: x > y |
| >= | greater than or equal | example: x >= y |
| <> | not equal | example: x <> y |

no sense. To properly define a real-valued function of two variables, you should use the arrow operator (or a procedure statement as discussed later). Thus, for example,

```
> f := (x,y) -> a*((x^2+y^2)^3-5*y)/(x^(-1)+y^(1/3));
```

defines such a function, which in this case also contains an unspecified parameter a. Various manipulations with f (like plotting its graph) will require that the parameter a be assigned a value first.

Vector-valued functions of one or more variables require a slightly different syntax in their definition. For example, the statement

```
> g := (x,y) -> [x-x^3*y, x^2*y^4, 1-x^2-y^2];
```

defines a vector-valued function $g : \mathbb{R}^2 \to \mathbb{R}^3$. *Maple*'s notation for a vector (in 3-D) is [a,b,c].

*Maple* has all the standard mathematical functions built into its library, and in most cases it uses the usual notation for these functions. Table C.3 shows some of the functions in *Maple*'s Library. Notwithstanding the standard functional notation, it has become customary to omit the parentheses in certain cases when using trig, hyperbolic, and log functions (as we have done throughout the text).

Table C.3.    Some of Maple's library functions

| | |
|---|---|
| exp | the natural exponential function |
| ln | the natural log function |
| sin, cos, tan, cot, sec, csc | the trig functions |
| arcsin, arccos, arctan | inverse trig functions |
| sinh, cosh, tanh, coth, sech, csch | the hyperbolic functions |
| sqrt | the square root function |
| abs | the absolute value function |
| Heaviside | Heaviside's function |
| BesselJ, BesselY | two kinds of Bessel functions |

Thus, for example, one writes $\sin x$ and $\ln x$ for the more correct expressions $\sin(x)$ and $\ln(x)$. Beware: in *Maple* this is *not* permitted. For example, you must use `sin(x)` for the value of the sine function at $x$. If you happen to use `sin x` in a *Maple* expression, it will not be correctly interpreted (*Maple* thinks it is a new name for something else).

The comparison operators in Table C.2 are used to build logical expressions that can be either true or false (depending on the values of x and y, in the examples shown).

## C.2.   Packages

A *Maple package* is a collection of commands to do specific tasks, like plotting, solving differential equations, and linear algebra. A basic set of commands is always present whenever you start *Maple*, but to take advantage of the additional commands available in a package, the package has to be loaded using the `with` command. Here are the three main examples:

```
> with(plots);
```
*animate, animate3d, implicitplot, implicitplot3d, contourplot, cylinderplot, spacecurve, . . .*

```
> with(DEtools);
```
*DEplot, DEplot3d, DEplot1, phaseportrait, . . .*

```
> with(linalg);
```
*rref, gausselim, matrix, augment, . . .*

The particular commands available in each package are shown in the output above. The ellipsis indicates that there are more commands in the package. The list contains some of those that are pertinent here. *Note*: it is possible to use a single command from a package without loading the whole package. For example, `plots[animate]` allows you to use the `animate` command without loading the whole `plots` package. See the next section.

## C.3.   Plotting and Visualization

*Maple* has many powerful features for plotting and constructing complicated graphical images in two and three dimensions. Table C.4 shows just a few of the basic commands that will suffice for doing the scientific visualization in this text. All of these commands, except `plot` and `plot3d`, are part of the `plots` package.

Table C.4.    A few commands for graphing and visualizing various structures

| plot | plots the graph of a function of a single variable |
|---|---|
| plot3d | plots the graph of a function of two variables |
| implicitplot | plots the graph of a curve given implicitly by an equation in two variables |
| implicitplot3d | plots the graph of a surface given implicitly by an equation in three variables |
| spacecurve | plots a curve in $\mathbb{R}^3$ that is given parametrically |
| contourplot | plots level curves for a function of two variables |
| cylinderplot | plots the graph of the polar coordinate version of a function of two variables |
| display | displays several 2-D plots simultaneously |
| display3d | displays several 3-D plots simultaneously |
| animate | simulates the motion of the graph of $u(x, t)$ |
| animate3d | simulates the motion of the graph of $u(x, y, t)$ |

(1) $\boxed{\texttt{plot(expr,ranges,options)}}$ This command can be used to plot the graph of a real-valued function $f : [a, b] \to \mathbb{R}$ or of a vector-valued function $g : [a, b] \to \mathbb{R}^2$, of a single variable $x \in [a, b]$. The latter is often called a curve or a parametrization for a curve in the plane, and mathematically, the plot will be of the set of points $g[a, b]$, often called the trace of the curve. The former is a special case of the latter, since $f$ determines a vector-valued function $g$, via $g(x) = (x, f(x))$.

*Note*: For the sake of explanation here, we use x for the independent variable name. You can use any name you wish (like s or theta) in applications. In the syntax for this command, expr can be either (i) an expression involving x, or (ii) something like f(x), where f is the name for a real-valued function of a real variable. In either case there must be no other variables or parameters in expr. The ranges consists of two ranges, one like: x=a..b for the independent variable and one like c..d or y=c..d for the dependent variable. The latter range is optional but convenient when f has vertical asymptotes in its graph. The options enable you to control the way the graph looks and consists of a list of items like color=black and numpoints=201 with commas separating the items in the list. Each option has a default value that is used if you do not specify that option. For example, numpoints has default value 49, which means that the plot is executed by evaluating expr at a *minimum* of 49 equally spaced points $\{x_n\}_{n=1}^{49}$ in the interval $[a, b]$, with $x_1 = a$ and $x_{49} = b$.

*Note*: *Maple* uses an adaptive method for this type of plot, so it may actually use more points than specified in numpoints.

You can consult the help menu to see what the other options are or you can, as I do, select most of them from the plot window after executing a basic plot with no options.

The plot command works differently (and a little unexpectedly) for parametrized curves in the plane $g(x) = (h(x), k(x))$, $x \in [a, b]$, where $h$ and $k$ are the component functions. The proper syntax for the plot is

```
plot([h(x),k(x),x=a..b])
```

Multiple plots in the same picture can be achieved by using a list of valid expressions {expr1,expr2,...,exprn} in place of the single expression expr. The ranges must be the same for each of these expressions. See pp. 90, 94.

(2) | plot3d(expr,ranges,options) | This command can be used to plot the graph of a real-valued function $f : [a, b] \times [c, d] \to \mathbb{R}$ or of a vector-valued function $g : [a, b] \times [c, d] \to \mathbb{R}^3$ of two variables $(x, y) \in [a, b] \times [c, d]$. The former is a special case of the latter, since $f$ determines a vector-valued function $g$, via $g(x, y) = (x, y, f(x, y))$. In the plot command syntax, expr can be either (i) a single expression involving two independent variables, say x and y; or (ii) something like f(x,y), where f is a name for a real-valued function of two variables; or (iii) an expression list of the form [h(x,y),k(x,y),l(x,y)], where h,k,l are names for real-valued functions of two variables; or (iv) something like g(x,y), where g is a name for a function of two variables with values in $\mathbb{R}^3$. The ranges part of the plot3d command has the form x=a..b,y=c..d (there is no optional z range). An example of one of the options in the plot3d commmand is grid=[51,61], which causes the rectangle $[a, b] \times [c, d]$ to be divided into $50 \times 60 = 3000$ equal subrectangles by choosing 51 equally spaced points $x_n$ in $[a, b]$ (starting at $a$ and ending at $b$) and 61 equally spaced points $y_m$ in $[c, d]$. The plot is executed by evaluating expr at all the points $(x_n, y_m)$, $n = 1, \ldots, 51$, $m = 1, \ldots, 61$. The default value is grid=[25,25]. Other possible options can be explored by consulting the help menu, or can be selected from the plot window once a basic plot is executed. I find it more convenient to do the latter, changing the style, selecting color and lighting, adding the axes, and selecting the view and perspective, after doing a basic plot with no options.

Multiple plots in the same picture can be achieved by using a list of valid expressions {expr1,expr2,...,exprn} in place of the single expression expr. The ranges must be the same for each of these expressions. See CD-ROM, Figures 6.4 (Example 6.1), 6.6, 6.8, 6.9 (Example 6.5).

(3) | implicitplot(eqn,ranges,options) | This command is used to plot an equation of the form $f(x, y) = c$, where $c$ is a constant and $f$ is a real-

valued function of two variables. The plot can be thought of as (a) the graph of a curve in the plane; (b) the graph of a level curve for the function $f$; or (c) the graphs of the various functions $y$ of $x$ defined implicitly by $f(x, y) = c$. In the syntax for this command, eqn is the equation, and `ranges` has the form x=a..b,y=c..d, which represents the viewing rectangle $[a, b] \times [c, d]$ for the plot. One commonly used option is grid, which allows you to obtain a finer resolution in the plot. The default is grid=[25,25].

(4) | `implicitplot3d(eqn,ranges,options)` | This command is used to plot an equation of the form $f(x, y, z) = c$, where $c$ is a constant and $f$ is a real-valued function of three variables. The plot can be thought of as (a) the graph of a surface in 3-space; (b) the graph of a level surface for the function $f$, or (c) the graphs of the various functions $z$ of $(x, y)$ defined implicitly by $f(x, y, z) = c$. In the syntax for this command, eqn is the equation, and `ranges` has the form x=a..b,y=c..d,z=m..n, which represents the viewing box $[a, b] \times [c, d] \times [m, n]$ for the plot. One commonly used option is grid, which allows you to obtain a finer resolution in the plot. The default is grid=[10,10,10]. See the CD-ROM, Figures 8.3 (Example 8.2), 8.10 (Example 8.4).

(5) | `spacecurve(expr,range, options)` | This command is used to plot the trace of a vector-valued function $g : [a, b] \to \mathbf{R}^3$, which is often called a curve or a parametrized curve in 3-space. In the syntax, expr can be either (i) an expression list of the form [h(x),k(x),l(x)], where h,k,l are real-valued functions of a single variable x, or (ii) g(x), where g is a function of a single variable with values in $\mathbb{R}^3$ (a vector-valued function). Some useful options are color=black and numpoints=200. The default value for numpoints is 50, and unlike the use of this option in the plot command, the number of points specified is the actual number of points used. Thus numpoints=3 produces a plot of three points connected by two straight line segments. See pp. 122, 399.

(6) | `contourplot(expr,ranges,options)` | This command is *not* very useful since you can accomplish the same thing with plot3d. It is mentioned here because contour plots are important. The command is used to plot a sequence of level curves $f(x, y) = c_n, n = 1, \ldots, N$, for a real-valued function of two variables. Mathematically, this is the same as slicing the graph of $f$ (a surface) by a sequence of horizontal planes $z = c_n$ and projecting the resulting curves on the $x$-$y$ plane. In the syntax, expr can be either (i) an expression involving two variables (say x,y) and numbers but no other variables or parameters, or (ii) an expression like f(x,y) where f is a real-valued function of two variables. ranges has the form x=a..b,y=c..d. As mentioned, you can get the same graphic as contourplot(expr,ranges)

by using `plot3d(expr,ranges)`, choosing `style=contour` from the plot window, and then rotating the view to one looking down on the $x$-$y$ plane. See the CD-ROM, Figure 7.13 (Example 7.4).

(7) `cylinderplot([r,theta,K(r,theta)],ranges,options)`    This command plots the graph of a real-valued function $S$ of two variables when the polar coordinate version $K$ of $S$ is given. Here

$$K(r, \theta) \equiv S(r \cos \theta, r \sin \theta).$$

Using `cylinderplot` in this way is equivalent to using `plot3d` to plot the vector-valued function

$$g(r, \theta) = (r \cos \theta, r \sin \theta, K(r, \theta)).$$

`ranges` has the form `r=a..b`, `theta=c..d`. An example of a useful option is `grid=[5,51]`, which because of the transformation embodied in $g$, causes the grid lines to be displayed on the surface in the form of four equally spaced concentric circles and 50 equally spaced radial lines. The default is `grid=[25,25]`, which often gives too many concentric circles for readability and too few radial lines for smoothness of the plot. See pp. 375, 399.

(8) `display3d(list)` This command is used to display the results of several 3-D plot structures in the same picture. In the syntax, `list` has the form {`p1,p2,...,pn`}, where each of the p's in the list is a (3-D) plot structure. A plot structure is a name to which has been assigned the output of one of *Maple*'s plotting commands. For example, the commands

```
p1:=plot3d(x^2+y^2,x=-1..1,y=-1..1)
line:=spacecurve([t,t+1,t+2],t=-1..1,color=black)
```

create two plot structures named `p1` and `line`. Then the command `display3d({p1,line})` will display the surface and the line in the same picture. *Note*: When creating a plot structure, you should end the assignment statement `p1:=...` with a colon before entering the command. This prevents the result from being written back to the monitor screen.

(9) `display(list)` This command works the same way as `display3d`, except that the `list` contains 2-D plot structures.

(10) `animate(u(x,t),ranges,options)` In this command, `u` is a real-valued function of two variables, and `ranges` has the form `x=a..b,t=c..d`. What is displayed in the animation window is a succession of plots of

$u(\cdot, t_1), u(\cdot, t_2) \ldots, u(\cdot, t_n)$ as functions of a single variable $x$. Here $t_i$, $i = 1, 2, \ldots n$, are equally spaced times, with $t_1 = a$ and $t_n = b$. Also, $n$ is the number of frames (an option). The default is `frames=8`. See the CD-ROM, Figures 11.7 (Example 11.2), 11.9 (Example 11.3).

(11) $\boxed{\texttt{animate3d(u(x,y,t),ranges,options)}}$ This command is entirely similar to `animate`, except that now u depends on three variables. See the CD-ROM, Figures 11.13 (Example 11.4), 11.16 (Example 11.5).

## C.4.   Plotting Flows

Suppose $v : \Omega \times I \to \mathbb{R}^3$ is a time dependent vector field on a domain $\Omega$ in $\mathbb{R}^3$. Here (as opposed to the Introduction) we denote the three spatial variables by $x, y, z$ and write the component expression for $v$ as

$$v(x, y, z, t) = \left( h(x, y, z, t), k(x, y, z, t), l(x, y, z, t) \right),$$

where $h, k, l$ are scalar functions (the components of $v$). In the special case where $v$ does not depend on $z$ and has values in $\mathbb{R}^2$ (i.e., is a *planar* vector field), it has the form

$$v(x, y, t) = \left( h(x, y, t), k(x, y, t) \right)$$

Physically, the flow generated by $v$ can represent heat flow lines, electric field lines, or fluid flow lines when $v$ is a heat flux vector field, a static electric field, or a velocity vector field for a fluid. Mathematically, the individual flow lines (curves) comprising the flow are solutions of the first-order system of DE's:

$$\frac{dx}{dt} = h(x, y, z, t),$$

$$\frac{dy}{dt} = k(x, y, z, t),$$

$$\frac{dz}{dt} = l(x, y, z, t),$$

which reduces to $dx/dt = h(x, y, t), dy/dt = k(x, y, t)$ when $v$ is a planar vector field. These are known as three-dimensional and two-dimensional systems respectively, and the flow lines are synonymous with what are known as integral curves or solution curves for the system of DE's (or for the vector field $v$).

*Maple* has many tools in the package DEtools for studying differential equations. For plotting the flow of $v$, i.e., a set of its integral curves, the commands DEplot, for 2-dimensional systems, and DEplot3d, for 3-dimensional systems, are useful:

(1) DEplot(deqn,vars,trange,initset,options) This command plots one or more (depending on the number of initial conditions specified) integral curves for a two-dimensional system of first order DE's. In the syntax, deqs is a list of the two DE's comprising the system and has the form

```
[D(x)(t)=h(x,y,t),D(y)(t)=k(x,y,t)]
```

*Note*: The equality operator = is used here, not the assignment operator : =. Also, there can be no parameters or symbolic constants (like $\pi$) involved in the functions h,k. The vars part of the command has the form [x,y] and just indicates what the two unknown functions are. The trange is typically t=a..b. The initial conditions for the plots are specified by initset, which has the form {[t1,x1,y1],...,[tn,xn,yn]}. This plots the $n$ integral curves that pass through the points $(x_i, y_i)$ at time $t_i$, for $i = 1, \ldots, n$. Finally, options is a sequence of options separated by commas. For example, stepsize=.05,arrows=thin is a sequence of two options you might want to use. Also, you can use x=a..b,y=c..d to set the viewing window.

I find it convenient to define the component functions $h$ and $k$, assign a name to the generic system of two DE's, and assign values to all parameters before executing a DEplot. Likewise, if the list of initial conditions is long, it is convenient to name it first. Thus, for example, you might want to do something like

```
> h:=(x,y) -> -y+x*(a-x^2-y^2);
> k:=(x,y) -> x+y*(a-x^2-y^2);
> sys := D(x)(t) = h(x,y), D(y)(t) = k(x,y);
> a:=0.8;
> init:= [0,2,2],[0,-2,2],[0,2,-2],[0,0.1,0];
> DEplot([sys], [x,y],t=0..20,{init});
```

See also p. 157.

(2) DEplot3d(deqs,vars,trange,initset,options) This command plots one or more (depending on the number of initial conditions specified) integral curves for a three-dimensional system of first order DE's. *Caution*: The command is valid only in *Release 4*. To do the same thing in *Release 3*, use DEplot with the interpretation of its arguments given here. Apparently, the designers of *Release 4* wanted to have the distinction in the commands

plot, plot3d also appear in the DE plotting commands. In the syntax, deqs is a list of the three DE's comprising the system and has the form

```
[D(x)(t)=h(x,y,z,t),D(y)(t)=k(x,y,z,t),
 D(z)(t)=l(x,y,z,t)]
```

The vars part of the command has the form [x,y,z] and just indicates what the three unknown functions are. The trange is typically t=a..b. The initial conditions are specified by initset, which has the form {[t1,x1,y1,z1],...,[tn,xn,yn,zn]} and has a similar interpretation as that in the DEplot command above. Finally, the options is a sequence of options separated by commas. For example, stepsize=.05,arrows=thin is a sequence of two options you might want to use. See the CD-ROM, Figure 8.4 (Example 8.2).

# C.5.  Programming

*Maple* has all the programming constructs typical of most programming languages. Table C.5 shows the few that we have used in this book.

(1) | if *condition* then *statements* fi |  This works pretty much as expected: If the *condition* is true, then the *statements* are executed. Otherwise the *statements* are skipped, and control passes to the next statement after fi. See p. 376.

(2) | if *condition* then *statements1* else *statements2* fi |  Similar to the above: If the *condition* is true, then the *statements1* are executed. Otherwise, the *statements2* are executed. In either event, control passes afterward to the next statement after fi. See p. 432.

(3) | for *name* to *expr* do *statements* od |  This is the basic "do loop" in *Maple* and works in a standard way: at the start, the variable named in *name* is assigned the value 1, and if *name* < *expr*, then *statements* are executed. After this, *name* is incremented by 1, and if *name* < *expr*, then *statements*

Table C.5.   A few programming constructs

| | |
|---|---|
| if..then..fi | single branching statement |
| if..then..else..fi | double branching statement |
| for..from..by..to..do..od | looping (repetition) statement |
| proc..end | procedure statement |

are executed. This pattern is repeated until *name* ≥ *expr*. The optional items
from and by can be included in the command in order to begin and increment
*name* with values other than 1. See: pp. 89, 92, 94, 121, 123, 376, 432.

(4) | proc( *params* ) *statements* end | Procedures in *Maple* are similar to sub-
routines in FORTRAN and other languages. Here, params is a sequence of
parameters that are passed to or returned from the procedure. The body of
the procedure consists of the *statements* (each separated by a semicolon)
contained between proc and end.

We have used procedures here mainly to define functions that are truncated
Fourier series. For example,

```
f:= proc(x,y,N) s:=0; for n to N do for m to N do
    s:=s+A[n,m]*sinh(n*Pi*x)*sin(n*Pi*y) od;od;end;
```

defines the function $f(x, y, N) = \sum_{n=1}^{N} \sum_{m=1}^{N} A_{nm} \sinh n\pi x \, \sin m\pi y$. The
parameters here are x,y,N (all of which are used to pass values to the
procedure), and the body consists of three statements. The variables s,n,m
are *local* to the procedure, i.e., they exist and have values associated with
them only when the procedure is executing. On the other hand, A is a *global*
variable (in this case a 2 × 2 array) and must have values assigned to it before
the procedure can execute correctly. *Maple* will automatically identify the
local variables for you, but good programming practice dictates that you
declare them in the definition of the procedure (see the help contents).

In this example, the procedure is invoked (or called) by writing something
like f(0.5,0.5,20), which should return the numerical value of the sum
of the 400 terms. In general, if no values are explicitly assigned to any of
the parameters of the procedure, then the last value calculated within the
procedure is returned. In this example, the last time through the do loop, the
procedure assigns a value to s, and this is the value returned. See pp. 89, 92,
94, 121, 123, 432.

As an example of a procedure that returns values through parameters,
consider the following:

```
linsolve:= proc(A,b,v) d:=det(A); if d=0
            then ERROR('det'=0) else
            v[1]:=(b[1]*A[2,2]-b[2]*A[1,2])/d;
            v[2]:=(-b[1]*A[2,1]+b[2]*A[1,1])/d;
            fi; end;
```

This procedure solves the linear system $Av = b$ for $v$, where $A$ is a given
2 × 2 array and $b$ is a given 2 × 1 array. The parameters A,b are used to pass
values to the procedure, and the procedure assigns a value to the parameter
v (which implicitly is a 2 × 1 array). A typical call to the procedure would

be something like  `>linsolve(M,c,w);`, where `M,c` are the arrays for a particular system $Mw = c$ you wish to solve, and `w` is the array where you want to store the results. Then the command `> evalm(w);` will display the results. *Note*: This example requires the linear algebra package.

# D Symbols and Tables

| | |
|---|---|
| $\nabla^2$ | Laplacian, p. 8 |
| $\nabla_v$ | covariant derivative, p. 7 |
| $\nabla$ | gradient operator, p. 4 |
| curl | curl operator, p. 4 |
| div | divergence operator, p. 4 |
| $n$ | unit normal to a surface, p. 443 |
| $\tau$ | unit tangent to a curve, p. 439 |
| $\nu$ | outward-directed normal to a boundary curve, p. 482 |
| $\int_S E \cdot n \, dA$ | surface integral of $E$ over $S$, p. 443 |
| $\int_C E \cdot \tau \, dL$ | integral of $E$ along $C$, p. 440 |
| $\int_\Omega f \, dV$ | volume integral of $f$ over $\Omega$, p. 445 |
| $E'$ | Jacobian matrix of $E$, p. 7 |
| $A^*$ | transpose of matrix $A$, p. 10 |
| $H_{a,p,b}$ | triangular pulse function, p. 61 |
| $K_{a,b,L}$ | parabolic pulse function, p. 319 |
| $\mu$ | mean curvature normal p. 199, 459 |
| $H^1(\Omega)$ | Sobolev space, p. 410 |
| $H^k(\Omega)$ | Sobolev space, p. 410 |
| $H^1_0(\Omega)$ | closure of $\mathcal{D}(\Omega)$ in $H^1(\Omega)$, p. 410 |
| $L^2(\Omega)$ | space of square integrable functions on $\Omega$, p. 409 |
| $\mathcal{D}(\Omega)$ | space of test functions on $\Omega$, p. 221, 409 |
| $\mathcal{D}^*(\Omega)$ | space of distributions on $\Omega$, p. 221, 409 |

Table D.1.   Table of integrals

$$2 \int_0^1 \sin n\pi x \, dx = \frac{2[1 - (-1)^n]}{n\pi}$$

$$2 \int_0^1 x \sin n\pi x \, dx = \frac{2(-1)^{n+1}}{n\pi}$$

$$2 \int_0^1 x \cos n\pi x \, dx = \frac{2[(-1)^n - 1]}{n^2 \pi^2}$$

$$2 \int_0^1 x^2 \sin n\pi x \, dx = \frac{2(-1)^{n+1}}{n\pi} + \frac{4[(-1)^n - 1]}{n^3 \pi^3}$$

$$2 \int_0^1 x^2 \cos n\pi x \, dx = \frac{4(-1)^n}{n^2 \pi^2}$$

$$2 \int_0^1 (x - x^2) \sin n\pi x \, dx = \frac{4[1 - (-1)^n]}{n^3 \pi^3}$$

$$2 \int_0^1 (x - x^2) \cos n\pi x \, dx = \frac{2[(-1)^{n+1} - 1]}{n^2 \pi^2}$$

$$\int x \sin ax \, dx = \frac{-x}{a} \cos ax + \frac{1}{a^2} \sin ax$$

$$\int x \cos ax \, dx = \frac{x}{a} \sin ax + \frac{1}{a^2} \cos ax$$

$$\int x^2 \sin ax \, dx = -(\frac{x^2}{a} - \frac{2}{a^3}) \cos ax + \frac{2x}{a^2} \sin ax$$

$$\int x^2 \cos ax \, dx = (\frac{x^2}{a} - \frac{2}{a^3}) \sin ax + \frac{2x}{a^2} \cos ax$$

The general formulas for the above particular cases are:

$$\int x^p \sin ax \, dx = -h_p(x, a) \cos ax + \frac{p}{a} h_{p-1}(x, a) \sin ax$$

$$\int x^p \cos ax \, dx = h_p(x, a) \sin ax + \frac{p}{a} h_{p-1}(x, a) \cos ax$$

In the above $h_p(x, a)$ is (by definition):

$$h_p(x, a) = \sum_{k=0}^{[p/2]} \frac{(-1)^k p}{(p - 2k)} \frac{x^{p-2k}}{a^{2k+1}}, \qquad \text{for } p \geq 1$$

*Note*: $h_0(x, a) \equiv \frac{1}{a}$,   and $[p/2]$ is the greatest integer $\leq p/2$

Table D.2.  Eigenvalues and eigenfunctions for the
Sturm–Liouville problem $X'' = kX$ with the BCs shown.
*Note*: The eigenvalues are $k = -\lambda_n^2$, with $\lambda_n$ as indicated

| BCs | $\lambda_n$ | $X_n(x)$ |
|---|---|---|
| $X(0) = 0$ <br> $X(1) = 0$ | $n\pi$ | $\sin n\pi x$ |
| $X'(0) = 0$ <br> $X'(1) = 0$ | $0$ <br> $n\pi$ | $1$ <br> $\cos n\pi x$ |
| $X(0) = 0$ <br> $X'(1) = 0$ | $\dfrac{(2n-1)\pi}{2}$ | $\sin \dfrac{2n-1}{2}\pi x$ |
| $X'(0) = 0$ <br> $X(1) = 0$ | $\dfrac{(2n-1)\pi}{2}$ | $\cos \dfrac{2n-1}{2}\pi x$ |
| $X(0) = 0$ <br> $X'(1) + X(1) = 0$ | $\tan \lambda_n = -\lambda_n$ | $\sin \lambda_n x$ |
| $X'(0) = 0$ <br> $X'(1) + X(1) = 0$ | $\cot \lambda_n = \lambda_n$ | $\cos \lambda_n x$ |
| $X'(0) - X(0) = 0$ <br> $X'(1) + X(1) = 0$ | $\tan \lambda_n = \dfrac{2\lambda_n}{\lambda_n^2 - 1}$ | $\sin \lambda_n x + \lambda_n \cos \lambda_n x$ |

Table D.3.  Eigenvalues and eigenfunctions for the Sturm–Liouville
problem $R'' + r^{-1}R' + (\lambda^2 - m^2 r^{-2})R = 0$ with the BCs shown. The
differential equation is Bessel's equation and the eigenvalues/eigen-
functions depend on $m = 0, 1, 2\ldots$, as the notation indicates. The first
three BCs are for a disk of radius 1, and the last two are for an annulus
with radii 1 and 2. For clarity we have explicitly pointed out that the
second and fifth BCs have zero for an eigenvalue. This occurs only for
$m = 0$, and the corresponding eigenfunction is $R_0(r) = 1$.
*Note*: The eigenvalues are $-\lambda_{mn}^2$, with $\lambda_{mn}$ satisfying the indicated equation.

| BCs | $\lambda_{mn}$ | $R_{mn}(r)$ |
|---|---|---|
| $R_m(1) = 0$ | $J_m(\lambda_{mn}) = 0$ | $J_m(\lambda_{mn}r)$ |
| $R'_m(1) = 0$ | $0$ <br> $J'_m(\lambda_{mn}) = 0$ | $1$ <br> $J_m(\lambda_{mn}r)$ |
| $R'_m(1) + R_m(1) = 0$ | $J'_m(\lambda_{mn}) + J_m(\lambda_{mn}) = 0$ | $J_m(\lambda_{mn}r)$ |
| $R_m(1) = 0$ <br> $R_m(2) = 0$ | $J_m(\lambda_{mn})Y_m(2\lambda_{mn}) =$ <br> $J_m(2\lambda_{mn})Y_m(\lambda_{mn})$ | $J_m(\lambda_{mn}r)Y_m(\lambda_{mn}) -$ <br> $J_m(\lambda_{mn})Y_m(\lambda_{mn}r)$ |
| $R'_m(1) = 0$ <br> $R'_m(2) = 0$ | $0$ <br> $J'_m(\lambda_{mn})Y'_m(2\lambda_{mn}) =$ <br> $J'_m(2\lambda_{mn})Y'_m(\lambda_{mn})$ | $1$ <br> $J_m(\lambda_{mn}r)Y'_m(\lambda_{mn}) -$ <br> $J'_m(\lambda_{mn})Y_m(\lambda_{mn}r)$ |

# References

[AM 78] R. Abraham and J. Marsden, *Foundations of Mechanics*, 2nd ed., Addison-Wesley, Reading, MA, 1978.

[AMR 82] R. Abraham, J. Marsden, and T. Ratiu, *Manifolds, Tensor Analysis, and Applications*, Addison Wesley, Reading MA, 1982.

[Arn 78] V.I. Arnold, *Ordinary Differential Equations*, MIT Press, Boston, 1978.

[AB 84] O. Axelsson and V.A. Barker, *Finite Element Solution of Boundary Value Problems*, Academic Press, New York, 1984.

[BM 94] N.R. Blachman and M.J. Mossinghoff, *Maple V Quick Reference*, Brooks/Cole Publishing, Pacific Grove, 1994.

[Boo 75] W.M. Boothby, *Introduction to Differential Manifolds and Riemannian Geometry*, Academic Press, New York, 1975.

[CM 79] A.J. Chorin and J.E. Marsden, *A Mathematical Introduction to Fluid Mechanics*, Springer-Verlag, New York, 1979.

[Fo 95] G.B. Folland, *Introduction to Partial Differential Equations*, 2nd ed., Princeton University Press, New Jersey, 1995.

[Fr 63] A. Friedman, *Generalized Functions and Partial Differential Equations*, Prentice-Hall, Engelwood Cliffs, New Jersey, 1963.

[Fu 65] Y.C. Fung, *Foundations of Solid Mechanics*, Prentice-Hall, Engelwood Cliffs, New Jersey, 1965.

[Gur 81] M.E. Gurtin, *An Introduction to Continuum Mechanics*, Academic Press, New York, 1981.

[HHR 96] K.M. Heal, M.L. Hansen, and K.M. Rickard, *Maple V: Learning Guide*, Springer-Verlag, New York, 1996.

[Ja 75]  J.D. Jackson, *Classical Electrodynamics*, 2nd ed. Wiley, New York, 1975.

[Kl 78]  W. Klingenberg, *A Course in Differential Geometry*, Springer-Verlag, New York, 1978.

[Ko 89]  H. Koçak, *Differential and Difference Equations through Computer Experiments*, 2nd ed., Springer-Verlag, New York, 1989.

[LC 62]  P. Lorrain and D.R. Corson, *Electromagnetic Fields and Waves*, W.H. Freeman, San Francisco, 1962.

[Mal 69]  L.E. Malvern, *Introduction to the Mechanics of a Continuous Medium*, Prentice-Hall, Engelwood Cliffs, New Jersey, 1969.

[MH 83]  J.E. Marsden and T.J.R. Hughes, *Mathematical Foundations of Elasticity*, Prentice-Hall, Engelwood Cliffs, New Jersey, 1983.

[Mon 96]  M.B. Monagan, K.O. Geddes, K.M. Heal, G Labahn, and S. Vorkoetter, *Maple V: Programming Guide*, Springer-Verlag, New York, 1996.

[Ne 67]  J. Nečas, *Les Méthodes Directes en Théorie des Equations Elliptiques*, Masson, Paris, 1967.

[ON 91]  P.V. O'Neil, *Advanced Engineering Mathematics*, 3rd ed., Wadsworth, Belmont, 1991.

[Per 91]  L. Perko, *Differential Equations and Dynamical Systems*, Sringer-Verlag, New York, 1991.

[Pin 91]  M. Pinsky, *Partial Differential Equations and Boundary-Value Problems with Applications*, 2nd ed., McGraw-Hill, New York, 1991.

[Red 94]  D. Redfern, *Maple Handbook: Maple V Release 3*, Springer-Verlag, New York 1994.

[RMC 93]  J.R. Reitz, F. J. Milford, and R.W. Christy, *Foundations of Electromagnetic Theory*, 4th ed., Addison-Wesley, Reading, 1993.

[RR 93]  M. Renardy and R.C. Rogers, *An Introduction to Partial Differential Equations*, Springer-Verlag, New York, 1993.

[Rh 60]  G. de Rham, *Variétés Différentiables*, Hermann, Paris, 1960.

[Rou 65]  H. Rouse, *Advanced Mechanics of Fluids*, Wiley, New York, 1965.

[Sch 66]  L. Schwartz, *Théorie des Distributions*, nouvelle ed., Hermann, Paris, 1966.

[Sch 67]  L. Schwartz, *Mathematics for the Physical Sciences*, Addison-Wesley, Reading, MA, 1967.

[Str 92]  W.A. Strauss, *Partial Differential Equations, an Introduction*, Wiley, New York, 1992.

[Te 77]  R. Teman, *Navier-Stokes Equations, Theory and Numerical Analysis*, North-Holland, Amsterdam, 1977.

[Tru 77]  C. Truesdell, *A First Course in Rational Continuum Mechanics*, Academic Press, New York, 1977.

[Ver 90]  F. Verhulst, *Nonlinear Differential Equations and Dynamical Systems*, Springer-Verlag, New York, 1990.

# Index